THE MECHANICS OF FROZEN GROUND

THE MECHANICS OF FROZEN GROUND

N. A. Tsytovich

Professor and Corresponding Member
of the U.S.S.R. Academy of Sciences
Distinguished Scientist and Engineer
of the RSFSR

edited by

George K. Swinzow

Geologist
U.S. Army Cold Regions Research and Engineering Laboratory

Gregory P. Tschebotarioff
Advisory Editor

Scripta Book Company
Washington, D.C.

McGraw-Hill Book Company

New York   St. Louis   San Francisco   Auckland   Düsseldorf
Johannesburg   Kuala Lumpur   London   Mexico   Montreal
New Delhi   Panama   Paris   São Paulo   Singapore
Sydney   Tokyo   Toronto

Library of Congress Cataloging in Publication Data

TSytovich, Nikolaĭ Aleksandrovich.
    The mechanics of frozen ground.

    Translation of Mekhanika Merzlykh gruntov.
    Bibliography.
    1. Frozen ground.   2. Soil mechanics.   I. Title.
TA713.T72813   624'.1513   74-32196
ISBN 0-07-065410-7

THE MECHANICS OF FROZEN GROUND

First published in the Russian language under the title *Mechanics of
Frozen Soil* in 1973 by Vysshaya Shkola Press, Moscow. Translated
by Scripta Technica, Inc.

1 2 3 4 5 6 7 8 9 0 K P K P 7 8 4 3 2 1 0 9 8 7 6 5

This book was set in Press Roman by Hemisphere Publishing
Corporation. The editors were Jeremy Robinson and Ezra Kohn;
the designers and compositors were Bernie Doenhoefer and Pat
Hopper. The printer and binder was The Kingsport Press.

To the glowing memory of

Academician Vladimir Afanas'yevich Obruchev

CONTENTS

FOREWORD TO THE ENGLISH TRANSLATION

Prior to the construction of the Alaska-Canada highway during World War II little attention was paid in the United States to permafrost problems. Because of the early development of vast areas covered by permafrost in Siberia the construction problems presented by them received earlier attention there from both a practical and a theoretical point of view. The late academician V. A. Obruchev, to whom Professor N. A. Tsytovich dedicates his book, is revered among Soviet engineers as the father of such modern studies there.

Dr. Tsytovich himself is a leading pioneer in such work and has received many honors for it in his country. For many years he has served as the President of the USSR National Society for Soil Mechanics and Foundation Engineering. Among Western theoreticians his best known original contributions are his studies of the role of unfrozen water in frozen soils and his quantitative description of the consolidation of thawing frozen ground. His practical contributions are also impressive. As early as 1928 he successfully introduced into construction practice the principle of preserving the frozen state of the ground under foundations by providing air spaces open to the wind under the floors of buildings.

Professor Tsytovich was always willing to share his accumulated knowledge with foreign colleagues. Thus, as a member of the Soviet delegation in the first USA–USSR official exchange in the field of soil engineering, he presented in 1959 a relevant paper at seminars held at U.S. universities (see *Highway Research Board, Special Report No. 60*, pp. 48–63, Washington, D.C., 1960) and discussed it with panels of American colleagues. In 1963 he took part in the First International Conference on Permafrost, which was held at Purdue University.

Dr. Tsytovich's book gives a summary of the present status of this branch of engineering and science in the Soviet Union, as it developed there over the past forty-five years. It should be of considerable interest in this country not only to specialists, but to anyone connected with the development of Alaska and of other regions in the far North. The appearance of its translation is most timely.

Much new and original work on frozen ground has been done in the United States since World War II. It is fortunate that the editing of the English translation of the present book could be undertaken by an active participant in this work here. At the suggestion of Professor Gerald A. Leonards, the McGraw-Hill Book Company entrusted this work to Dr. George K. Swinzow, who ably carried out his assignment for them.

Gregory P. Tschebotarioff
Consulting Engineer, P.E.

INTRODUCTION TO THE ENGLISH TRANSLATION BY THE AUTHOR

This book sets forth the basic laws established by the author for frozen soil mechanics as a science, presents a number of solutions obtained for problems in the mechanics of frozen ground, and deals with the practical application of investigations performed in the last decades, in most cases by the author himself and his co-workers.

The Russian edition of this book was published prior to the Second International Conference on Permafrost, which was held in Yakutsk (USSR, July 13-28, 1973). Hence, the materials of this conference have not been included. Two sessions of this conference—the 4th Session (Section 1: "Physics and Mechanics of Frozen Ground," where the state-of-the-art report was presented by the author) and the 7th Session (Principles Controlling Cryogenic Processes During Construction in Permafrost Regions)—were devoted to a certain extent to problems dealt with in this book. These include the improvement of investigation techniques, the devising of methods for obtaining predesigned changes in the mechanical properties of permafrost, the investigation of the mechanism of deformation of frozen ground, studies of experience gained in construction on frozen ground, etc. All these data, of course, require analyzing in further investigations of the mechanics of frozen ground.

In writing this book, the author has tried to present the analytical solutions, obtained for many problems in frozen soil mechanics, as clearly and concretely as possible so that they can be efficiently and extensively employed in practice for the development of permafrost regions. The solutions have been illustrated by the addition of numerical examples, while the test results are interpreted as the quantitative characteristics of certain cryogenic processes and as the indices of the mechanical properties of frozen and permafrost ground.

N. A. Tsytovich

PREFACE

Based upon analysis of the results of research performed over many years on the physicomechanical properties of freezing, frozen, and thawing soils and the cryogenic processes that take place in them, the present book establishes laws of frozen-ground mechanics as a science and their fundamental practical applications.

During the 40 years that have passed since publication of the first paper reporting a study of the mechanical properties of frozen soils (the author's article in Collection No. 80 of the KYePS AN SSSR [Permanent Commission for the Study of Natural Productive Forces of the USSR, USSR Academy of Sciences], 1930), and especially during recent decades, a series of solid monographs and many articles bearing on specific problems of frozen-ground mechanics have been published. Thus, the following well-known monographs have made a recognized contribution to the emergence and development of the mechanics of frozen soils as a science: the first book on frozen-soil mechanics, "Fundamentals of the Mechanics of Frozen Ground" (N. A. Tsytovich and M. I. Sumgin, USSR Academy of Sciences Press, 1937); "Principles of the Mechanics of Frozen Ground" (N. A. Tsytovich, USSR Academy of Sciences Press, 1952), which was awarded a State Prize in 1950 "For Development of the Fundamentals of Frozen-Ground Mechanics"; Fundamentals of the Mechanics of Freezing, Frozen, and Thawing Ground (N. A. Tsytovich *et al.*, vol. II, chap. 3 of the collective publication "Fundamentals of Geocryology," USSR Academy of Sciences Press, 1959); "The Rheological Properties and Bearing Strength of Frozen Ground" (S. S. Vyalov, USSR Academy of Sciences Press, 1959); and "A Study of the Deformations of Frozen Ground" (part II of Tsytovich's doctoral dissertation, 1940).

Finally, note should be taken of Yu. K. Zaretskiy's analytical work on the rheology and consolidation of frozen and thawing ground (Frozen Soils Mechanics Group of the NIIosnovaniy [Scientific Research Institute of Foundations and Underground Structures]), which he published in his doctoral thesis "Problems in the Theory of the Creep and Consolidation of Soils and Their Practical Applications" in 1971) and in various journal articles.

G. V. Porkhayev has published a highly important work on the thermophysics of frozen and permafrost soils, whose solutions are absolutely indispensable for prediction of the thawing depths of frozen soils and, from them, the settling of thawing soils: "Thermal Interaction of Buildings and Structures with Permafrost" (Nauka Press, 1970).

Interesting material will also be found in collections of Moscow State University papers: "Frost Investigations" (8 issues, 1961–1968); "Materials of the Eighth All-Union Conference on Geocryology" (8 issues, Yakutsk, 1966);

"Materials of the Conference-Seminar on Construction on Permafrost Soils" (20 issues, Krasnoyarsk Promstroy-NIIproyekt [Scientific Research and Planning Institute for Industrial Construction], 1964); the Collections of the Institute of Permafrost Studies, Siberian Division, USSR Academy of Sciences; the transactions of VNII-1 [All-Union Scientific Research Institute-1] (Magadan, 1961, 1963), and many separate articles in engineering journals, which have run to more than 5,000 printed pages over the past decade alone.

Finally, 1967 saw the publication of a section of official "Construction Norms and Rules" (SNiP II-B. 6-66, more than 52 pp., under the title "Bases and Foundations for Buildings and Structures on Permafrost Soils") with a supplementary "Design Handbook" (compiled under the direction of S. S. Vyalov and G. V. Porkhayev), within a volume of 133 pp.

To be sure, the limited size of this book prevented the full encompassing of the extensive problem circle connected with frozen ground mechanics. It is therefore natural that certain problems are not illuminated to the extent the author would have liked to provide, and not all the useful conclusions possible have been extracted from them.

However, the author feels that the time has come to analyze the fundamental results of many years of research done by both Soviet and other investigators on mechanics of frozen ground in the light of the most recent experimental and theoretical data, and to draw general conclusions after establishing the theoretically and practically important laws of frozen-soil mechanics.

The author formulates only the fundamental laws and relationships that enable us to comprehend the most complex problems of frozen-ground mechanics and legitimize methods for stable construction on permafrost; here, for the most part, he sets forth the fundamental aspect of the calculations and devotes much attention to the physical side of the mechanical processes that take place in freezing, frozen, and thawing soils and their analytical description; the construction of foundations and detailed elaboration of actual construction methods are treated in the textbook "Bases and Foundations."

The systematic exposition of frozen-soil mechanics will aid readers, scientific workers, engineers, and students in familiarizing themselves with the enormous literature on the mechanics of frozen soils—a matter of great importance for engineering geologists engaged in field surveys for construction in permafrost regions, and especially for construction engineers who use data from frozen-ground mechanics in their practice.

Knowledge of the laws of frozen-ground mechanics is also necessary for evaluation of the physicogeological phenomena and processes taking place during freezing, frozen, and thawing strata; treatment of these processes is beyond the scope of the present work since they constitute one of the problems of general geocryology (permafrostology), which is the subject of specialized works (see, e.g., B. N. Dostovalov, V. A. Kudryatsev, et al., "General Permafrostology," Moscow State University Press, 1967), they are taken into consideration in the present study but not expounded on in detail.

The present volume consists of two parts: Part One, general experimental and theoretical, which sets forth laws of frozen-ground mechanics that can be used both in study of geocryological processes in rock massifs and in evaluation of frozen rocks as bases and environments for various types of engineering structures, and Part Two, the applied, which is devoted to the development of a scientific basis for stable construction of various types of structures (residential and industrial buildings, highways, and hydrotechnical structures) on perma-frost.

The work does not pretend to be an exhaustive coverage of the problems of frozen-ground mechanics, but merely summarizes and generalizes the results of experimental and analytical investigations over approximately the past 40 years, which even now permits the successful use of the results in practical engineering survey and construction work under permafrost conditions.

N. A. Tsytovich

THE MECHANICS OF FROZEN GROUND

INTRODUCTION

Difficulties in the erection of structures on permafrost. There are extensive regions in the northern and northeastern USSR, covering an area of about 11 million square kilometers, and regions of nearly equal total area in other countries (northern Canada, Alaska) in which the ground and the rocks in general thaw only to a shallow depth (about 1-3 meters) in summer, and are permanently (eternally) frozen at greater depths (down to 50-1000 meters).

Structures built on permafrost without observing special and extraordinary measures and methods are subject to totally unacceptable deformations that make it difficult to use them and may result in their total destruction.

Because of the presence of ice-cement bonds, frozen and permafrost soils are quite strong and stable natural formations as long as their temperatures remain below freezing. But as the temperature rises and falls (and even if it remains in the subfreezing range), significant changes take place in the properties of the soils, giving rise to property instability of frozen soils; when interstitial ice melts, ice-cement structural bonds are catastrophically lost, substantial deformations arise, and high-ice permafrost soils with silty and clayey composition turn to liquified masses.

In the case of relatively dry (low-ice content) frozen soils, thawing of the interstitial ice lowers structural stability by breaking ice-cement bonds between particles, increases substantially the compressibility and water permeability of the thawed soils, and lowers their strength to a significant degree. On the other hand, when icy and high-ice (ice-supersaturated) soils thaw, their structure disintegrates catastrophically with an abrupt change in porosity, and rapid local (where thawing takes place) settlement occurs, often accompanied by extrusion of liquified masses of thawed soils away from loaded surfaces (e.g., from beneath the foundations of structures).

Settling is the principal cause of the unacceptable deformation of buildings erected on permafrost soils without special measures to prevent their thawing under the structures.

Construction in the permafrost regions, especially in the far north, appears to be even more complicated due to supersaturation of the upper soil layers with moisture or, in the frozen state, with ice. Figure 1 shows a cut opened in the course of dirt road construction in the Aldan region which is characterized by moisture supersaturation of the upper soil layers resulting in great excavation difficulties. Besides that, ice-supersaturated layers observed generally at the depth of thaw penetration, on top of permafrost, turn upon thawing into liquified masses.

Many years of practical construction work on permafrost soils has taught us that when such soils thaw under buildings and structures, significant and often

Fig. 1. Dirt road construction in a permafrost
region.

totally unacceptable deformations arise in the foundations and superstructures.
Thus, according to P. D. Bondarev, who investigated buildings erected on
permafrost soils in the city of Vorkuta and its district, it was found that about
80 percent of the buildings had unacceptable deformations, among them: 78
percent severely deformed stone (brick) buildings (full destruction of walls in 6
percent of the buildings, presence of gapping crevasses in 67 percent); 75 percent
of wood-log buildings and 90 percent of wood-frame buildings. Among the stone
buildings investigated, 30 percent had deformations of catastrophic character
and required immediate major repairs. Of the wooden buildings, 50 percent
needed major repairs and restoration.

Figure 2 shows the deformation of a slag-concrete boilerhouse and its leaning
smokestack (as a consequence of nonuniform settling of the thawed base).
Figure 3 shows a cross section through the building and the ground under it with
permafrost boundaries indicated in its original position (1945) and after two
years (1947) of operation of the boiler installation without provisions for
removal of the heat which penetrated the ground. The improper foundation
design (without heat removal) resulted in substantial deformations of the
building—in two years, the boilers settled 40 centimeters, the concrete floors
sagged, and the concrete foundation of the smokestack settled unevenly and
caused the stack to lean toward the building, which was the subsequence of the

Fig. 2. Deformations of slag-concrete boilerhouse building.

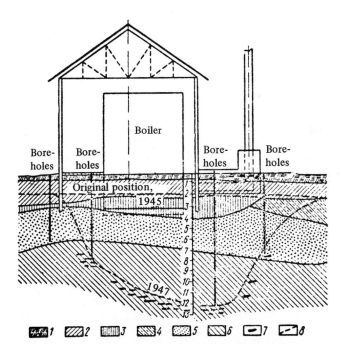

Fig. 3. "Thaw bowl" in frozen soils under the boilerhouse: (1) fill; (2, 3, 6) various loams; (4) sandy loam; (5) fine sand; (7) ice lenses; (8) permafrost boundary, original and after two years.

formation of an assymetrical "thaw bowl" in the ground under the boiler-house.

An example of strongly nonuniform settling, which caused large cracks in the floor of a building, appears in Fig. 4, which shows the interior of one of the buildings of a fish-processing plant in the settlement of Anadyr', which had been built without consideration for the properties of the high-ice permanently frozen soils. The deformations were so large that their further growth lead to the full destruction of the building.

Figure 5 illustrates differential settlement in the footings and walls of a wooden structure in the town of Khal'mer-Iu (north from Vorkuta) after one year in use. Deformations appeared as a consequence of settling and thaw of ice-rich ground at the bottom of the foundation.

Nonuniform high-magnitude frost heaving of the upper soil layers and buckling of foundations during freezing of the soils surrounding them, which is especially severe in cases when there is water influx from the outside or in the presence of the frequently observed over-moistened strata at the upper boundary of the permafrost, is another highly important cause of substantial deformations in structures built on permafrost soils. Thus, Fig. 6 shows a wooden yard gate in Yakutsk differentially heaved by frost action. The forces may develop an enormous magnitude and unless measures are taken to eliminate or control them they may cause nonuniform foundation buckling and severe deformations of structures. The magnitude of frost-heaving forces can be appreciated from Fig. 7, which shows a rather thick larch tree trunk split by a crack in a frost mound near Igarka.

It would be possible to cite many more examples of structures on permafrost that have been deformed as a result of uneven settling of thawing soils or sagging

Fig. 4.   Cracks in floor of bakery at Chukotka.

Fig. 5. Sagging of foundation and walls of apartment building in Khal'-mer-Yu settlement.

Fig. 6. Yard gate in Yakutsk tilted by nonuniform heaving of soils.

of soils with high ice content during thawing, or nonuniform foundations buckling by frost heave; such instances were common in permafrost regions until recently.

It is apparent that the development of theoretical premises and practical methods for stable construction on permafrost must be based on consideration of the characteristics of the permafrost regions, detailed study of the properties of freezing, frozen, and thawing soils, investigation of the mechanical processes

Fig. 7. Larch split by crack in a frost mound (photograph by A. P. Turtikov).

taking place under the influence of natural factors, their interaction with the structures, and a search for ways and means to adjust the properties of the soils in desirable directions.

However, it must be noted that it is possible even now to establish scientific bases for stable construction on permafrost on the basis of solutions that the mechanics of frozen ground has obtained by generalizing many years of experimental and theoretical research on the physicomechanical properties of freezing, frozen, and thawing soils; part 2 of the present work is devoted to this possibility.

We note that even the earliest domestic work (that of the present author from 1928 to 1930) on the design of foundations for structures to be built on permafrost, which was the first to be reported in either the domestic or the foreign technical literature, made it possible to map out a correct general approach and correct basic principles for construction on permafrost—a beginning that has since been fully vindicated in practice.

Thus, the use of subfloor spaces open to the wind, which the author first quantitatively substantiated back in 1928, made possible the development of a method now in widespread successful use: that in which structures are built on permafrost on the principle of preserving the frozen state in the bases.

One of the first structures built on this principle, with the author as consultant was the building of the Naryn high-altitude weather station in 1930 on the Petrov Glacier (Altay) and has given many years of excellent service with no structural deformation whatever (Fig. 8).

In 1931–1933, for the first time in construction practice, a large industrial building—the Yakutsk Central Thermal Power Station (YaTsES)—was constructed to plans based on the author's design formulas and in consultation with him on the principle of preserving the frozen state of its soil bases (Fig. 9). Since 1933 (i.e., for about 40 years), this building has been in operation with no

Fig. 8. Naryn high altitude weather station built on Petrov Glacier (1930).

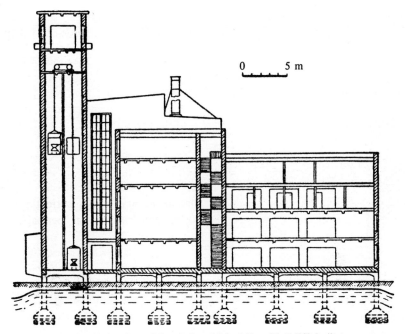

Fig. 9. Sectional drawing of first industrial building (YaTsES) built to preserve the frozen state of the ground bases.

inadmissible strains (with a rather high temperature maintained inside the building); moreover, the upper boundary of the permafrost under the building (according to examination data by the Yakutsk Scientific Research Station of the USSR Academy of Sciences) has not sunk, but has risen from 0.8 to 1.2 meters above its original level, thus increasing the stability of the foundations.

There are a number of other examples of successful completion of buildings and structures that are still in existence and were designed to allow for settling of the frozen soils on thawing, an effect that can be predicted on the basis of solutions from the mechanics of frozen soils, as we shall see later on.

In conclusion, we include a photograph of the main street in the subpolar city of Noril'sk–V. I. Lenin Prospekt–(Fig. 10), whose buildings were designed and constructed to preserve the frozen state of the base ground. These buildings are centrally heated, with hot and cold running water and all modern conveniences. We see from the illustration that even under the harsh conditions of the subarctic permafrost, these buildings, which were designed in conformity with the scientific fundamentals of the mechanics of frozen soils to take advantage of the useful properties of permafrost (when kept below the freezing point), are in good condition.

Principal concepts and definitions. The basic criterion of the freezing of soils, rocks, and particulate solids of all other types is the crystallization (at the

Fig. 10. V. I. Lenin Prospekt in Noril'sk, which was built to preserve the frozen state of the ground underneath.

appropriate negative temperatures) of ice in their pores. The crystallization of ice in the pores of soils is accompanied by a whole series of extremely complex physicochemical phenomena and processes: migration of water, congelation of mineral particles of the soil, an increase in the concentrations of the solutions in the pores, etc. These phenomena and processes form frozen soils with new properties different from those observed before freezing, with prime importance going to the process of cementing (cohesiveness) of the mineral particles by ice to one degree or another.

However, the pore water in soils and other particulate solids comes under the influence of the surface electromolecular forces of the mineral particles (this applies at least to the first few layers of water directly at the particle surface), with the result that not all of the water turns to ice (crystallizes) as the soil is cooled, but only those water layers the forces of whose interaction with the mineral particle surface are weaker than the crystallization forces of the ice at the particular subfreezing temperature.

If no ice forms in the pores of a soil as it is cooled, there is no reason for calling it a frozen ground. We shall therefore apply the terms frozen earth, frozen rocks, and frozen soils to earth, rocks, soils, and dispersed materials at negative or freezing temperatures when at least part of the water is frozen, i.e., turned into ice and mineral particles are cemented together. Rocks at subfreezing or freezing temperatures in whose pores ice has not crystallized will be referred to as chilled (or frosted) and supercooled if they contain strongly bound or mineralized water that does not freeze at the particular subfreezing temperature. All other rocks (ground, soils, etc.) that are at above-freezing temperatures will be called unfrozen, but if they were previously frozen for some time and then thawed out, we shall call them thawed rocks.

Thus, all soils ("uncohesive rocks") and other rocks (e.g., fissured and broken rocks) will be treated in three classes, depending on the phase composition of their pore water: (1) frozen rocks, (2) chilled rocks, and (3) unfrozen rocks.

We shall devote all of our attention below to investigation of the mechanical properties of frozen and permanently frozen ground and the mechanical processes that take place in them, i.e., to an exposition of the problems of frozen-soil mechanics.

We shall adhere to the following nomenclature of frozen ground (which we shall also apply to frozen fissured and broken rocks and soils) depending upon the time of their existence.

Term Used	*Time of Existence*
Permafrost	Centuries and millenia
Perenially frozen	From a few years (but no fewer than three) to several decades
Seasonally frozen	From one to two seasons
Briefly frozen	From a few hours to a few days (not occurring regularly)

If the time of existence of the frozen state of a rock (soil, etc.) is not to be reflected in the term used, we shall simply call them frozen (naturally frozen or artificially frozen).

Thus, in recognition of the firmly rooted popular notion of the "eternal frost," we shall call ground and other rocks that have been in the frozen state for a long time (centuries) permafrost ground and, in general, permafrost rocks. Rocks will be called perenially frozen if they have existed for many years—at least for one human generation. Finally, soils, topsoils, etc. that have frozen during one season (winter) and remain frozen for one to two years will be referred to as seasonally frozen. In the latter case, two-season frozen soils and other rocks are often called pereletok.*

We shall treat the permanently frozen state of rocks (soils, etc.) as a geological phenomenon (thus, the permafrost soil strata of northeastern Siberia range in age from 10,000–12,000 to 280,000 years), the perenially frozen state as a modern phenomenon, and the seasonally frozen state as a regular phenomenon that recurs from year to year, and, finally, the briefly frozen state as a phenomenon of irregular seasonal freezing that persists for an insignificant time.

*Term also accepted in English.

In our general description of frozen and permafrost soils, we stress once again that the quantity of prime importance in evaluation of their mechanical properties will be the degree to which the mineral particles are cemented together by ice (their cohesion), which can be evaluated for the most part in terms of the iciness of the frozen soils and the quantitative content of pore water that is not frozen at the particular subfreezing temperature in the frozen soils.

When frozen ground and permafrost are used as bases or as materials for various types of structures, it is important to establish at the very outset to which of the ice content and physical-state categories they belong.

Categories of Frozen Soils

By iciness

High-ice (ice content above 50%);

Low-ice (ice content below 25%);

Icy (ice contents from 25 to 50%).

By physical state

Hard-frozen (low-temperature) soils rendered firmly cohesive by ice and practically incompressible.

Plastic-frozen (high-temperature) soils with high unfrozen-water contents and relatively low compressibilities in the frozen state

In addition to the above categories of frozen soils, another category, that of friably frozen soils, is distinguished (in SNiP II–B, 6–66); these are uncohesive and coarse fractured soils that have not been cemented by ice even though their temperatures are below the freezing point and have very small water contents in their pores. These soils do not exhibit the specific properties of frozen soils, since there are practically no changes in their mechanical properties when the temperature rises above the freezing point. The author feels that it would be more correct to classify these soils not as frozen, but as chilled (frosty) uncohesive grounds.

Let us briefly describe the properties of the various soil categories.

High-ice content grounds, among which we include frozen ground and permafrost with volumetric ice content (the ratio of the volume of the ice present in the frozen soil to the total volume of the soil) above 50 percent and appropriate granulometric composition (clay, loam, or fine silt), acquire a fluid, fluid-plastic, or plastic state on thawing which often is connected with slumping. These soils are characterized by very low bearing capacity in the thawed state and by high compressibility (the usual coefficients of relative compressibility $a_r \geqslant 0.05$ cm^2/kgf).

Low-ice content grounds (with volume iciness below 25 percent) contain insignificant amounts of excess ice, and with clayey, loamy, or silty compositions thaw to a stiff plastic or semisolid consistency, with relatively low compressibility (relative compressibility coefficients on the order of $a_r \leqslant 0.01$ cm^2/kfg).

Icy grounds (volumetric ice content from 25 to 50 percent) have properties imtermediate between those of the two soil categories described above.

Hard-frozen grounds are firmly cemented (cohered) by pore ice and have temperatures low enough so that most of the water present in them has frozen. The approximate temperature limits of the hard-frozen state are listed in the SNiP: $-0.3°C$ for silty sands, $-0.6°C$ for sandy loams, $-1.0°C$ for loams, and $-0.5°C$ for clays; more precise values will be established in the exposition to follow. The last-mentioned limit ($-1.5°C$) applies (according to our figures) only for clayey soils whose particles are not too fine (and usually with fine particles having kaolinite mineralogical composition); for very fine clays (and especially those with the montmorillonite mineralogical composition), on the other hand, the stated limit should be lowered to -5 or even to $-7°C$.

Soils in the hard-frozen state are characterized by brittle fracture and are practically incompressible under loads smaller than 5–10 kgf/cm^2 (relative compressibility coefficients $a_r < 0.0001$ cm^2/kgf).

Plastic-frozen soils exhibit viscous properties in virtue of their high water contents (often amounting to more than half of the total pore water) and are characterized by relatively high compressibility in the frozen state. All high-temperature frozen clayey and loamy soils at subfreezing temperatures higher than the value characterizing the hard-frozen state of the soil are classified as plastic-frozen soils. This group also includes highly compressible ($a_r > 0.01$ cm^2/kgf) frozen soils that are completely saturated with ice and water, and also less compressible ($a_r \approx 0.01$–0.001 cm^2/kgf) high-temperature frozen soils with total water saturation (including all forms of water) below 0.8. Soils of this category may cause considerable foundation-base settling, even when the bases are kept frozen.

As we noted above, the greatest difficulties encountered in the erection of various types of structures in the North and Northeast come from beds of permafrost with especially high ice content in the plastic-frozen and high-temperature states.

Permafrost soils occupy about 49 percent of the area of the USSR (more than 10.5 million square kilometers), range in thickness from a few meters to 1.5 km (in the Vilyuy River valley), and are encountered at depths of 0.5 to 1 meter in the north of the permafrost region and at depths of up to 3.0–4.5 meters in the south.

The flagged line on the permafrost distribution map of the USSR (Fig. 11) indicates the southern extent of the permafrost, i.e., the boundary to the south of which, as a rule, islands of permafrost are no longer encountered, although

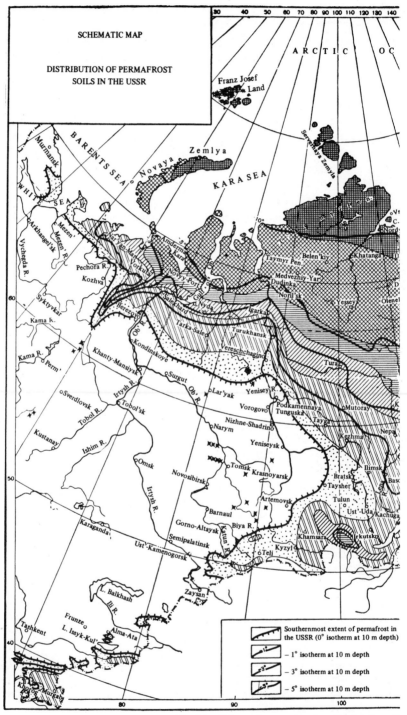

Fig. 11. Schematic map showing distribution of

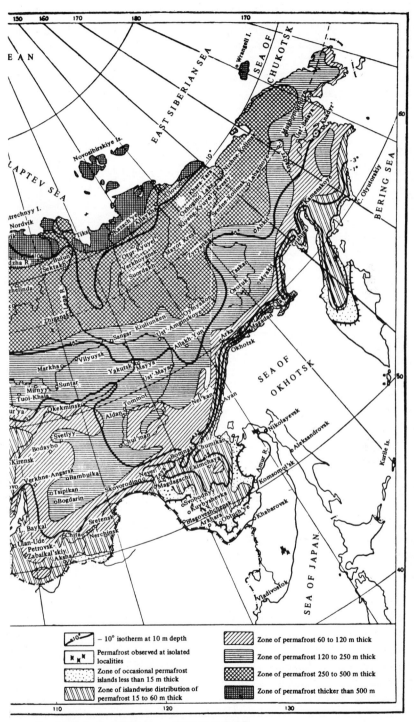

permafrost soils in the USSR (from SNiP–II–B 6-66).

occasional anomalies (e.g., those due to terrain relief and the direction in which cold air masses move) may still occur (see Fig. 11).

The map indicates, in addition to permafrost zones with various maximum thicknesses of the frozen layer (identified by shading differences at 60-500-meter intervals), also isolines of subfreezing temperatures at a depth of 10 meters below the surface. In determining the depths of soil thawing and in other thermal calculations for the bases of structures to be erected on permafrost soils, these temperatures are taken as established initial working negative temperatures at the depth corresponding approximately to zero temperature amplitudes (to within ± 0.1°C).

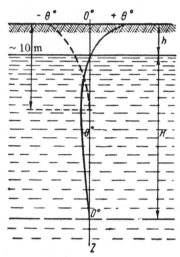

Figure 12 shows a schematic temperature section through the layer of permafrost soils. We see from the vertical temperature profile that the temperature does not remain constant down to the depth h corresponding to the seasonal thaw, but varies from positive to negative as the seasons change, and that the temperature amplitude in the seasonally frozen layer is higher the more continental the particular area of the permafrost region that is considered. Below the upper boundary of the permafrost layer (the maximum depth to which the summer thawing of the soil extends), the ground temperatures are always at or below zero centigrade, and the vertical variations of this temperature extend to 10 meters and somewhat more without leaving the subfreezing range (accurate to ± 0.1°C).

Fig. 12. Schematic temperature section through permafrost bed.

Finally, we observe, beginning at a certain depth, a gradual rise of the temperature of the frozen bed to 0°C, after which the temperature curve enters the positive range. We should note that in the lower zone of the temperature curve, a certain layer of soils will not be frozen even though it has a (slight) negative temperature because of the heavy pressure exerted upon it by the overlaying layers, i.e., it will not contain ice in its pores because the freezing temperature of water is depressed by a rise in pressure. This layer of soils will contain only supercooled water that does not freeze at the particular temperature.

The distance H from the top to the bottom of the permafrost bed determines its thickness.

The depth of the summer thaw (h in Fig. 12) for the permafrost-soil region and the corresponding depth of winter frost h' for regions outside the limit to which permafrost extends are highly important characteristics of the zone of seasonal soil-temperature variations from positive to negative and vice versa, i.e., the zone that is subject to periodic freezing and thawing and (as will be discussed in detail in later chapters) a whole series of attendant complex physical and physicomechanical processes and phenomena (sometimes of enormous intensity), effects that determine, among other things, the foundation depths specified for structures and some of the design features of the foundations. In engineering practice, the layer of annual winter frost and summer thaw is known as the active soil layer.

Origins and development of the mechanics of frozen soils in the USSR. Investigation of the mechanical processes that occur in the active layer as it freezes and thaws and in the permafrost layer under the influence of external disturbances and in its upper layers in particular, study of the strength, stability, and deformability of freezing, frozen, and thawing soils and the stress-strain interaction of the structures with permafrost are all within the problem complex of the mechanics of frozen soils. The mechanics of frozen soils grew out of the practical necessities of opening up the North and Northeast of the USSR, with expansion of highway, residential, industrial, hydraulic-engineering, and other types of construction in the permafrost regions.

Although the phases in the genesis and development of frozen-soil mechanics in the USSR can be demarcated only somewhat arbitrarily, this is done here on the basis of landmarks that are, in our opinion, important: establishment of fundamental relationships or publication of scientific papers illuminating various problems in frozen-soil mechanics, generalizing earlier research and summarizing results that have been obtained and have an important bearing on the solution of practical problems. In reviewing the individual periods, we shall draw attention to the development of the various concepts, propositions, and ideas that have contributed to further progress in the mechanics of frozen ground.

Five periods in the genesis and development of frozen-ground mechanics in the USSR can be distinguished.

The first period (before 1927) also includes prerevolutionary times. During this period, it was merely observed that study of the mechanical properties of frozen and permanently frozen ground was necessary, but in the absence of any data from direct experiments, the strengths of frozen soils were equated to the average compressive strength of river ice (N. S. Bogdanov, 1912; V. N. Pinegin, 1924, and others).

The second period (from 1927 through 1937) was one of systematic experimental investigation of the mechanical properties of frozen and permanently frozen ground, growing out of the USSR's development of capital construction in the permafrost region: the Petrovsk-Transbaykal metallurgical plant, the Baykal-Amur trunk railroad and other facilities; here the rule was to use known testing methods that have been developed in the theory of materials strength, regarding the frozen soils as continuous single-component solids.

During this period, the first data on the compressive, tensile, shear, and torsional strengths of frozen soils were obtained as functions of their composition, moisture content, and magnitude of negative temperature; the first experiments were performed to study the total settling of frozen soils on thawing with lateral confinement and under local loads (the author's experiments at the Leningrad Construction Engineering Institute (LISI) and the USSR Academy of Sciences). The need to consider phase transformations of water in freezing and thawing soils in thermal foundation design was also demonstrated (S. S. Kovner, 1933, and others).

The end of this period saw the publication, on the basis of a generalization of many experimental studies of the physicomechanical properties of frozen soils, of the monograph "Fundamentals of the Mechanics of Frozen Soils" by N. A. Tsytovich and M. I. Sumgin (vol. 27, 537; USSR Academy of Sciences Press, 1937).

The third period (1937-1947)* was a period of generalization of experimental studies and establishment of physical foundations for the mechanics of frozen soils.

Here we should note first the monograph of V. P. Veynberg (1940), which generalizes many years of research on the physical and mechanical properties of ice, that basic component of frozen soils, and the monograph published by a group of authors headed by M. I. Sumgin in 1940 on the general natural-history aspects of permafrostology.

During the same period, the highly important proposition that any frozen soil, regardless of how far below freezing its temperature is, will always contain a certain (sometimes significant and sometimes very small) amount of unfrozen water than remains in this state indefinitely as long as external conditions remain constant was first established on the basis of an analysis of a multitude of tests of the mechanical properties of frozen soils, study of the nature of their strength, and specifically designed experiments (N. A. Tsytovich, 1940).

In 1945-1947, this author proposed a theory of dynamic equilibrium between unfrozen water and ice in frozen and permanently frozen ground which explained and generally permitted estimation of the mechanical property changes of frozen soils under the influence of external disturbances; this theory is one of the fundamental physical premises of modern mechanics of frozen ground.

A series of monographs that generalized previous experiments in the erection of various types of structures on permafrost was published at the end of this period on the basis of results from many years of field observations and numerous publications in the technical journals (N. I. Bykov and P. I. Kapterev, 1940; A. V. Liverovskiy, 1941; N. A. Tsytovich, 1941, and others).

The fourth period (1947-1959) was one in which results were summarized with revision of premises on a new physical basis (the theory of the equilibrium state of water and ice in frozen soils, structural glaciology, etc.).

---

*Here we have made minor changes in the chronology of frozen soil mechanics in the USSR that we proposed earlier. (*Trudy S.O. NIIOSPa*, no. 3, 1967.)

At the beginning of this period, the migration of water not only in freezing topsoils, but also in frozen ground (under the influence of a temperature gradient) was established under field conditions in the interesting doctoral thesis of I. A. Tytynov (1947). This confirmed the theory of the equilibrium state of water and ice in frozen soils that had been advanced earlier.

In 1952, a monograph by N. A. Tsytovich (USSR Academy of Sciences Press) stated the fundamental principles of frozen-ground mechanics and, in particular, drew attention to the importance of research on the fluidity* of frozen soils under load.

The fluidity of frozen soils, their rheological properties, and their bearing capacity were the subjects of S. S. Vyalov's doctoral dissertation (1957) (which was later published as a separate book entitled "The Rheological Properties and Bearing Strengths of Frozen Soils," USSR Academy of Sciences Press, 1959).

The publication, during the same period, of P. A. Shumskiy's monograph, "Fundamentals of Structural Glaciology" (USSR Academy of Sciences Press, 1955) made possible a new approach to the study of the mechanical processes in freezing soils and, in particular, to the formation of underground ice in frozen soils, a matter of some practical importance.

At the end of this period, the scientific staff of the V. A. Obruchev Institute of Permafrostology of the USSR Academy of Sciences devoted a great deal of attention to the compilation of the joint work "Fundamentals of Geocryology (The Study of Frost)"; vol. 1, "General Geocryology," and vol. 2, "Engineering Geocryology" (USSR Academy of Sciences Press, 1959), devoted to applied problems in geocryology, such as prediction of thermal interactions between structures and permafrost soils, the hydrothermal melioration of permafrost soils, problems in the mechanics of soils, bases, and foundations, the laying of underground pipelines and roadbeds, mining operations, etc.

Publications generalizing experience gained in the construction of foundations appeared during the same period (N. A. Tsytovich's monograph, "Bases and Foundations on Frozen Soils," USSR Academy of Sciences Press, 1958, a series of article collections, etc.).

In the fifth period (1959-1970), the materials accumulated previously, chiefly by the staff of the Obruchev Institute of Permafrostology of the USSR Academy of Sciences (prior to its rebasing in 1961), were put to use in practice and new regional research programs were set up (at Yakutsk, Noril'sk, Igarka, Krasnoyarsk, and Magadan).

This period saw an expansion of research on the rheology of frozen soils, which was published in the collections "Strength and Creep of Frozen Soils" of the Central (1962) and Siberian (1963) Institutes of Geocryology.

Also during this time, a number of reference aids were compiled on the construction (Yu. Ya. Velle, V. V. Dokuchayev, and N. F. Fedorov) and design of bases and foundations (V. V. Dokuchayev) and also a new edition of the

---

*Defined on next page as "relaxation and creep."—Ed.

"Structural Norms and Rules" (SNiP II-6-66, 1967 edition) and its attached "Handbook of the Design of Bases and Foundations for Buildings and Structures on Permafrost" (which was compiled at the NIIosnovaniy under the supervision of S. S. Vyalov and G. V. Porkhayev with the participation of the author and others, 1969).

The projected future development of the mechanics of frozen soils will proceed through the development of new research based on the most recent advances in the general mechanics of continuous and multiphase bodies and related disciplines, with the use of modern precision techniques of experimental physics, with a trend to more profound study of the physical nature of the mechanical processes and their progress in time (compaction, creep, flow, heaving, failures of strength and stability, etc.) in freezing, frozen, and thawing soils under the influence of external factors and the soil interaction with structures.

Fundamental principles. As the theoretical basis of engineering geocryology (engineering frost science), the mechanics of frozen soils provides a physical basis for understanding the phenomena and processes that unfold in freezing, frozen, and thawing soils and is therefore of considerable practical importance.

We submit the following as the fundamental elements of frozen-ground mechanics, which we shall substantiate in detail below on the basis of analysis of the enormous factual material presented by the results of many years of research:

1) the dynamic equilibrium between unfrozen water and ice in frozen soils

2) migration of water on freezing of soils as a result of disturbance of phase equilibrium

3) the fluidity (relaxation and creep) of frozen soils under prolonged action of loads

4) mechanical properties instability of frozen soils,

5) the compactibility of high-temperature frozen soils under load

6) structural instability, compressibility, and settling of icy frozen soils on thawing, and

7) nonlinearity of the strength changes of thawed weak clayey soils on compaction under load

The individual elements of frozen-ground mechanics will be discussed not in isolation from one another, but in their interaction as the properties of the soils in their various states are characterized (in the process of freezing, in the frozen state, and in the thawing process).

The mechanics of frozen ground will be built on the basis of these elements and will, in turn, serve as a base for the establishment of scientific methods for the stable construction of various types of facilities on permafrost and the use of artificial ground freezing in construction and mining.

Part One

EXPERIMENTAL AND THEORETICAL

Chapter I

THE NATURE OF FROZEN GROUND;
ITS PECULARITIES AND
PHYSICAL PROPERTIES

1. Frozen Ground as a Multicomponent Multiphase System of Interconnected Particles

Frozen and permanently frozen ground are highly complex multiphase natural formations consisting of components that differ in their properties, are in a variety of phases (solid, ideally plastic, liquid, gaseous), and are bound to one another. Such soils can be regarded as single-component (continuous) bodies only under certain conditions, e.g., when no redistribution of the individual phases occurs with time in a given volume of frozen soil.

As was noted in the introduction, frozen soils always contain, in addition to solid mineral particles, ice as a rock-forming mineral and one or another quantity of unfrozen water and gases.

While the classical mechanics of particulate bodies is the mechanics of the single-phase particle system (the influence of air in communication with the atmosphere is neglected), the mechanics of soils that are fully saturated with water that is for the most part free, unbound, and hydraulically continuous, and without gas content (the so-called ground mass) is the mechanics of a two-phase system, and the mechanics of soils whose pores contain vapor and gases (in addition to water) is the mechanics of a three-phase system, the frozen ground mechanics deals with an even more complex system of particles that is, at least a four-phase system which contains: (1) solid mineral particles, (2) ideally plastic ice inclusions (cementing ice and interlayer ice), (3) water in the bound and liquid states (Fig. 13), and (4) gaseous components—vapors and gases. All of these constituents are related to one another in ways that depend both on the properties of the individual phases and on the levels of external disturbances.

2. Basic Constituents of Frozen Soils

The following must be regarded as the basic constitutents of frozen soils: solid mineral particles, viscoplastic ice inclusions, liquid (unfrozen and strongly bound) water, and gaseous inclusions (vapors and gases).

The solid mineral particles of frozen soils exert an essential influence on the properties of the soils, which depend not only on the sizes and shapes of the

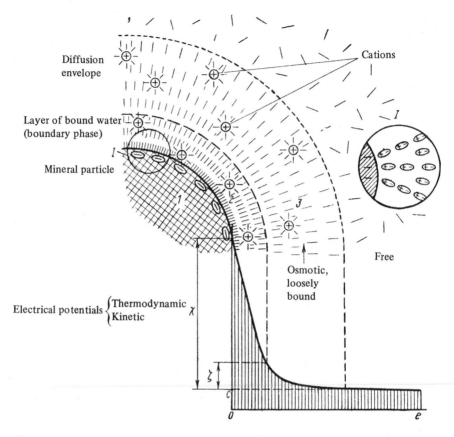

Fig. 13. Scheme of electromolecular interaction of the surface of a mineral particle with water: 1) mineral particle; 2) bound water; 3) loosely bound (osmotic) water.

mineral particles, but also on the physicochemical nature of their surfaces, which is determined chiefly by the mineralogical composition of the particles and the composition of cations that they have absorbed.

Frozen soils are classified on the basis of granulometric composition in the same way as unfrozen soils (SNiP II–B.1–62),* i.e., the following frozen soils are distinguished: coarse-fragment, sandy, and clayey (clay, loams, sandy loams).

But the properties of frozen soils are strongly influenced not only by the sizes of the mineral grains, but also by the shapes of the mineral particles. The magnitude of the local forces transmitted to the frozen soil from an external load depends on the shape of the solid grains. For example, an external pressure remains nearly untransformed at the points of contact between flat mineral

---

*The only difference from the case of unfrozen soils is that the word "silty" is prefixed to the usual term for the soil in the case of frozen soils that contain more than 50 percent of particles in the 0.05–0.005 mm (silt particles) size range.

particles of the kind sometimes observed in micaceous sands,* while it may reach enormous values in the case of sharp-angular granules (mountain sands). Thus, in an example cited in our book (see N. A. Tsytovich and M. I. Symgin, "Fundamentals of the Mechanics of Frozen Soils," Academy of Sciences USSR Press, 1937), the force at the contact of a rounded quartz grain 1 mm in diameter with a flat ice interlayer (at elastic moduli of 3×10^5 kgf/cm^2 for the quartz and 3×10^4 kgf/cm^2 for the underlying ice interlayer), calculated from the Hertz formulas known from elasticity theory, was about 1170 kgf/cm^2 under an external pressure of 2 kgf/cm^2; on the other hand, for a contact between two hard mineral grains of the same diameter (1 mm), the calculation shows that the forces would be several times larger. Such pressures can occur only in the elastic stage of deformation (e.g., under short-term loads); with the passage of time, the contact area increases and the forces at the contacts decrease as a result of creep of the contacting materials and especially of the ice interlayers, which "flow" even under very light pressures. However, the appearance of substantial pressures between mineral particles and ice and at points of contact between mineral particles no doubt has its effects on the properties of frozen ground, and in particular on their contents of unfrozen water (since ice melts under even a very light pressure), affects the structure of the soils, and increases friction between its particles, which in turn increases their shear resistance, etc.

The sizes of the mineral particles also influence the properties of frozen soils, chiefly with respect to the direction taken by physicochemical surface effects, whose rates also depend on the particle specific surface area, which, in turn, depends on the mineralogical composition of the soils. For example, the particles of kaolin clay have a specific surface area of 10 m^2/g, while those of montmorillonite clay have areas ranging up to 800 m^2/g, i.e., the particles in 1 g of a fine soil may have a specific area ranging from a few square meters to several hundreds and even thousands of square meters. Some minerals (for example, quartz, feldspar, and certain others) interact less actively with pore water, and others (montmorillonite, attapulgite, and others) much more weakly; there are also variations in the nature of the interaction, since the number of centers at which soil particles interact with their environment depends on their mineralogical composition and surface inhomogeneity.

The part played by the mineral component of soils is governed by the enormous energy of the chemical bonds between the surfaces of the mineral particles and the medium surrounding them, e.g., with the pore water and interstitial ice.

Unlike solid soil particles, *ice*, which is an inevitable component of frozen ground, represents a monomineralic cryohydrate rock with highly unique

---

*In construction-site excavations in the Transbaykal region (settlement of Tarbagatay), the author once observed dry sands consisting almost exclusively of minute micaceous plates, which endowed these sands with properties (compressibility, shear strength) similar to those of plastic clays.

physicomechanical properties that sharply differ from those of other rocks. Frozen soils may also contain other cryohydrate minerals (minerals that exist only at subzero temperatures), such as sodium carbonate (Na_2CO_3, freezing temperature $-2.1°C$), magnesium chloride ($MgCl_2$, freezing temperature $-3.9°C$), and others.*

All solid modifications of water are called ice, irrespective of their crystalline or amorphous state.

We presently recognize one amorphous modification of ice (which is formed on quick, very "deep" freezing), three modifications of ice (I, II, III) that exist at subzero temperatures and appropriate pressures, and modifications of ice (crystalline water) formed at pressures of several tens of thousands of atmospheres, which can exist not only at subzero, but also at above-zero temperatures.

According to Tamman's experiments (our source here is V. P. Vaynberg's monograph *Ice*, Gostekhteorizdat, 1940), ordinary crysdalline ice I turns into ice II (which is heavier than water) as its temperature is lowered and the pressure is raised up to 2,200 atm, and then to ice III at a pressure 2,236 atm and temperatures of -34 and $-64°C$. These transitions involve sharp changes in volume and absorption of tremendous amounts of heat.

The modification of ice that is most common in nature—ordinary ice I—exists at ordinary pressures and not very low temperatures (at least down to about $-100°C$).

Ice I is the most important component of frozen soils; its highly peculiar properties are responsible in large measure for the mechanical properties of the frozen soils. Ice I is a crystalline solid in the hexagonal system and exhibits conspicuous property anisotropy. It has its maximum viscoplastic deformation in the direction perpendicular to the principal optical axis, while the manifestation of rheological properties in the parallel direction is so weak that brittle fracture occurs right after elastic deformation. At the same time, ice always has viscoplastic strains (flow strains) under load, even at low stress levels. Appropriate experiments show that its viscosity can vary through a factor of 100 and more, depending on the direction of the forces. Ice exhibits elastic properties only under instantaneous loads, and its elastic limit (according to Vaynberg) is so low that the range of pure elastic strain is of no practical importance.

A special structural property of ice I is the mobility of the hydrogen atoms in its crystal lattice, which varies incessantly under the influence of external disturbances (variations of subfreezing temperature, pressure, etc.). Upon lowering the temperature, the mobility of the hydrogen atoms decreases and the ice acquires a more organized (denser and stronger) structure; at $-78°C$, the crystal lattice of ice enters a stable state (according to B. A. Savel'yev), and at

---

*The author speaks here about cryohydrates of sodium carbonate, magnesium chloride, etc.—Ed.

temperatures below $-70°C$, it changes from the hexagonal to the cubic system. With rising temperature, the activation energy of its molecules increases, accelerating their regrouping with weaker intermolecular bonds, the net effect of which is a decrease in the strength properties of the ice.

It follows from the above that under natural conditions, when there are always certain variations of thermodynamic conditions (temperature, pressure, etc.), the properties of ice (its structure, viscosity, etc.) may undergo considerable variation. These changes result in instability of the properties of both the ice and the frozen soils in response to any change in natural conditions. We note also that the surface electromolecular bonds of ice are considerably stronger than the molecular bonds of free water, so that free water is absorbed by the surface of the ice.

Water in the liquid phase is always present in one quantity or another in frozen and permanently frozen ground as unfrozen water at ordinary subzero temperatures (at least down to about $-70°C$), as was shown on the basis of theoretical considerations back in 1939[1a,b] and has since been fully confirmed by the results of direct experiments under both laboratory and field conditions.

We note that incomplete freezing of water in clay was observed in laboratory experiments by E. Young in 1932, but this phenomenon did not receive the attention that it merited.

Unfrozen water can exist in two states in frozen ground and permafrost:

1. Water strongly bonded by the surfaces of the mineral particles (with an activation-energy excess), in which case, because of the enormous electromolecular forces of the surface, the water cannot form the hexagonal crystal lattice of ice even at very low temperatures;

2. Loosely bound water of variable-phase composition (according to B. N. Dostovalov and V. A. Kudryavtsev[2] with an activation-energy deficiency), capable of releasing heat of crystallization, observable by calorimetry, and freezes at temperatures below $0°C$. The thinner the layers of loosely bound water, the stronger will be the effect of the soil mineral particle surfaces upon it and the lower will be its freezing temperature. According to Dostovalov and Kudryavtsev,[3] the following formulation would be more correct: "Bound water (loosely bound, of variable phase composition) freezes at the lower temperatures the more rapid the diminution of its bonds as compared with free water and the more rapid the formation of high-mobility zones, as a result of the opposing action of contiguous structures." The freezing temperature of loosely bound water is depressed because there forms, between the layer of bound water and the free water, a layer of less strongly bound and more mobile water that is, as it were, "warmer" than the free water, requiring more energy and a lower temperature for its crystallization.

The amount of unfrozen water in frozen ground and permafrost decreases as the soil temperature falls farther below zero; each soil type is characterized by a specific curve of unfrozen-water content.

The contents of unfrozen water in frozen ground and permafrost are determined on soil samples with a sensitive calorimeter, and the results of calorimetry are processed by considering that only frozen water (ice) releases its latent heat of fusion ($\sim$ 80 cal/g) on thawing, while unfrozen water has no latent heat of fusion.

The unfrozen-water content in frozen soils and its variations under the influence of external disturbances determine in many respects the physico-mechanical properties of frozen ground and permafrost and are of enormous importance in the physics and mechanics of frozen soils.

The gaseous components of frozen soils are water vapor (when the frozen soil is not fully saturated by water and ice) and gases.

The water vapor in frozen soils may be important in some cases, since, as we know, it is displaced from points with higher vapor pressure (determined, for the most part, by temperature) to points with lower vapor pressure and may, in water-unsaturated soils, be the basic cause of moisture redistribution during temperature variations and freezing of the soils.

As for the gases, their role in frozen soils is reduced to the mere generation of porosity and, if closed gas vacuoles are present, to an elasticity increase.

All the individual components listed in the present paragraph—solid mineral particles, ice, unfrozen and strongly bound water, vapor, and gases—endowed with their own specific properties, interact with the others in frozen soils in ways that are determined primarily by the force fields of the mineral-particle surfaces and the ice with water in various states. The strength of the interaction depends on the specific surface area and physicochemical nature of the solid soil components and the composition of their exchange cations, as well as external disturbances (temperature, pressure, etc.).

3. Peculiarities of Pore-water Freezing in Soils

The freezing of pore water in soils has certain features that arise out of the interaction of the water with the surface of the soil's mineral particles and the presence of a certain quantity of salts dissolved in the water.

As we know, free unbound water has a freezing point of $0°C$ at standard atmospheric pressure, but pore water, which is situated in the force field of the mineral-particle surfaces, especially in thin films of water (film water), will have a lower freezing temperature.

The freezing temperatures of soils (i.e., the crystallization temperatures of the pore water that they contain) are of enormous importance for determination of the depth of freezing and thawing of soils, i.e., for establishment of the zone in which various important physicomechanical processes unfold under natural conditions and in interaction with structures: frost weathering with changes in the internal bonds of the soils, cracking, frost heaving, etc. Because of these factors, the layers of the annual freeze and thaw are unreliable as bases for structures, and it is important to regard the ground freezing depth as one of the most important parameters determining the depth below grade of structural foundations (outside of the permafrost region).

Different soils have different freezing temperatures (from -0 to $-2.5°C$ and slightly lower); here the term applies to the stable freezing temperature of the pore water (after the temperature jump if supercooling took place) which is accompanied by an increase in the volume of the soil, segregation of ice, cohesion of part of the soil by freezing, etc.

Let us consider the process in which the temperature of a soil is lowered until the pore water in it begins to crystallize, the freezing process of the soil, and further intensification of the freezing and chilling of the frozen soil.

We should note that the freezing of soils also depends on external conditions: the magnitude of the negative temperature, the rate at which it is lowered, etc. Thus, for example, the author established in experiments with stiff plastic clays (of the Kinelian stratum) that they began to freeze or frost-heave only at a certain initial gradient ($\Delta\theta \approx 5$ deg/cm), which is sufficient to overcome the internal resistance to migration of the water as it moves toward the freezing front.

P. I. Andrianov[4] made a detailed study of the freezing temperatures of various soils with disturbed structure under laboratory conditions; he showed that as a soil specimen is cooled from all sides (the soil was frozen in a test tube inserted in a Dewar flask filled with a freezing mixture), there is an initial supercooling $-\theta_s^o$ and then a sharp rise in temperature—a temperature jump (Fig. 14) resulting from release of the latent heat of ice formation—and finally, after the temperature has reached a certain value (the freezing temperature), freezing of the entire specimen occurs. The highest and most persistent temperature $-\theta_f^o$ observed at the temperature jump corresponds (according to P. I. Andrianov) to the freezing temperature of the soil.

For soils with various mechanical compositions and approximately the same moisture content (about 20 percent by weight), freezing temperatures from -0.03 (for a soil with 0.21 percent hygroscopicity) to $-1.56°C$ (hygroscopicity 4.62 percent) were obtained on cooling from -4.5 to $-7.5°C$.

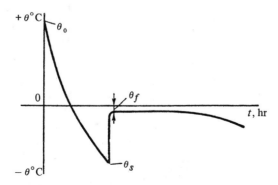

Fig. 14. Cooling and freezing curve of soil.

A later study of soil freezing temperatures in the USSR was organized under the supervision of the author by A. P. Bozhenova[5] at the Central Permafrostology Laboratory of the USSR Academy of Sciences; Bozhenova used the same technique as Andrianov, but the entire process of chilling and freezing was recorded automatically on photographic paper with the aid of a mirror galvanometer. We present some of the cooling and freezing curves that were obtained for characteristic soil types.

Figure 15 shows the freezing curve of a water-saturated Lyubertsy quartz sand. This curve can be regarded as consisting of several segments. The first segment I corresponds to cooling and supercooling of the soil sample, i.e., only temperature lowering of soil without ice formation.

In both this and other similar cases, this segment of the cooling curve is convex toward the temperature axis. The largest supercooling temperature in this experiment was $\theta_s = -3.0°C$. We note that the supercooling temperature depends on the conditions of the experiment, and chiefly on the overall heat budget of the soil sample being frozen.

Later, as soon as the pore water has begun to freeze in the soil sample, a significant amount of latent heat of ice formation is released and the temperature of the soil rises sharply (segment II in Fig. 15). In this particular case (water-saturated sand), this rise reached 0°C ($\theta_f \approx 0°C$). At this temperature, all the free water present in sand freezes; the time of freezing depends on the moisture content of the sand, the cooling rate, and the dimensions of the sample. The experiments also show that the freezing temperature, i.e., the stable temperature reached at the temperature jump, is near 0°C for all moist and water-saturated sands (segment III on the curve of Fig. 15).

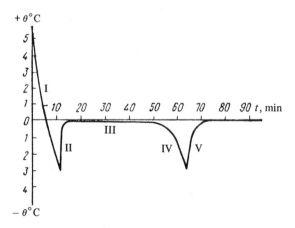

Fig. 15. Cooling and freezing curve of sand (moisture content $W = 19.6$ percent, temperature of the refrigirant $-10°C$).

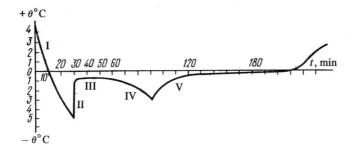

Fig. 16. Chilling and freezing curve of fine bentonite clay (moisture content $W = 80.5$ percent; cooling temperature $-10°C$).

If freezing of the soil continues, its temperature begins to drop after a certain time has elapsed—at first along a curve (to -0.5 to $-1°C$ for water-saturated sand) whose convexity faces away from the temperature axis (and not toward it, as during chilling). This indicates that a certain amount of the latent heat of ice formation is still being released on this segment, with freezing of loosely bound water (of variable phase composition). But at temperatures below $-1°C$ (for this particular experiment), cooling proceeds linearly (segment IV on the curve of Fig. 15), i.e., the sand may be regarded as frozen, with practically all of its water having turned to ice, and even if a certain insignificant amount of unfrozen water has remained, it no longer has any practical effect on the heat budget.

As the temperature is raised (segment V on the curve of Fig. 15), the temperature varies linearly at first, but then in curvilinear fashion (for sands at temperatures from approximately -0.5 to $-1°C$), i.e., absorption of the latent heat of ice formation begins even before the thawing temperature of the soil has been reached.*

As for the value of the supercooling temperature ($-\theta_s$ in Fig. 14), it is not a constant for a given soil and is observed in soil samples (in laboratory experiments) and in the upper layers of natural soils only when chilling is gradual and the soil contains no ice crystals ("ice seeds," e.g., snowflakes). But if precipitation of ice has begun at some point, supercooling will no longer occur, but the freezing temperature of the soil (which is, generally speaking, below zero degree) will depend on the mechanical and mineralogical composition of the soil, the saturation of its particles by various cations, its natural porosity density, etc.

All of the above is fully applicable to clayey soils as well (clays, loams, and sandy loams), but these soils will freeze differently in that a lower temperature (on the order of -0.1 to $-2.5°C$ or slightly lower, depending on the properties of the soil) is established after the temperature jump when they are cooled, and the curvilinear segment along which the already frozen soil is further chilled and

*The reader may find that certain observations by the author hold true only for the particular framework of the experiment.—Ed.

frozen (segment IV in Fig. 16) extends considerably farther, and complete freezing (the practically frozen state) may be attained in fine clays at very low temperatures (on the order of several tens of degrees below zero), when the temperature curve has become a straight line, i.e., cooling of the now frozen soil occurs with no appreciable release of latent heat of ice formation.

Table 1 shows certain results from the described laboratory experiments to determine the freezing temperatures of soils.

We note also that the freezing points of salt soils also depend on the salt concentration K_z in the pore water, which is equal to the ratio of the weight of the dissolved salts to the weight of the pore solution (including the salt content), and, according to LenmorNIIproyekt can be determined from empirical formulas; the average values, e.g., for salt loams and sandy loams, will be approximately as follows:

$$At\ K_z = 0.01, \text{freezing point } \theta_f = -0.5°C$$
$$At\ K_z = 0.05, \text{freezing point } \theta_f = \ \ 2.2°C$$

The freezing temperatures of the soils will be even lower at higher salt concentrations in the pore water. Thus, the freezing temperatures of various soils (or, more precisely, the temperature during the first main period of freezing) with definite physicochemical and physicomechanical properties (natural degree of compaction, moisture contents, consistencies, etc.) are quite well defined, and this must be remembered in freezing calculations for soils: the freezing point will be near 0°C for water-saturated sandy, fluid, and fluid-plastic clayey soils; it will average −0.1 to −1.2°C for plastic clayey soils, depending on compaction; it will range from −2 to −5°C for semihard and hard clays. The above data must be adjusted for specific conditions on the basis of results from direct experiments and field observations.

Data on the freezing temperatures of soils are needed for thermal-engineering calculations related to the freezing of soils, and especially with determination of the freezing depth during a given time interval at a given subzero temperature of the outside air.

Table 1. Soil freezing temperatures

Designation of soil	Compaction, kgf/cm²	Temperature of cooling mixture, °C	Soil moisture content, wt-%	Temperature, °C	
				Super-cooling	Freezing
Medium-grain sand	1	− 10	20.5	− 3.0	0.0
Loam (20.8% content of particles < 0.001 mm)	0.5	− 10	32.2	− 4.0	− 0.1
The same	10	− 10	19.6	− 3.3	− 0.9
Clay (bentonite)	1	− 10	80.5	− 4.9	− 0.7

As we know (see SNiP II-B.1-62), the maximum seasonal freezing depth of the soil under natural conditions determines the minimum burial depths of foundations for residential and industrial buildings and, outside of the perma-frost region, the thickness of the so-called active layer, i.e., the layer in which a whole series of physical and physicomechanical processes capable of influencing significantly the strength and stability of structural foundations come actively into play.

For permafrost regions, the active layer will be determined not by the maxi-mum freezing depth of the soils, but by their maximum thawing depth. The latter is the depth of complete melting of the ice present in soils of the annual frost layer overlying the permafrost massif.

The temperature of complete thawing of frozen soils will be near $0°C$ unless the soils are saline. The soil thawing depth and, consequently, the thickness of the active layer is determined for permafrost by the depth of penetration of above-zero and zero temperatures into the frozen soil.

The thawing depth can be determined by a thermal-engineering calculation or, in approximation, from isoline charts of seasonal soil thawing (Figs. 17 and 18),[6] which indicate thawing depths near the maxima for sandy soils with mois-ture contents (in fractions of unity) $W_d = 0.05$ and for clayey soils at $W_d = 0.15$. The thawing depth will be less at higher moisture contents, as determined from the graphs of Figs. 17 and 18 by multiplying the thawing depth by $K_w = f(W_d)$.

For permafrost regions, as we noted earlier, the maximum thaw depth will also correspond to the thickness of the active layer, which has the following typical values (see Figs. 17 and 18):

For the Far North

For sandy soils . . . . 1.0–1.8 m
For peat bog and clayey soils . . . . 0.4–1.2 m

For the Southern Region

For sandy soils . . . . 2.5–4.5 m
For clayey soils . . . . 1.0–2.5 m

It follows from the above data that the freeze and thaw of soils (especially disperse clays) are extremely complex physical phase change processes of water in the frozen soils, with a strong influence on the properties of freezing, frozen, and thawing soils and therefore require special, more detailed analysis in the exposition to follow.

4. Characteristic Ranges of the Phase Transitions of Water to Ice, and the Unfrozen-water Contents in Frozen Soils

On freezing of soils, and especially fine soils (e.g., clayey soils), by no means all of the pore water change to ice at the soil's freezing temperature, but only part of it. With further lowering of the negative temperature, phase transitions of

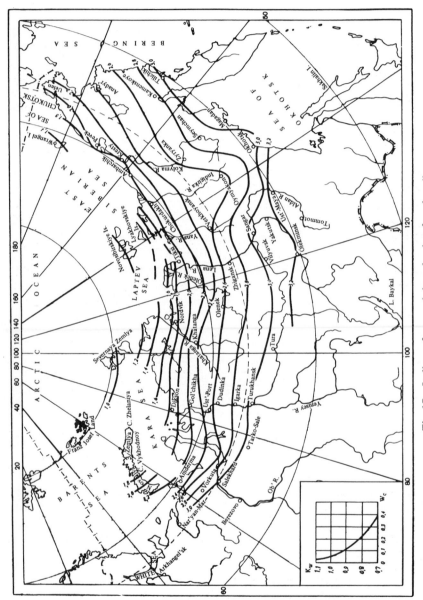

Fig. 17. Isolines of seasonal thaw depths of sandy soils.

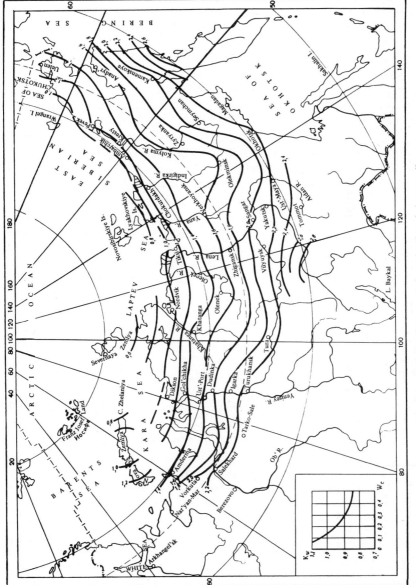

Fig. 18. Isolines of seasonal thaw depths of clayey soils.

the water continue, but at steadily decreasing rates, with the quantity of freezing water dependent both on the magnitude of negative temperature (the main factor) and on the specific surface area of the mineral particles, the composition of absorbed cations, pressure, etc.

As indicated above, the freezing temperature of pore water falls due to opposing effects of attraction forces to the zone of strongly bound water and water of variable phase composition, since (according to B. N. Dostovalov and A. A. Ananyan) a layer with more mobile molecules—a "warmer" layer that requires a lower freezing temperature, appears between them.

Studies of unfrozen-water content as a function of degrees of negative temperature, carried out mainly in the Laboratory of Frozen Soil Mechanics of the Obruchev Permafrostology Institute, USSR Academy of Sciences, under the author's supervision, provide the means to distinguish three principal phase-transition ranges for the water in frozen soils:[7]

1) a range of significant phase transitions, in which the amount of unfrozen water W_u changes by 1 percent or more for every $1°C$ (with respect to the dry weight of the soil);

2) a transitional range, in which the variations of unfrozen-water content are smaller than 1 percent but larger than 0.1 percent;

3) the range of the practically frozen state, in which the phase transitions of the water to ice do not exceed 0.1 percent per $1°C$.

In the range of significant phase transitions of the water in frozen soils, i.e., on the initial regions of the graphs of unfrozen-water content W_u plotted against the negative temperatures $-\theta°$ of the soils (Fig. 19), all of the free water (that in large pores and capillaries) freezes, along with some of the loosely bound water; in the transitional region, on the other hand, the water of variable phase composition freezes (the loosely bound water), and the range of the practically frozen state corresponds to the

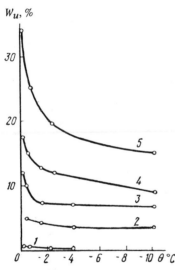

Fig. 19. Graph of unfrozen-water content in frozen soils as a function of negative temperature: (1) quartz sand; (2) sandy loam; (3) loam; (4) clay; (5) clay containing montmorillonite.

content of strongly bound water only, which closely approaches the maximum hygroscopicity of the soil. But at $-70°C$ and below, the overwhelming majority of soils are practically completely frozen (the liquid phase of the water has been converted entirely to ice).

Calorimetric experiments performed to determine the unfrozen-water contents of frozen soils also showed that for each type of soil there is a

Table 2. Unfrozen-water contents of salt free soils vs. degrees of negative temperature*

Designation of soil	Amount of unfrozen water, in percent of weight of dry soil, as a function of temperature in °C					
	$-0.2-$ -0.5	$-0.5-$ -0.5	$-1.0-$ -1.5	$-2.0-$ -2.5	$-4.0-$ -4.5	$-10.0-$ -11.0
Sand	0.2	0.2	–	–	0.0	0.0
Sandy loam	–	5.0	4.5	–	4.0	3.5
Loam	12.0	10.0	7.8	–	7.0	6.5
Clay	17.5	15.0	13.0	12.5	–	9.3
Clay (containing montmorillonite)	34.3	25.9	–	19.8	–	15.3

*According to the experiments of Z. A. Nersesova (see N. A. Tsytovich and Z. A. Nersesova, Physical Phenomena and Processes in Freezing, Frozen, and Thawing Soils, Collection No. 3, "Reports on Laboratory Studies of Frozen Soils," N. A. Tsytovich, ed., Academy of Sciences USSR Press, 1957.

characteristic graph of unfrozen-water content, i.e., a curve of the unfrozen-water content W_u against negative soil temperature $-30°$ (see Fig. 19 and Table 2), and that it has the same importance for estimation of the physical and mechanical properties of frozen ground and permafrost as the compression curve has for unfrozen soils.

We see from Table 2 and the graphs of Fig. 19 that the finer (more clayey) the soil, the larger the amount of unfrozen water it contains at a given subzero temperature. This is understandable when we realize that finer soils have larger specific surface mineral-particle areas and, consequently, have a greater capacity for the binding of pore water.

According to research done at the Moscow State University (T. A. Litvinova), the unfrozen-water contents of extremely fine soils depend not only on their porosity, but also on their microporosity (the internal porosity of their mineral particles).

Detailed studies made by Z. A. Nersesova,[8] who used a sensitive calorimeter to determine unfrozen-water contents in frozen soils, showed that the amount of unfrozen water is practically independent of the total soil moisture content for a given soil (if, of course, the soil moisture content exceeds its molecular moisture capacity.–N.T.)* and is determined principally by its negative temperature (this is also confirmed by the experiments of P. Williams).[9]

This proposition has an important practical application: by determining the total moisture content of the soil and its negative temperature under field conditions (as is mandatory for all engineering cryological surveys) and the amount of unfrozen water in a permanent laboratory on a moistened sample of the same soil with disturbed structure, it is possible to calculate the amount of

*Here and elsewhere, N.T. means author's note.

unfrozen water and the iciness of the soil in its natural state as functions of temperature. These characteristics are needed for determination of the physical properties and evaluation of the state of frozen soils to find whether the SNiP recommendations for construction on permafrost soils can be applied.

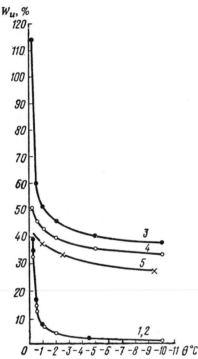

Fig. 20. Amount of unfrozen water in clays as a function of exchange-cation composition and negative temperature ($-\theta°$): (1) Ca-kaolin; (2) Na-kaolin; (3) Na-allophane; (4) Ca-allophane; (5) Fe-allophane.

The studies also showed that in fine clays, in which exchange cations have a strong influence on particle size, water properties, etc., the content of unfrozen water at a given subzero temperature depends on the saturation of the clays with various exchange cations, e.g., saturation with the Na ion endows montmorillonite clay (allophane) with the finest particle sizes and the highest contents of unfrozen water. Figure 20 shows graphs of unfrozen-water content in single-mineral soils when their exchange complexes are saturated with various cations: Na, Ca, and Fe.[10]

We should note that when allophane is saturated with iron (Fe), in which case coagulation and aggregation are most conspicuous and the ultraporosity is lowest, the amounts of unfrozen water at various temperatures will be substantially smaller than when allophane is saturated with the sodium ion.

The above data indicate that the amounts of unfrozen water are different when the absorbing complex of the allophane is saturated with sodium and iron, even in the range of the practically frozen state. For fine kaolin, saturation of the absorbing complex with sodium or calcium ions has practically no effect on the unfrozen-water content.

The influence of water-soluble compounds in soils is manifested both in a freezing-point depression and in a decrease in the thickness of the water envelopes on the colloidal soils particles, depending on the composition of the water-soluble compounds.

The complexity of the phase transitions of water in soils stems both from the complexity of the composition and structure of fine soils and from the structural features of the film water and ice.

5. Dynamic Equilibrium of Unfrozen Water and Ice in Frozen Soils

Study of the mechanical properties of frozen soils began back in the 1930's, but until 1945 (when the principle of dynamic equilibrium of water and ice in frozen soils was formulated),[11] no general theory capable of explaining the origin of various mechanical-property changes of frozen soils and the effects of external disturbances (temperature, pressure, etc.) upon them had been proposed. It was usually assumed that severe frosts "welded" the frozen soil and ice together more strongly, but exactly what constituted this "welding" was unclear. It was also assumed that to freeze something meant to produce a "petrified" material that did not change over very long times under ordinary variations of temperature, pressure, etc.

However, investigation of the properties of frozen soils showed that they are extremely sensitive to changes in subzero temperature and undergo abrupt changes on thawing. These changes are so extensive that it is absolutely necessary to reckon with them for practical purposes.

Thus, according to our experiments, the increments in the short-term compressive strength of frozen clay per $1°C$ were 9.6 kgf/cm^2 in the range from -0.3 to $-1.0°C$, 4.5 kgf/cm^2 from -1.0 to $-5°C$, and 3.8 kgf/cm^2 from -5 to $-10°C$. Exactly the same picture is observed for the case of sandy loam, where the short-term compressive strength changed by amounts ranging from 16.0 to 4.8 kgf/cm^2 per $1°C$, depending on the temperature range.

In determining the compressive strength of frozen soils, the author found that when samples of frozen sand are kept frozen for one day, they had a compressive strength of 48 kgf/cm^2 at $-5°C$; but after three and five days at the same temperature, the figures were 59 kgf/cm^2 and 64 kgf/cm^2, respectively.

Thus, even at constant temperature, the time in the frozen state also influences the mechanical properties of frozen soils, especially immediately after freezing.

Another factor that strongly influences the properties of frozen soils is the time of action of the load: the more rapidly the load is applied, the stronger they are, and vice versa. Under slow loading with holding until the strains from the particular degree of load have become constant, the strength of frozen soils drops off considerably, an effect that we shall examine in detail below.

We note that the rigidity of frozen soils decreases with increasing external pressure, with a strong effect on the normal elastic moduli of the materials.

It is obvious from the above examples that frozen soils must be regarded as materials that are extremely sensitive to changes in external influences.

To establish the physical causes of a change in the mechanical properties of frozen soils and the laws that control these changes, it is necessary to immerse ourselves in study of the nature of frozen soils, with prime emphasis on the variations of their ice content and unfrozen-water content under the influence of temperature and other external influences.

A certain amount of liquid-phase (unfrozen) water is invariably present in any particulate frozen soil, even at very low temperatures (at least down to $-70°C$). The appropriate experiments indicate that any change in the amount of unfrozen water in frozen soils causes quite distinct and sometimes even major changes in their mechanical properties. Even in frozen sands containing insignificant amounts of unfrozen water, a change in this water content on depression of the temperature is manifested in a change in strength properties.

This is because the unfrozen water is concentrated in extremely narrow chinks, capillaries, and contact points between mineral grains, and any change in the amount of this water strongly affects cohesion between the mineral particles.

On the other hand, the quantity of unfrozen water present in a given frozen soil (and, consequently, its iciness and the cementing of the particles by ice) is not constant, but varies with variations of external factors.

Negative temperature variations and variations in the external pressure have a particularly strong influence on the amount of liquid-phase water present in a frozen soil and, consequently, also on its ice content. In addition, there are also other, often considerably weaker effects, such as migration of film water in frozen soils, the time since freezing, etc. The influence of film-water migration on the unfrozen-water content of a frozen soil must be considered only if the soil is subject to temperature gradients that persist over very long times, since film migration is a very slow process.

Temperature of the frozen soil. As we indicated under the previous heading, the lower the temperature, the smaller will be the unfrozen-water content of a soil, and the higher will be its ice content. Thus, according to experiments that we published earlier, Moscow-area topsoil (with a moisture content of about 30 percent) contains 74 percent unfrozen water (and was accordingly 26 percent ice saturated) at $-1.6°C$; at $-17.4°C$, the unfrozen-water content was 57.4 percent (ice saturation 42.6 percent). Similarly, a medium-grain sand had only 2 percent of unfrozen water at $-0.5°C$ and less than 0.2 percent at $-10°C$ (percentages referred to the weight of the dry soil).

Without dwelling in detail on analogous data, we note only that all of our experiments and those of numerous later investigators indicate that the unfrozen-water and ice contents of a frozen soil vary as functions of negative frozen-soil temperature, following curves that are quite well defined for each soil type (see, for example, Fig. 19).

However, lower negative temperatures not only lower the unfrozen-water content of a soil, but also its composition changes, since the amount of ice in the soil increases due to attachment of molecules of pure water to ice crystals, thus increasing the salt concentration in the unfrozen water and depressing its freezing temperature even further. The physical properties of the unfrozen water also change as the temperature goes down, e.g., its viscosity increases. Thus, according to Dorsey, water has a viscosity of 1.91 poises at $-2°C$, 2.14 at $-5°C$, and 2.60 poises at $-10°$; at $+20°C$ the viscosity of water is one poise.

All of the above indicates that both the amount of unfrozen water in a frozen soil and its composition and properties change as the temperature is lowered through the negative range.

The author has shown[12] that any rise in the subzero temperature thaws part of the ice in any frozen soil, even in the negative temperature range.

Young's[13] results from a study of the hysteresis of water content in a clayey soil as it was frozen and thawed lead to similar results. The clays he investigated did not freeze at all at 0°C, but as the temperatures were lowered to −19°C, from 35.04 percent (at a total moisture content of 13.3 percent) to 89.48 percent (at a total moisture content of 33.1 percent) of their water froze. On thawing, on the other hand, the curve of the ice content of the frozen clay coincided almost perfectly with the curve for freezing, i.e., the experiments indicated that part of the pore ice thaws with rising temperature even in the subzero range.

The problem of ice thawing in soils at subzero temperatures was investigated in detail by Z. A. Nersesova[14] at the Obruchev Institute of Permafrostology of the USSR Academy of Sciences.

Table 3 presents the results of Nersesova's experiments to determine the iciness i (the ratio of the weight of ice to the total weight of the water present in the frozen soil) for three typical soils; these data are plotted as graphs in Fig. 21.

The data of Table 3 and Fig. 21 not only confirm the previously established fact that ice melts in frozen soils as the temperature rises through the subzero range, but also indicate differences in the ice-content variations as the negative temperatures rise toward the temperature of complete thawing. Thus, in sand, where practically all of the water is free, practically no subzero thawing is observed (Fig. 21); in loam, thawing of pore ice begins at about −2° and is complete at 0°C; in the clay (whose water-extract residue amounted to 1.4

Table 3. Variations of relative ice content on freezing and thawing of soils

Temperature, °C	Sand, W_{tot} = 11–12%		Loam, W_{tot} = 21–22%		Clay, W_{tot} = 30–31%	
	Frozen	Thawed	Frozen	Thawed	Frozen	Thawed
− 0.3	0.94	0.97	0.19	0.42	0.0	0.0
− 0.5; 20.6	0.95	0.97	0.31	0.48	0.0	0.12
− 0.85; − 1.0 ..	0.97	0.99	0.43	0.55	0.0	0.21
− 2.1	0.97	0.99	0.53	0.62	0.20	0.28
− 3.0	−	−	0.55	0.63	0.30	0.37
− 4.4; − 4.5 ...	0.97	1.0	0.56	−	0.33	−
− 10.5	0.99	−	0.60	−	0.43	−
− 15.2	1.0	1.0	0.66	0.70	0.49	0.52

Note: W_{tot} is the ratio of the weight of water to the total weight of the soil (wet sample weight).

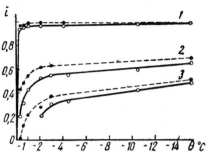

Fig. 21. Variation of soil ice contents during freezing and thawing: (1) sand; (2) loam, (3) clay. Solid line: freezing; dashed line: thawing.

percent, indicating some degree of salting), thawing of the ice began at $-3°$, and the clay had thawed out completely at $-0.3°C$.

Thus, the ice- and unfrozen-water contents of frozen soils are in dynamic equilibrium with the effective negative temperature.

We note that similar data indicating partial thawing of pore ice with rising subzero temperature have also been obtained for soils with natural undisturbed structure. As we noted earlier, the temperatures of the onset of freezing and thawing of frozen soils depend on the initial temperature gradient, i.e., the unfrozen-water content at a given point in time may also depend on the effective temperature gradient.

The external pressure may, as we noted earlier, affect the unfrozen-water contents of frozen soils. The author designed several special experiments to establish the fact that external pressure influences unfrozen-water content.

Two perfectly identical samples of a fine clayey soil were compressed until no further compaction occurred under the particular load, and were then placed for freezing in a refrigeration chamber, where one of the samples was frozen under no load and the other under the same pressure at which it had been compacted.

Table 4. Effects of external pressure on unfrozen-water content in frozen Moscow-area toploam (clay-particle content 20.8%) according to experiments of N. A. Tsytovich

Preliminary compaction, kfg/cm²	Temperature, °C	Moisture content (on wet specimen), %	Unfrozen-water content, %	
			Under load	Without load
2	-24.2	22.7	–	50.2
2	-26.9	24.4	61.5	–
10	-28.6	21.5	–	72.7
10	-27.8	19.0	74.2	–

The results of these experiments (Table 4) indicate that external pressure strongly increases the unfrozen-water contents of frozen soils, and that this effect is larger the higher the pressure under which the soil is frozen. The external-pressure effect will be much less conspicuous in very dense, fine clayey soils that contain water only in the form of thin films and insignificant amounts of free water, since the layers of film (loosely bound and bound) water may experience such strong electromolecular attraction from the surfaces of the mineral particles that the action of an external pressure may not be sufficient to overcome it.

The above experimental data indicate that even at low temperatures, external pressure affects unfrozen-water content (and, consequently, also the iciness) of a frozen soil.

The relationship established by the author (in 1940) between the unfrozen-water contents of frozen soils and the effective external pressure has also been confirmed in the very recent (1969) experiments of P. Hoekstra,[15] according to which the following variations (ΔW_u) of the unfrozen-water content (expressed as a fraction of the weight of the clay) were obtained for lower (weak) London clays on an increase in the external pressure by 100 kgf/cm²:

At temperature $-1°C$ change $\Delta W_u = 0.071$
At temperature $-1.2°C$ change $\Delta W_u = 0.067$
At temperature $-1.5°C$ change $\Delta W_u = 0.026$
At temperature $-3.0°C$ change $\Delta W_u = 0.018$
At temperature $-10.0°C$ change $\Delta W_u = 0.005$

Hoekstra also confirms our tenet to the effect that the creep of frozen soils is strongly related to the presence of unfrozen water and its movements under the influence of the stress gradient.

Thus, on the basis of all of the above, we arrive at the following proposition, which the author regards as one of the fundamental principles of the mechanics of frozen soils:[16] the quantity, composition, and properties of the unfrozen water and ice present in frozen soils are not constant, but change with variations of external factors, with which they are in dynamic equilibrium.

This principle, which has come to be known as the principle of the equilibrium state of water and ice in frozen soils, establishes a dynamic equilibrium between the amounts of unfrozen water and ice in frozen soils and the values of external factors: temperature, pressure, etc.

The principle of the equilibrium state of water with ice in frozen soils also explains the physical nature of the basic disturbances responsible for changes in the physicomechanical properties of frozen soils.

In the opinion of this author, this principle is the tenet that can be applied successfully in investigation of frozen and permanently frozen soils. Thus, frozen-soil mechanics is quite familiar with such facts as the strength increase of frozen soils with decreasing negative temperature, the decrease in the normal elastic moduli of frozen soils with increasing external pressure (which we shall discuss in detail later on), the dependence of the plastic properties of frozen soils on negative temperature, etc. All of these phenomena (and many others) are explained basically by the principle of the equilibrium state of water and ice in frozen soils, especially if changes in physicomechanical properties occur in the range of intensive phase transformations or in the transitional range.

We note that the principle of dynamic equilibrium of water and ice in frozen soils should be regarded only as a natural conformity reflecting the most important interactions in the phase compositions of frozen soils, such as those with temperature, pressure, etc. In separate cases, however, it is necessary to consider the effects of other factors on the phase composition of the water and the mechanical properties of frozen soils, e.g., migration of film moisture, variations of the mobility of hydrogen atoms in the ice crystal lattice, the viscosity of the pore water, its dissolved salt content, etc. However, it is our view that these influences will not affect the basic dependence established by the principle of the equilibrium state of water and ice in frozen soils.

6. The Physical Properties of Frozen Soils

As we noted earlier, frozen soils are four-component systems of interrelated particles (solid mineralic particles, plastic-ice, liquid-unfrozen-water, and gaseous). Thus, while a single characteristic—specific gravity—is sufficient for determination of the basic physical properties of a single component (such as massive rocks) or quasi-single–component (e.g., loose materials, if the air communicating with the atmosphere is neglected) systems, two-component systems (gas free soils material) require two indicators—specific gravity and moisture content—and three-component systems require three characteristics—specific gravity, bulk density, and moisture content; frozen ground and permafrost, as four-component particulate systems, require determination of a minimum of four fundamental indicators for experimental evaluation of their physical properties and states:

1) The bulk density of the frozen soil with natural undisturbed structure* γ in gf/cm^3;

2) The total moisture content W_d of the soil (expressed as a fraction of unity);

3) The specific gravity γ_{sp} of the solid particles in gf/cm^3;

4) The unfrozen water content W_u (expressed as a fraction of unity) or, as a substitute, the relative ice content i, which is the ratio of the weight of ice to the total weight of water present in the frozen soil.

The bulk density γ of the frozen soil, that most important characteristic of the natural degree of compaction of frozen soils, is determined on solid blocks of frozen soil that have been taken out with a special coring tool, or by

---

*American researchers use the word texture.—Ed.

hydrostatic weighing of a piece of frozen soil in chilled kerosene, or, finally, by the method that we regard as best of all: by digging a small pit to a given depth (for example, to foundation depth below grade) with careful collection of all of the excavated soil and subsequent weighing of this soil (for example, in a box or barrel) and measurement of the side lengths of the excavated parallalepiped with a rule. Needless to say, the bulk densities of frozen soils must be determined at negative air temperatures.

Then, for any of the above methods, the bulk weight of the frozen soil is

$$\gamma = \frac{g_{so}}{V_{so}} \tag{I.1}$$

where g_{so} is the weight of the frozen-soil sample with undisturbed structure and V_{so} is the volume of the same soil sample.

The total moisture content W_d is best determined by the "groove" method: a pick is used in the excavation to open a trench of frozen soil in depth to the total thickness of the soil layer whose total moisture is to be determined, and the total moisture content W_d is determined on the material from that trench. Continuous ice interlayers exposed in plan (not ice lenses) are measured separately if they are thicker than 0.5 cm.

We note that SNiP II-B.6-66 and its attached "Handbook" recommend differentiated determination of the moisture contents in frozen ground and permafrost: moisture due to ice inclusions W_i; the moisture of interlayers of frozen soil between layers of ice W_s; moisture due to pore ice cementing mineral particles W_c; moisture of unfrozen water W_u, and, finally, the total moisture content W_d, which will be equal to

$$W_d = W_i + W_s = W_i + (W_c + W_u) \tag{I.2}$$

The moisture W_s is determined from carefully selected samples of the mineral soil (that between the ice interlayers), and the ice-inclusion moisture W_i from test measurements on the ice interlayers or from (I.2) if the values of W_s and W_d have been determined.

It is our view that these recommendations are valid only for scientific research studies, since it is extremely difficult to follow the instructions for experimental determination of moisture in mineral interlayers of frozen soil situated between thin ice inclusions and lenses of ice, i.e., W_s.

However, the SNiP permits equating the value of W_s for clayey soils to the moisture content at the plastic limit, i.e., $W_s \approx W_r$.

Concerning the unfrozen-water content W_u in frozen soil, it is always necessary, in our view, to determine this quantity in survey work by the thoroughly developed calorimetric method. As we indicated above, this determination may be made in stationary laboratories on soil samples with disturbed structure at the natural subzero temperature.

SNiPII-B.6-66 includes a tabulation of coefficients that can be used to determine the unfrozen-water content W_u approximately from the negative temperature $-\theta°C$ of the frozen soil, its plasticity index W_p, and the moisture content of the soil at the plastic limit W_r, using the formula

$$W_u = k_u W_r \tag{I.3}$$

where the coefficient $k_u = f(W_p, -\theta°)$, i.e., it is a function of the plasticity index and soil temperature.

As for the specific gravity γ_{sp} of the solid particles of frozen soils, it is determined by the conventional pycnometer method on soil samples with disturbed structure.

Thus, the following quantities must be determined to characterize the physical properties of frozen soil and permafrost: for the undisturbed structure, the bulk density γ of the frozen soil and its total moisture content W_d and, on samples with disturbed structure, the moisture W_u due to unfrozen water which corresponds to the natural temperature $\theta°C$ of the frozen soil and the specific gravity γ_{sp} of its mineral particles.

The most important phase-composition characteristic of frozen and permanently frozen soils is their ice content at the natural ground temperature.

The relative gravimetric ice content i (referred to the weight of the dried soil) is determined by the expression

$$i = \frac{W_d - W_u}{W_d} \tag{I.4}$$

Volumetric ice content, i.e., the ratio of the volume of ice present in the frozen soil (except for the thicker continuous interlayers, which are measured separately) to the volume of the frozen soil, is a significant characteristic of such soils.

Since the weight of water in a unit volume of soil equals the gravimetric moisture content of the soil multiplied by the bulk weight of the particles of the soil skeleton, we readily obtain the following expression for the volumetric ice content:

$$i_{vol} = \frac{\gamma}{\gamma_i} \times \frac{(W_d - W_u)}{(1 + W_d)} \tag{I.5}$$

where γ_i is the specific gravity of the ice.

We have recommended this very simple expression for the volumetric ice contents of frozen soils for extensive use in practice.

We should note that the SNiP recommend determination, as a classification index for frozen soils, of the so-called inclusion volumetric ice content I_{vol} due solely to ice inclusions (except for pore ice), which is defined as

$$I_{vol} = \frac{\gamma_{sp} W_i}{\gamma_i + \gamma_{sp} (W_d - 0.1 W_u)} \tag{I.6}$$

When formula (I.6) is used, it is difficult to make a sufficiently accurate determination of W_i, the moisture due to ice inclusions, i.e., lenses and interlayers of ice; in (I.5) and (I.6) γ_i is assumed equal to 0.9 g/cm^3 = 0.0009 kgf/cm^3.

In determining the total moisture of frozen soils, it is often more informative and more convenient to calculate not the gravimetric moisture content W_d (for a dry sample), but what we call the "total moisture" W_{tot} referred to the weight of the whole (undried) soil, since this makes it possible to avoid awkward frozen-soil moisture contents greater than 100 percent. For example, the frozen-soil gravimetric moisture content W_d = 200 percent will correspond to a total moisture of 66 percent, i.e., water and ice of all categories will account for 66 percent of the total weight of the frozen soil.

Since the total moisture content W_{tot} equals the ratio of the weight of the water to the weight of the whole soil, and recognizing that the weight of the dried soil (for 1 cm^3 of soil) will equal the weight γ_{sk} of its skeleton, and that the latter is, as we know, equal to $\gamma_{sk} = \gamma/(1 + W_d)$, we obtain

$$W_{tot} = \frac{W_d}{1 + W_d} \tag{I.7}$$

For a unit volume of frozen soil (1 cm^3), we shall have the weights of the following constituents:
for solid particles

$$g_{sk} = \gamma(1 - W_{tot}) \tag{I.8}$$

for ice

$$g_i = \gamma W_{tot} i \tag{I.9}$$

for liquid-phase water

$$g_w = \gamma W_{tot}(1 - i) \tag{I.10}$$

Needless to say, the sum of the weights of all constituents will equal the weight of a unit volume of the frozen soil, i.e.,

$$g_{sk} + g_i + g_w = \gamma$$

The above four indicators of the basic physical properties of frozen soils enable us to determine, in addition to the gravimetric and volumetric ice contents and the contents by weight of the constituents of the frozen soils, many other physical-property indicators: porosity coefficient, coefficient of water saturation (percentage moisture content), volume of gases per unit volume of soil, etc.

Formulas for calculation of the listed frozen-soil characteristics and determination of certain average values of these characteristics (excluding continuous ice interlayers of substantial thickness) are given in Table 5.

We note that when they are carried out without knowledge of the above properties of frozen soils, engineering calculations concerned with the use of frozen ground and permafrost as bases and settings for structures lose all

Table 5. Interrelation between indicators of fundamental physical properties of frozen grounds

Quantity determined experimentally	Quantity calculated from formulas
Bulk weight γ of natural undisturbed structure	Total moisture $$W_{tot} = \frac{W_d}{(1 + W_d)}$$
Specific gravity γ_{sp} of solid particles	Relative ice content (gravimetric) $$i = \frac{W_d - W_u}{W_d}$$
Total gravimetric moisture content (on dry sample) W_d	Volume ice content $$i_{vol} = \frac{\gamma}{\gamma_{ice}} \times \frac{W_d - W_u}{1 + W_d}$$
Moisture W_u due to unfrozen water (expressed as a fraction of the weight of the dry soil)	Bulk weight of soil skeleton $$\gamma_{sk} = \gamma (1 - W_{tot})$$ Void ratio of frozen soil $$\epsilon_f = \frac{\gamma_{sp} - \gamma_{sk}}{\gamma_{sk}}$$ Coefficient of water saturation (in SNiP, the degree of moisture G) $$I_w = \frac{W_d \times \gamma_{sp}}{\epsilon_f \gamma_w}$$ where γ_w is the bulk weight of water: $$\gamma_w = 0.001 \text{ kg/cm}^3$$ Weights of constituents (in 1 cm³) of frozen soil: solid particles $g_{sk} = \gamma(1 - W_{tot})$ ice $g_{ice} = \gamma W_{tot} i$ unfrozen water $g_w = \gamma W_{tot} (1 - i)$ volume of gases in 1 cm³ of frozen soil $$V_a = \frac{\epsilon_f}{\gamma} = \frac{W_d}{\gamma_w} \gamma_{sk}$$

relation to the actual soil conditions and yield results that are not sufficiently reliable.

Heat and mass transfer coefficients (according to A. V. Lykov[17-19] form a second group of physical characteristics of frozen soils, namely:

1) the thermal diffusivity

$$a = \frac{\lambda}{C\gamma_{sk}} \text{ m}^2/\text{hr} \tag{I.11}$$

which can be determined separately for the frozen and unfrozen (thawed) states of the soils with consideration of the phase transitions of water (here λ is the thermal conductivity in kcal/m $\cdot$ hr $\cdot$ deg, C is the specific heat in kcal/kg $\cdot$ deg, and the values of λ and C for frozen and thawed soils are given in Table 10 of SNiP II–B.6–66);

2) the coefficient of potential conductivity

$$a' = \frac{\lambda'}{C'\gamma_{sk}} \text{ m}^2/\text{hr} \tag{I.11'}$$

which is also determined separately for thawed and frozen soils (here λ' is the moisture conductivity in kg/m $\cdot$ hr $\cdot$ potential unit and C' is the specific "moisture capacity" of the soil in reciprocal potential units).

Values of the potential conductivity can be obtained only by performing special experiments to plot vertical profiles of moisture and ice content at various points in time after the onset of freezing (Fig. 22).

The following expression can be used to calculate the potential conductivity a':

$$a' = \frac{(\Delta\Phi_{\Delta t})_m}{\left[\dfrac{\Delta f(x)}{\Delta x}\right]_{\Delta t}}$$

where $\Delta\Phi_{\Delta t}$ is the area enclosed between two neighboring moisture-distribution curves (Fig. 22a) or ice-distribution curves (Fig. 22b), m is the number of cross

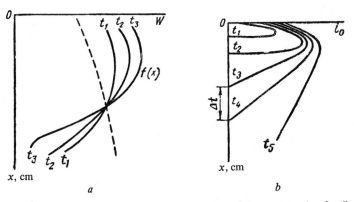

Fig. 22. Time curves of moisture content W and ice content i_v of soil after various time lapses since the beginning of freezing (according to Institute of Permafrostology laboratory experiments, 1956).

sections, $\Delta f(x)/\Delta x$ is the average moisture gradient during the time from t to $t + \Delta t$, which equals their arithmetic mean value.

Knowledge of the thermal diffusivities a is necessary for prediction of the depth of thaw or permafrost under heated buildings, while the potential conductivity a' is needed to calculate the migration of moisture during freezing.

7. Forms and Significance of Ice in Frozen Ground and Permafrost

We shall give more detailed consideration to the types of ice in frozen ground and permafrost as the basic component determining their frozen state.

In P. A. Shumskiy's classification,[20] ice is formed in frozen ground by three basic processes: (1) on freezing of moistened strata—constitutional ices; (2) on filling of cavities by ice—wedge and recurrent—wedge ices; (3) when snow and ice are buried—buried ices.

1. Constitutional ice is the most important one in shaping the structure and texture of frozen ground and permafrost, as a structure-forming factor for frozen soils.

This ice forms both interstitial (pore) ground ice and interlayers of pure ice encountered at depth in frozen soils.

It has been established by observations under natural conditions and by special laboratory experiments that thick ice interlayers and lenses are forming in freezing ground when the freezing level is held for some time at a certain depth, so that there is enough time for freezing of the film and capillary water, which is slowly drawn toward the freezing front. Such conditions are observed, for example, during thaws or when the freezing boundary fluctuates in a certain area in the soil and the ground receives an influx of ground water.

But if freezing occurs very quickly with large temperature gradients, as in severe frosts, there is insufficient time for the water in fine soils to migrate toward the freezing front, so that formation of continuous ice interlayers and lenses is difficult and only pore ice, which bonds the mineral particles firmly together, is formed.

If no appreciable expansion of the soil's mineral skeleton is observed on the formation of pore ice, the ice is called cement ice, to distinguish it from the other, considerably larger ice formations, which are known as inclusion ice or, to use Shumskiy's terminology, as segregational ice.

Depending on the severity of the frost (the temperature gradient) and the boundary conditions (i.e., whether freezing is strictly unilateral (one-dimensional problem) or the soil is frozen from several sides), the presence of water seepage and delays in the advance of the freezing boundary or their absence, a peculiar cryogenic (frost) texture forms in frozen soils during freezing which essentially determines their properties.

At least three basic types of frozen ground texture are distinguished (Fig. 23): fused (massive), laminar, and cellular (network).

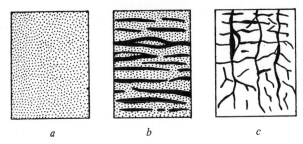

$$a \qquad\qquad b \qquad\qquad c$$

Fig. 23. Fundamental types of cryogenic textures of frozen soils: (a) fused; (b) laminar; (c) cellular.

The fused (massive) cryogenic structure* of frozen soils forms only on rapid freezing and is characterized by practically uniform distribution of the ice crystals (pore ice) in the frozen soil. Frozen soils with the massive texture exhibit the strongest resistance to external forces in the frozen state and settle least of all on thawing.

Frozen soils with laminar and cellular textures exhibit lower mechanical strength in the frozen state and settle most on thawing, with cellular (network)-textured frozen soils settling more on thawing than frozen soils with the laminar texture (at the same ice content).

In the case of frozen soils with cellular (network) texture, it is practically impossible to determine with any accuracy the moisture content of the frozen soil between the ice inclusions, i.e., W_s, or even the moisture content due to lenticular and interlayer inclusions of ice W_i and from it the ice content I_i due to the ice inclusions; it is therefore more advantageous to classify permafrost not on the basis of inclusion ice content (I_i), but on the basis of volumetric ice content (i_{vol}), which also includes the pore ice, so that it is necessary to increase slightly the ice-content limits recommended by SNiP II–B.6–66, assuming

for $I_i = 0.2$ correspondingly $i_{vol} = 0.25$
for $I_i = 0.4$ correspondingly $i_{vol} = 0.50$
for $I_i > 0.03$ correspondingly $i_{vol} > 0.05$

This proposal greatly simplifies determination of ice contents, since the permafrost soils are classified on the basis of a very simple relationship (expression (I.5)) that requires determination only of the bulk density γ of the frozen soil, its total moisture content W_d, and the moisture due to unfrozen water W_u, which are always determined in engineering surveys.

But if a permafrost soil contains areas extended into continuous ice interlayers that are so thick that they can be measured with sufficient accuracy with a millimeter rule (i.e., more than 5–10 mm in thickness), reference should

---

*The author uses "structure" and "texture" as synonomous terms.–Ed.

be made to the "Note" in Sec. 2.6 of SNiP II–B.6–66, and the ice contents I_i due to ice inclusions should be determined roughly from the results of direct measurements made on the ice inclusions in the process of the engineering-geological study of the soils.

In determining the volumetric ice content i_{vol}, on the other hand, full account is taken of thin interlayers, lenses, streaks, and other similar ice inclusions. Then the total thickness of the significantly large continuous ice interlayers is added to the value for the volumetric ice content of the frozen soil in the general evaluation of their ice content.

2. Wedge and recurrent–wedge ices are formed, as has been shown by research made over 50 years, on recurrent winter cracking of the upper layers of soil; for certain regions in the far north of the USSR, they sometimes account for 50 percent and more of the volume of the top 20 meters of soil, leaving a strong imprint upon the local relief and creating special construction difficulties in these regions.

Wedge and recurrent-wedge ices often occur in frozen ground and permafrost in the form of vertical *seams*, usually wedge-shaped and tapering down with increasing depth; these wedges are formed as a result of recurrent frost cracking of the soils with subsequent filling of the cracks by water, which then freezes.[21]

Recurrent-wedge ice is widely distributed over an enormous territory in the north of the USSR (Fig. 24).[22]

It has now been demonstrated (by numerous investigators, but primarily by P. A. Shumskiy in the USSR, R. Black in Alaska and Canada, and others) that major accumulations of underground ice are of ice-wedge origin.

Recurrent-wedge ices are most strongly developed in alluvial soils in river terraces, and may extend vertically for several tens of meters and be as wide as 10 meters across the top. These types of ice differ in composition from others in their abundance of impurities, for the most part soil, which arises out of the process by which they were formed—filling of frost cracks by ground water and mud. According to Shumskiy, the amount of solid mineral impurities in recurrent-wedge ices can range up to 3–5 percent of the total weight of the ice, and the volume of the cavities filled by gas up to 4–6 percent of the total volume. However, recurrent-wedge ices are only weakly mineralized and have the composition of the surface water.

In plan view, recurrent-wedge ice forms a polygonal lattice with side lengths varying over a broad range—from a few meters to 100 meters and more—and with 3 to 6 angles on a polygon, forming a frost-crack polygon terrain relief.

In the northern permafrost regions, recurrent-wedge ice is responsible for thermokarsts, a common phenomenon in the far north in which sink holes (karsts) are formed as a result of thawing of recurrent-wedge ices.

As a rule, thermokarst first appears at points of intersection of ice wedge polygons where the ice crops out of the tundra moss. For this reason,

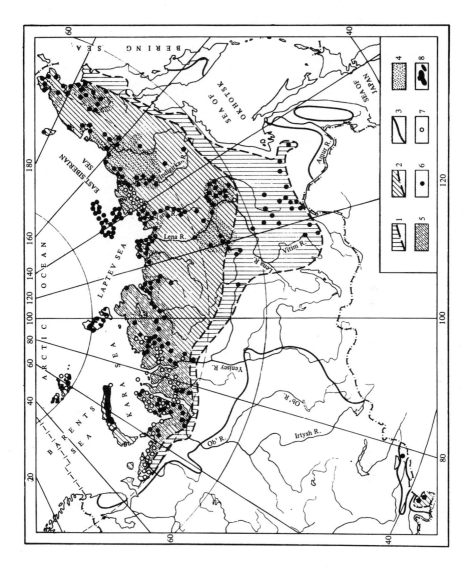

Fig. 24. Map showing distribution of recurrent-ice wedges in the USSR (prepared by P. A. Shumskiy and B. I. Vtyurin). (1) Range and southern boundary of fossil recurrent-wedge ices; (2) range and southern boundary of fossil and modern recurrent-wedge ices; (3) southern boundary of region of perennially frozen rocks; (4) regions of distribution of thick layers of icy (containing wedge ices) silty strata; (5) regions with silty icy deposits of shallow depths; (6) deposits of wedge ice according to surface investigations; (7) deposits of wedge ice according to aereal observations of ice wedge polygon relief; (8) glaciers.

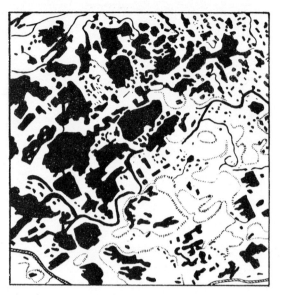

Fig. 25. Plan-view diagram of thermokarst lakes in
Anadyr' tundra.

thermokarst lakes formed by cave-ins are in many cases (according to S. P. Kachurin[23] of a partially rectangular shape (Fig. 25) resulting from the polygonal arrangement of the recurrent-wedge ices.

Engineering studies made in Anadyr' have established that thermokarst lakes originating from recurrent-wedge ices expand rapidly in plan and migrate if the terrain is level, remaining shallow, since the ice wedges taper rapidly out with depth. Constant erosion of sagging masses of soil and thawing of the wedge ice result in rapid retreat of the higher shore; the resulting deposit creates a new incompetent, which, after the water level falls, is broken up by recurrent-wedge ice into polygons, and the process of thermokarst formation is repeated.

Special experiments performed by S. V. Tomirdiaro[24] have shown that the following two situations must prevail simultaneously if a thermokarst is to form regardless of the severity of the Arctic conditions:

1) ground ice must be exposed to the sun, and
2) the melt water must remain in a depression in the ice.

Drill cutting samples from the thawed beds of thermokarst lakes have shown that the previously ice-saturated soils have undergone considerable self-compaction and consolidation. The soils of tundra regions are subject to large-scale cryogenic reworking by thermokarst lakes moving across the tundra after having been generated by the thawing of recurrent-wedge ices.

The surfaces of flatlands in certain regions of the permafrost range are subject to such reworking, and a locality will sometimes become unrecognizable within a few years; fine soils are subject to far-reaching changes.

Thus, the formation of frozen ground and permafrost is influenced not only the formation of ice interlayers, lenses, and wedges, but also by thermokarst phenomena.

3. Buried ices are formed when firm ice and snow are gradually covered by deposition of unconsolidated rock material or by debris carried by spring-thaw runoff water. In their disintegration stage, they are characterized by thermokarst sinkholes, funnels, cirques, etc.

Before the 1930's, the literature sometimes gave too much significance to buried ices ("firn fields") in the formation of underground ices. But the most recent data, e.g., those from a study made in central Yakutiya, have shown that even in this region, the underground ice is for the most part recurrent-wedge formations.

The brief description that we have given of the types of underground ice that form under permafrost conditions indicates that its study is important both in its scientific aspect (as a means of establishing the genesis of the entire encompassing stratum of permafrost) and especially for practical purposes, since the greatest difficulties encountered in the erection of structures under permafrost conditions are due to nothing other than the underground ice.

References

1a. Tsytovich, N. A.: "Issledovaniye deformatsiy merzlykh gruntov" (A Study of the De-formations of Frozen Soils), vol. 2 of Doctoral thesis, pp. 45, etc., Leningrad, 1940.

1b. Tsytovich, N. A.: "Construction Under Permafrost Conditions," abstracts of papers at Conference of USSR Academy of Sciences Council for the Study of Productive Resources (SOPS), USSR Academy of Sciences, 1941.

2. Dostovalov, B. N. and V. A. Kudryavtsev: "Obshcheye merzlotovedeniye" (General Permafrostology), Moscow State University Press, 1967.

3. *Ibid.*

4. Andrianov, P. I.: "The Freezing Temperatures of Soils" (SOPS), USSR Academy of Sciences, 1936.

5. Bozhenova, A. P.: "Supercooling of Water on Freezing in Soils and Subsoils," Reports on Laboratory Studies of Frozen Ground under the Supervision of N. A. Tsytovich, Collection 1, USSR Academy of Sciences Press, 1953.

6. "Handbook of the Design of Bases and Foundations for Buildings and Structures on Permafrost." Scientific Research Institute of Foundations and Underground Structures (NIIOSP). Stroyizdat, 1969.

7. Tsytovich, N. A.: Certain General Methodological Problems of Research on the Physicomechanical Properties of Frozen Soils, in "Reports on Laboratory Studies of Frozen Soils," no. 2. USSR Academy of Sciences Press, 1954.

8. Nersesova, Z. A. and N. A. Tsytovich: Unfrozen Water in Frozen Soils, papers at International Conference on Permafrostology, USSR Academy of Sciences Press, 1963.

9. Tsvetkova, S. T., N. L. Brattseva, et al.: "The Present Status of Geocryological Research Abroad," Publishing House of the All-Union Scientific Research Institute of Hydrogeology and Engineering Geology (VSEGINGEO), 1966.

10. Nersesova and Tsytovich, *op. cit.*

11. Tsytovich, N. A.: On the Theory of the Equilibrium State of Water in Frozen Soils, *Izv. Akad. Nauk SSR, Ser Geog Geof*, no. 5–6, vol. 9, 1945.

12. Tsytovich, N. A.: On Unfrozen Water in Loose Rocks. *Izv. Akad. Nauk SSSR, Ser. Geog.*, no. 3, 1947.

13. Tsytovich, N. A.: "Principles of the Mechanics of Frozen Soils," USSR Academy of Sciences Press, 1952.
14. Nersesova, Z. A.: Melting of Ice in Soils at Negative Temperatures, *Dok. Akad. Nauk. SSSR*, vol. 29, 1951, No. 3.
15. Hoekstra, P.: The Physics and Chemistry of Frozen Soils, *Highway Research Board Spec. Rep.* 103, pp. 78–80, 1969.
16. Tsytovich, N. A.: A New Principle of Frozen-Soil Mechanics, USSR Academy of Sciences Journal *Merzlotovedeniye,* vol. 1, No. 1, 1946.
17. Lykov, A. V.: "The Theory of Heat Conduction," Gosteroizdat, 1954.
18. Lykov, A. V.: "The Phenomenon of Transport in Capillary-Porous Bodies," Gostekhizdat, 1954.
19. Lykov, A. V. and Yu. A. Mikhaylov: "The Theory of Energy and Mass Transport," Belorussian Academy of Sciences Press, Minsk, 1959.
20. Shumskiy, P. A.: "Fundamentals of Structural Glaciology," USSR Academy of Sciences Press, 1955.
21. Popov, A. I.: Frost Cracks and the Problem of Fossil Ices, *Transactions of the Permafrostology Institute,* vol. 16, USSR Academy of Sciences Press, 1960.
22. *Papers at International Conference on Geocryology*, N. A. Tsytovich (ed.), USSR Academy of Sciences Press, 1963.
23. Kachurin, S. P.: "Thermokarst on the Territory of the USSR," USSR Academy of Sciences Press, 1960.
24. Tomirdiaro, S. V.: The Physics of Lake Thermokarst in the Polar Lowlands and in Antarctica, and Cryogenic Reworking of Soils, *Kalyma*, nos. 7 and 8, 1965.

Chapter II

ON MIGRATION OF MOISTURE DURING FREEZING OF SOILS, AND FROST-HEAVING FORCES

1. Present State of the Problem of Moisture Migration During the Freezing of Soils

As we noted earlier, the freezing of moist particulate soils is accompanied by a number of physical, physicochemical, and physicomechanical phenomena and processes. Research done over recent decades both by Soviet (M. I. Symgin and N. A. Tsytovich, M. N. Gol'dshteyn, N. A. Puzakov, I. A. Tyutyunov, Z. A. Nersesova, A. M. Pchelintsev, V. O. Orlov, and many others) and by foreign scientists (Bouyocos, Taber, Beskow, Jumikis, Penner, and others) have shown that the basic process in freezing soils is a redistribution of moisture as a result of migration of water during freezing and deposition of ice with all of the consequences that follow from these processes.

Generalization of materials on the migration of water in freezing soils and the resulting frost heaving of the soils would require an extensive monograph, and this is beyond the scope of the present work. Detailed surveys of the literature on this problem can be found in the following sources: "Fundamentals of Geocryology," by a team of authors of the V. A. Obruchev Institute of Permafrostology of the USSR Academy of Sciences, vols. 1 and 2 (USSR Academy of Sciences Press, 1959); V. O. Orlov's "Cryogenic Heaving of Fine-Particle Soils" (USSR Academy of Sciences Press, 1962); "Thermophysics of Freezing and Thawing Soils," by a group of authors, Porkhayev, G. V., ed. (USSR Academy of Sciences Press, 1964); A. M. Pchelintsev's "The Structure and Physicomechanical Properties of Frozen Soils" (Nauka Press, 1964); N. S. Ivanov's "Heat and Mass Transfer in Frozen Rocks" (Nauka Press, 1969), and others.

When water freezes in soils, there is not only a sharp change in the properties of the soils themselves (cohesion increases by many times, and with it the resistance to external forces), but also a substantial increase in the volume of the frozen soil, an increase that is not, as a rule, distributed uniformly through the ground.

It has been established that freezing may create conditions such that the increase in soil volume due to migration of moisture toward the freezing front and its freezing may range into the tens and even hundreds of percent.

The problem of moisture migration during the freezing of soils is of very great importance; numerous investigators have occupied themselves with this problem since the end of the Nineteenth Century. However, because of the extreme complexity of the moisture-migration process during the freezing of soils and the effects of various factors on this process, the physics of moisture migration in soils has not yet been adequately studied. Several theories of moisture migration during soil freezing have been proposed to conform to various observed phenomena, but no mathematical theory capable of yielding quantitatively accurate results in the general case has as yet been developed.

Let us review very briefly the basic theories of moisture migration during the freezing of soils and outline the ranges of their validity, at least roughly (to the extent that this is possible).

We should note that the theories of migration are set forth in detail in the literature cited in this section,* so that there is no need to go into details here.

One of the first theories advanced in Russia was the "capillary theory of freezing pores" (V. I. Shtukenberg, 1885), in which it was first recognized that frost heaving is caused by migration of moisture toward the freezing front as a result of the rise of water in capillaries formed by cracks and "freezing pores." However, the formation of the "freezing pores" and the presence of capillary menisci at the freezing boundary were not confirmed by experiments, and this theory was not developed further.

On the basis of their observations of hummock formation in permafrost, K. O. Nikiforov (1912) and M. I. Sumgin (1929) and others advanced the "pressure-head theory," which is now viewed as applicable only for the case in which water-saturated soils freeze in sealed (closed) systems.

The very important theory of the film migration of water in moist particulate soils was developed during the 1920's (A. F. Lebedev, 1919; G. Beskow, 1935, and others); it is applicable both to unfrozen and, with certain refinements, to freezing particulate soils.

The "theory of crystallization forces" was proposed to account for the processes that occur when ice crystallizes in freezing soils (Bouyocos, 1923; Taber, 1929, and others); it supplements the theory of film migration for the case of freezing soils, but does not offer an explanation for the migration of water in soils at above-freezing temperatures, and is therefore not a general theory.

The "theory of osmotic pressures" (M. N. Gol'dshteyn, 1948) is similar to the theory of the film migration of moisture in freezing soils, but, as has been shown by A. P. Bozhenova,[1] it contributes practically nothing to determination of the moisture-migration mechanism in freezing soils, since perceptible results of osmosis are observed only at rather high salt concentrations in the pore water.

The generalized "theory of absorption forces" (R. Ruckli, 1943; N. A. Puzakov, 1960, and others) has been rather widely recognized in recent decades; in this theory, suction forces are regarded as a kind of theoretical equivalent of the net effect of migration forces, but the nature of the latter is not considered.

---

*See, e.g., V. O. Orlov's monograph. USSR Academy of Sciences Press, 1962, etc.

We note that other theories of moisture migration in freezing soils have also been advanced on the basis of different premises: the theory of chemical potential (I. A. Tyutyunov, 1959, and others), according to which the migration of moisture in freezing soils is a function of the free surface energy of the mineral particles or of their isobaric potential; the molecular-kinetic theory, which is based on concepts of the structural properties of water in thin films and capillaries (A. A. Ananyan, 1959); the theory of adsorption pressures of fully water-saturated soils, with use of the Tammann-Bridgeman empirical equation to relate pressure and the melting point of the ice (Kh. R. Khakimov, 1957); the theory of pore vacuum in freezing soils (S. I. Gapeyev, 1956), and others.

The adsorption-film theory, which is based on the combined action of several processes, now enjoys the widest currency (Beskow, 1936; A. P. Bozhenova, 1957; V. O. Orlov, 1962; Jumikis, Hoekstra, 1960, and others). According to present-day conceptions, the properties of water in films of various thicknesses, which are governed by the influence of active centers on the mineral-particle surfaces, set the water in motion from thick films with more mobile molecules toward thin films.

Finally, phenomenological theories of the migration of moisture in soils have recently come into rather widespread use; they are based on mathematically rigorous differential equations of heat and mass transfer derived by A. V. Lykov (Chap. I[17] papers by I. A. Zolotar', 1958; N. A. Puzakov, 1960; G. M. Fel'dman, 1963; N. S. Ivanov, 1969, and others).

We note that I. A. Zolotar' was the first to solve the problem of moisture conductivity in freezing soils with consideration of a specified law of motion of the freezing front in the case of a deep water table.

N. A. Puzakov[2] assumed in elaborating a computation of moisture accumulation in freezing soils that the pressure gradient in the bound water is directly proportional to the gradient of unfrozen water content, which, in turn, is proportional to the negative value of the temperature, and, working from hypotheses adopted as to the net effect of adsorption-crystallization, diffusion, and wedging forces, determined a "theoretical suction force" and, from it, the influx of water into the freezing soil layer. However, we should note that the method used to determine certain characteristics for the zone of water phase transformations, e.g., the moisture conductivity coefficient, have not been developed sufficiently, and that this makes it difficult to use the solutions obtained.

G. M. Fel'dman[3] solved the problem of moisture migration in freezing soils on the basis of Lykov's theory of heat and mass conduction for rhythmic motion of the freezing front (such as correspond to the freezing phenomena observed under natural conditions, e.g., in the experiments of R. Martin, A. M. Pchelintsev, and others) and, introducing certain simplifications which did not, according to estimates of the significance, introduce unacceptable errors, carried the calculations through for several of the practically most important cases, obtaining numerical values that were quite consistent with the results of field observations.

It follows from the above very concise description of the basic theories of moisture migration in freezing soils that water migration in soils is a highly complex natural phenomenon that exhibits great consistency in some respects and has certain aspects that can be described successfully by the various relationships that have been derived, although no integrated theory of moisture migration in freezing soils has as yet been elaborated in general form. In the next section, therefore, we shall examine the results of direct measurements of moisture redistribution in freezing and frozen soils and special experiments set up to study the migration of moisture during freezing.

2. Distribution of Moisture in Freezing Soils

Studies of the distribution of moisture in freezing soils at various time intervals after the onset of freezing were first instituted by Russian agronomists back at the end of the Nineteenth Century. Thus, G. Ya. Bliznin[4] cites experimental data on the redistribution of moisture in soil to a depth of 110 cm, reporting that the moisture content increased by 1.9-11 percent in the upper layers of the soil during one month (from January 27 through February 28), while it decreased by 0.3-1.6 percent in deeper layers.

Observing the distribution of moisture in soil during freezing and thawing, F. V. Chirikov and A. Malyugin[5] established that the moisture content increased to nearly double at depths of 0 to 5 cm and by a factor of 1.5 at 5 to 10 cm as compared with the initial moisture content (before the onset of freezing); here they concluded that the transfer of moisture from deeper levels occurred in the vapor state. Thus, the migration of moisture toward the freezing surface in frozen soil was recorded during the very first phase of the research. But in 1927, N. A. Kachinskiy[6] stated on the basis of his field observations that moisture is fixed by the frost on freezing. He writes, "As the frost sets in, it mechanically impedes further progress of the water, fixing it in the pattern that it occupied at the last instant before freezing."

The specially defined experiments of N. I. Sumgin[7] have proven beyond all doubt that moisture migrates and is not fixed in freezing soils.

S. Taber[8] also observed considerable migration of water toward the cooling surface in experiments in which cylindrical samples of clay were frozen; after freezing, rather thick interlayers of ice formed at the tops of the specimens, while the moisture contents at the bottom were down by an average of 25.5 to 16.8 percent. On the basis of a study of moisture redistribution during the freezing of soils, Taber proposed that two classes of freezing system be distinguished: closed (or lored) systems, in which the existing moisture is redistributed, and open systems, in which there is an influx of water from the outside. This subdivision of freezing soil systems has been generally accepted.

Subsequently, many investigators in the USSR studied the migration of moisture in freezing soils (A. Ye. Fedosov, A. P. Bozhenova, V. I. Moroshkin, N. V. Ornatskiy, S. L. Bastamov, A. Ya. Tulayev, N. A. Puzakov, M. N. Gol'dshteyn, and others). We note here only results most important for further elaboration of the problem.

Thus, A. Ye. Fedosov[9] established the fact of intravolume shrinkage (compression) of freezing clayey soils and suggested that the moisture content of aggregates between ice interlayers be assumed equal to the lower plasticity limit (to the moisture content at the plastic limit).

Continuing Sumgin and Taber's experiments, A. P. Bozhenova[10] observed different water-migration effects in clays and sands: the moisture contents of the clays increased in the direction toward the freezing front (the top of the specimen had a moisture content of 80.9 percent after freezing, at an average moisture content of 63.9 percent in the clay sample, while the content at the bottom became 36.8 percent), while the moisture content of medium-grain sand decreased (from 18.6 percent for the bottom of the specimen to 16.9 percent for the top at an average moisture content of 17.3 percent).

This last fact required clarification, since it was very important for practical purposes to establish whether ice accumulates or water is squeezed off when sandy soils are artificially frozen by wells, and under what conditions these effects occur. The author[11] made a detailed study of this problem under order from the Committee for the Promotion of Hydroelectric Power Station Construction of the USSR.

It was established that in water-saturated sands with free drainage of water in at least one direction, water does not migrate toward the freezing front, but is squeezed out, with the result that the porosity of frozen water-saturated sands remains practically the same (variations less than 0.2 percent. But when no drainage is provided, as when the material is frozen from all sides, an increase in the porosity of the frozen water-saturated sand was observed, ranging up to 4 percent at certain levels.

For fine clayey soils, the water pressure that appears as the ice crystals grow decays at a very short distance from the freezing front owing to the high internal resistance to filtration, but the crystallization-film mechanism of water migration continues to operate.

A special study made under the supervision of the author[12] in the laboratory of the USSR Academy of Sciences Institute of Geocryology was devoted to the problem of the basic mechanism of migration and estimation of the significance of the various other mechanisms of water migration in particulate soils during freezing.

The results showed that the adsorption forces of the mineral skeleton, which govern the "crystallization-film mechanism" of moisture transport toward the freezing front, are of basic importance in the migration of water in freezing soils.

Cylinders divided into two sections by a thermally nonconductive vertical partition were used to study the migration of water during the freezing of soils with various granulometric compositions:[10] the soil to be studied was placed in one compartment and fine glass tubes were inserted into the other at various heights and covered with fine cork. During freezing of water-saturated soils, the water was pressed out into the glass tubes, and the volume of migrating water could be measured. This experiment produced another literally obvious proof that water in freezing particulate soils migrates preferentially in the liquid phase. The experiments also showed that the water in fully water-saturated soils always migrated from course-granular soils to soils with finer particles, again for the most part in the liquid phase. For soils that were moist but not saturated, the migration of the water was determined entirely by the crystallization-film mechanism and by its transport on the vapor state.

The experiments of S. L. Bastamov[13] showed that when soil pores are partly filled with water, especially when $W < W_p$, the migration of vapor-phase moisture occurs as a persistant phenomenon. But although the accumulation of moisture in the direction toward the cooled surface takes place very slowly (up to 20-30 days), it may cause substantial moistening of a freezing water-unsaturated soil.

Finally, we cite data on the migration of unfrozen water in frozen soils, another observation from the experiments of G. Ya. Bliznin (1889).[4]

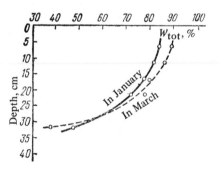

Fig. 26. Distribution of moisture in the upper layer of a forzen silt-mud loam, according to experiments of the Anadyr' Scientific Station of the USSR Academy of Sciences.

Layer-by-layer determinations of moisture content in topsoil made by I. A. Tyutyunov in January and March on Anadyr' tundra[14] showed that the total moisture content W_{tot} of the soil in its upper layers increased significantly from January through March, while the moisture in the lower layer (at depths of 32-33 cm) decreased slightly (Fig. 26); here migration of water in the vapor state was excluded due to oversaturation of the unfrozen soil with water.

It was found for another locality of the Anadyr' region that the total moisture content (with respect to the total weight of the soil) of a silt-mud loam at a depth of 35 cm below the surface was 25.9 percent in December, 31.1 percent in January, and 41.2 percent in April. These data indicate that the moisture redistribution in the frozen loam took place chiefly through the migration of unfrozen water.

In 1950, to check the possibility of migration of unfrozen water of frozen soils, the author designed a special laboratory experiment: a block of frozen Moscow-area loam was subjected for more than 24 hours to the constant action of a subfreezing-temperature gradient of about 1.7 deg/cm. The surface temperature of a specimen of heavily moistened loam frozen at $-4.4°$ was held at -4.2

Table 6.   Moisture distribution in a frozen Moscow-area loam (average moisture content $W_{tot} = 58.3\%$) after prolonged cooling from one side

Layer no.	Time of experiment, hr	Surface temperature of specimen		Moisture content of layer, % of average	
		Lower	Upper	Initial	After cooling from below
1	31	–	From – 4.2 to – 21.2	99.4	100.4
2		From – 4.2 to – 4.8		100.6	100.0
3			–	99.9	99.6

to $-4.8°$ on one side and at -20.2 to $-21.2°$ on the other. Table 6 shows the moisture distribution in the soil sample before freezing and 31 hours after unilateral exposure to the temperature gradient.

These results prove unequivocally that even in the transitional range of the phase transitions of water to ice, where the change in the content of unfrozen water is not so significant, migration of unfrozen water takes place in particulate frozen soils at subfreezing temperatures under the action of a temperature gradient. This is also confirmed by experiments on electroosmosis in frozen soils (A. A. Ananyan, 1952, Hoekstra and Chamberlain, 1964).[13a] This established fact is of both theoretical and practical importance; thus, it may be assumed that the moisture contents of permafrost change (if very slowly) with changes in temperature conditions in their upper layers and, consequently, that further increase in moisture content of the upper, already frozen layers occurs when soils freeze under natural conditions. Also indicative in this respect are the universally observed high moisture-content and ice-content values in the upper layers of frozen soils, which extend down to 10–15 meters, with a subsequent substantial decrease at greater depths.

Z. A. Nersesova[15] made a detailed analysis of foreign research on the migration of moisture in freezing soils and the conclusions drawn from them.

Summarizing her survey of the foreign literature, Nersesova writes, "The world literature has accumulated a large amount of factual material characterizing the influence of particle size, mineralogical composition, exchange bases, and capillarity (although the latter, according to Prof. Jumikis, is of very limited importance, since no capillary menisci were observed in his special experiments on the freezing of water-saturated soils.–N.T.) on the migration of water in freezing soils, on the formation of ice and on heaving in these soils, but due attention is not given to studies that treat the interaction of water and the aluminosilicate surfaces systems based on crystal-chemical conceptions and solid-state physics."

Within this last conception, I. A. Tyutyunov[15] proposed a new chemical-potential theory based for the most part on study of the interaction between the surfaces of soil mineral particles and pore water.

The theory of moisture migration in soils is now being developed on the basis of molecular-kinetic conceptions as to the structural features of water in thin films (A. A. Ananyan, 1959)[17] and the crystal chemistry of silicates (G. B. Bokiy, 1961).[18]

A study begun by Z. A. Nersesova[19] in the laboratory of the Obruchev Institute of Permafrostology of the USSR Academy of Sciences under the author's supervision is devoted to establishing the role of the physicochemical properties of soil surfaces on the migration of moisture and ice segregation. The processes of water migration, deposition of ice, and heaving during freezing were studied under open-system conditions in three loams and two monomineral clays in the natural state and with saturation by various exchange cations:

$$Al^{3+}, Fe^{3+}, Ca^{2+}, H^+, Na^+, \text{ and } K^+$$

Experiments were performed on a kaolinite clay (Glukhov kaolin), a montmorillonite clay (allophane), and the following loams: a Moscow-area loam from a depth of 80 to 100 cm, a (silty) Vorkuta loam from 50 cm deep, and a Yakutsk loam ("Trubka Mira" region) from a depth of 50 cm.

It was established that the composition of the exchange cations invariably has a strong influence on water migration during freezing. Thus, Fig. 27 shows vertical sections through specimens of Glukhov kaolin saturated with the cations

$$Fe^{3+}, Ca^{2+}, H^+, Na^+, \text{ and } K^+$$

We see from Figs. 27 and 28, which plot the variations of moisture content by weight in kaolin specimens after freezing, that the strongest migration of water during freezing, deposition of ice, and heaving is observed in the Fe kaolin

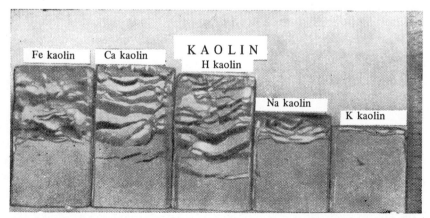

Fig. 27. Specimens of kaolin with various exchange-cation compositions (after freezing).

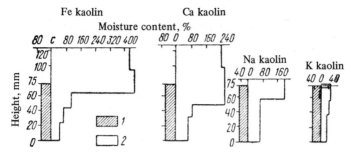

Fig. 28. Variation of height and moisture content of kaolin specimens after freezing: (1) before freezing; (2) after freezing.

(69.3 percent of the initial specimen height) and Ca kaolin (70.7 percent), and that the Fe kaolin showed somewhat stronger heave as compared to the Ca kaolin at the start of the experiment; on the other hand, the weakest water migration, icing, and heaving (to 8 percent) occurred in the K kaolin.

Freezing experiments with allophane, and a montmorillonite clay, showed that the strongest migration of water during freezing and the strongest heaving were observed when the allophane was saturated with multivalent cations (up to 91 percent for Al^{3+}), and that these phenomena were least pronounced for saturation with univalent cations: Na and K (to 8–10 percent) (Fig. 29).

From the results of these experiments, Nersesova concluded that montmorillonite clay becomes highly frost susceptible when univalent absorbed bases are replaced by multivalent bases.

Experiments with the heavy Moscow-area loam confirmed the view that the nature of the exchange cations is highly important for loams, although substantial separation of ice was observed only in the Fe-saturated Moscow-area loam.

Z. A. Nersesova also performed very interesting experiments with individual fine fractions of a pure quartz sand (particles from 0.1 to 0.005 mm and smaller

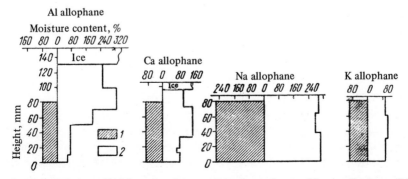

Fig. 29. Variation of height and moisture content in specimens of frozen allophane: (1) before freezing; (2) after freezing.

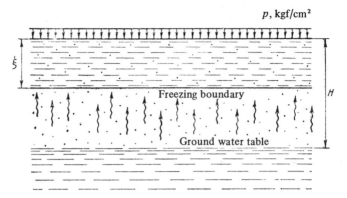

Fig. 30. Schematic section through frozen soil.

than 0.001 mm), which showed that moisture also migrated during freezing in the fine sand fractions (sizes below 0.001 mm) in the presence of water influx (Fig. 30), i.e., water also migrates toward the freezing front in a fine quartz sand when the area of its mineral particles is increased sharply and, consequently, it has a higher surface energy and increased capillarity, just as in the case of clayey soils.

The above highly interesting from the scientific and practical standpoints conclusions, are of importance not only for the creation of a general physicochemical theory of moisture migration in freezing soils and ice segregation, but also for the development of practical physicochemical countermeasures against the frost heaving of disperse soils.

3. The Motive Forces of Water Migration

Let us briefly summarize the purely experimental facts that determine the migration of water in freezing soils and the course that this process takes in time.

We note first of all that the migration of water in freezing and frozen soils can occur in three physical states: vapor, liquid, and solid.

This last type of water migration, i.e., migration in the solid state (in the form of ice) obviously can occur only as a result of purely plastic flows of the ice. However, "migration" (more precisely, the redistribution of ice by its conversion to water and immediate refreezing) has been observed in frozen and permanently frozen soils only under the action of an external load (experiments of S. S. Vyalov and others at Igarka),[20] and it may occur under natural conditions in ice-saturated lower layers of permafrost, which are under considerable pressure from the weight of the overlying soil layers. In fact, the corresponding calculations indicate that substantial local stresses (pressures) arise at the points of contact between mineral particles and ice and may initiate plastic flow of ice, i.e., they may be responsible for its redistribution in frozen soils and permafrost.

But this effect is not typical for freezing soils and we merely mention it here.

The migration of water in the liquid and vapor states occurs for the most part in freezing soils.

As we noted above, migration in the vapor state is important for soils with low degrees of water saturation, when the pores are only partially filled with water—a case investigated in detail by A. F. Lebedev[21] for unfrozen soils—but the conclusions drawn from these studies are fully applicable to freezing soils that are not saturated with water. Lebedev demonstrated with appropriate experiments that the migration of water in moist but unsaturated soils is due to the transport of water vapor from sites with higher, to points with lower vapor pressure. When the soil has a very low moisture content (below the maximum hygroscopicity), moisture can be transported only in the vapor state.

The conditions under which moisture migrates in freezing soils differ from the corresponding conditions in unfrozen soils, firstly in that in the former, water vapor not only condenses to liquid, but is also converted directly to ice, increasing the ice content of the soil at the freezing front and, secondly, in that vapor can migrate at one and the same temperature, since the vapor pressure over ice is lower than the vapor pressure over water. The migration rate of moisture vapor is proportional to the difference between the vapor pressures at the points in the soil under consideration, and this depends mainly (according to the Clausius-Clapeyron equation) on the absolute temperature of the vapor.

While it has some importance for undersaturated soils, vapor migration is not a determining factor in water-saturated soils, where, as we pointed out above, the liquid phase will be the dominant phase of the migrating water.

Generalizing from the foregoing experimental work results on water migration in freezing soils, we enumerate the basic established facts.

1. In all cases (with the very minor exception of quick-freeze—up to very low temperatures— below $-70°$), the fact of migration rather than fixation of moisture has been established (M. I. Sumgin et al.).

2. When fine clayey soils freeze, we observe intravolume shrinkage (consolidation) and aggregation of the soil between interlayers of ice (A. Ye. Fedosov, M. N. Gol'dshteyn, A. M. Pchelintsev, and others).

3. Different migration effects were established in the freezing of coarse sands—squeezing of water away from the freezing front—and fine clayey soils—inflow of water and an increase in the ice content at the freezing front (A. P. Bozhenova, N. A. Tsytovich, and others).

4. The rhythmical and stepwise nature of ice segregation and heaving of soils has been established under both laboratory and field conditions (R. Martin, A. M. Pchelintsev, and others).

5. Differences in the manner in which moisture is redistributed and ice accumulated in closed and open systems have been determined; the maximum ice separation (which frequently reaches tens and hundreds of percent in fine clayey soils of appropriate composition) occurs in open systems, while in closed

systems we observe redistribution of soil moisture and ice increasing toward the cooled area and dewatering of soil layers at a certain distance from the surface of the freezing front.

6. The fact of migration of unfrozen water in both freezing and frozen soils has also been observed and demonstrated experimentally (I. A. Tyutyunov, N. A. Tsytovich, A. A. Ananyan, and others).

7. It has been established that the migration of water in freezing soils occurs primarily in the liquid phase in the form of films (A. P. Bozhenova and others).

8. It has been demonstrated that the nature of the exchange cations strongly influences the rate of water migration and the amount of heaving in disperse soils, and that the greatest ice segregation is observed at saturation with multivalent cations and the least in the case of univalent cations (Z. A. Nersesova and others).

9. Moisture can migrate in freezing soils in all of the phases of water (gaseous, liquid, and solid) only if their equilibrium state is disturbed (A. F. Lebedev, N. A. Tsytovich, I. A. Tyutyunov, and others).

Water migrates in freezing soils under the influence of various forces, which determine the prevalence of one moisture-transfer mechanism or another. Let us enumerate the most important of these mechanisms on the basis of the foregoing data, indicating the conditions under which they predominate. We regard the following as such forces:

1) water vapor pressure forces (more important in soils with low-moisture contents)

2) pore-water capillary forces (a secondary effect of the action of adsorption forces and in the process of water migration in freezing soils, but without independent importance, govern the capillary-film mechanism at attainable water table depths shallower than the depth of maximum capillary rise)

3) external and internal pressures, including vacuum (causes pressure flow of soil waters and, under appropriate conditions, substantial soil heaving and hummocking)

4) osmotic forces (become dominant and determine migration only when the pore water contains substantial amounts of solutes)

5) ice-crystallization forces (they become important in the process of segregation and accumulation of ice at the freezing front, continually adding new portions of migrating water to the ice deposits that have already formed and strongly enhancing the adsorption-film mechanism of migration)

6) adsorption forces of the organomineral skeleton of the soil and ice, arising under the influence of the free surface energy of the mineral particles, are of prime importance for any freezing particulate soil and govern the rate of migration and the extent of frost heaving depending on the nature of the exchange cations and the capillary properties of the soil

Migration forces cause motion of pore water in freezing soils only when phase equilibrium is disturbed and conditions are created for the appearance of various gradients: moisture, temperature, adsorption-film, osmotic and other pressures and, finally, a gradient of the isobaric potential of mineral particle free energy.

In the general case, the migration water flux i_{mig} can be written in the form (see, e.g., Ref. 22)

$$i_{mig} = -k \operatorname{grad} F \qquad (II.1)$$

where k is a proportionality coefficient characterizing the specific resistance offered by the soil system to motion of the moisture and F is a generalized motive force.

The general equation (II.1) determines the flux of migration water along the generalized-force gradient, to which it is directly proportional.

Any of the migration forces enumerated above can be taken as the generalized force. The cases in which a generalized or particular migration force should be used must be established purely by experiment. There are now four conceptions that are used most often in choosing the generalized migration force F.

In the first conception, the so-called soil absorption force is taken as a generalized migration force (F. Ruckli, M. N. Gol'dshteyn,[23] N. A. Puzakov,[24] and others); it is governed by the combined action of a number of molecular disturbances and is to be determined purely by experiment. According to this conception, the migration flux depends on the difference between the "absorbing force" and the pressure produced by a load and by growing ice crystals (G. Beskow, 1935; M. N. Gol'dshteyn, 1947; N. A. Puzakov, 1948; B. I. Dalmatov, 1957, and others). The first conception has been used for the most part in roadbuilding.

In the second, thermodynamic conception, the generalized motive force of migration, based on the theory of heat and mass transfer in capillary-porous colloidal media (A. V. Lykov, 1954; N. A. Puzakov, 1948; I. A. Zolotar', 1958, N. S. Ivanov, 1969, and G. M. Fel'dman,[22] 1964) is the so-called mass-transport potential of the film moisture. Calculations of ice accumulation and heaving in frozen soils based on the second conception are used in roadbuilding and in residential and industrial building construction.

According to the third, physicochemical conception, the isobaric potential of the free energy of the mineral particles or thin films of water enveloping the soil particles is taken as the basic motive force of migration (on the basis of research by I. A. Tyutyunov, Z. A. Nersesova,[16-19] and others); this potential depends for the most part on the particle size of the soil, its capacity for absorption, and exchange-base composition. This conception is now used in the practical elaboration of physicochemical countermeasures against migration of moisture and frost heaving of soils.

The fourth conception regard the mobility of water molecules and structural features of water in thin films as the fundamental motive force behind migration (A. A. Ananyan).[25]

4. The Principle of Migration

Based on the foregoing exposition, and considering the experimentally established effects of various factors on the migration process and proposals for theoretical explanation of these effects with prerequisites for quantitative determination, we arrive at a definite general conclusion: all of the motive forces of migration are functions of electromolecular forces of the soils. The mechanism of water migration during freezing of various soils may take a very wide variety of forms (depending, generally speaking, on many factors), but it reduces in the final analysis and in all cases to the action of molecular adsorption-film, film-crystallization, and similar surface molecular forces of the mineral soil particles and the ice.

At an earlier time (1950-1952) (Chapter I) we formulated a general relationship that determines the moisture-migration process in freezing soils, calling it the "migration principle," namely: "The migration of water in freezing moist soils is a result of the action of molecular forces (surface tension, adsorption, osmotic pressures, stresses. . . ,)" attaching special importance in evaluation of the motive forces of migration to temperature gradients caused by the nonequilibrium state of the phases of the soil; this principle remains valid even today.

In the light of important experimental results that have been obtained over the last twenty years and have been described briefly above, our enunciation of the principle of moisture migration in freezing soils can now be expanded to some degree, but the general relationship described by this principle remains the same.

We formulate the principle of migration of moisture in freezing soils as follows: the migration of water in freezing moist soils is a process of moisture transport that operates consistently whenever the equilibrium state of the soil's phases is disturbed and external factors change (the presence of temperature, moisture content, pressure, mineral particle surface energy, mobility of molecules in water films, gradients, etc.).

This general formulation of the relationship that determines the migration of moisture in freezing soils provides a physical explanation for the various moisture migration mechanisms and incorporates all of the basic conceptions in a definition of generalized migration forces. Thus, the process of moisture migration can begin only on a disturbance of the equilibrium state of the phases of the soil, since migration of moisture will not occur in soils when they are perfectly homogeneous and under isothermal conditions in the absence of external disturbances.

All of the factors that have been identified in special experiments and field observations as influencing moisture migration in freezing soils are explained quite satisfactorily by the migration principle stated above. Thus, for example, the redistribution of moisture in freezing soils acquires a lucid physical explanation in the light of the migration principle, since the equilibrium of the phases is invariably disturbed on freezing. The consistently observed migration of water toward the freezing front in particulate soils (or toward cooled and frozen surfaces) also becomes quite understandable in accordance with the principle enunciated above, since points with lower ground temperatures will have lower vapor pressures, larger soil-skeleton adsorption forces, larger crystallization forces of the ice, and lower molecular mobilities in the films, i.e., gradients of the various molecular forces in the direction to the heat sink will exist and cause migration of liquid water in this direction. The fact that excess separation of ice occurs only in hydrophylic moistened soils when they freeze also becomes perfectly clear, since these soils have bed water films.

The relation between the migration rate and the cooling rate determines the number and thickness of ice interlayers in soils that freeze with seepage of water from the outside. The higher the freezing temperature is held at a given level, the thicker will be the ice interlayers that form there, since there is more time for sufficient water to be drawn in for growth of the ice crystals.

Finally, the migration of water in frozen soils, which is possible only in the presence of unfrozen water retained by adsorption on the surfaces of soil mineral particles and the ice, is subject to the same laws of film-water motion that were established for soils at positive temperatures, but has its own peculiarities because of the additional effects of ice crystallization forces; in this case, however, the migration process of the film water will be very slow.

We note in conclusion that we regard the migration principle as the basic relationship by which we must be guided in study of physicomechanical properties in freezing soils.

Concerning certain phenomena not reflected in the general formulation of the migration principle, we consider that they may call for certain corrections of secondary importance, but will not change the basic relationship, which covers the process of moisture migration in freezing and frozen soils quite thoroughly.

5. Prediction of Moisture Migration and Heaving in Freezing Soils

The migration principle formulated above and its concrete applications for calculation of ice accumulation in freezing soils now make it possible to take a quantitative account of the moisture-content increase in freezing soils and the frost (cyrogenic) heaving that it causes. The latter is a local increase in soil volume (generally nonuniform) that results from freezing of moisture present in the soil and moisture drawn in in the migration process.

There are now two basic methods for determination of the migration flow: the first is purely empirical, and the second theoretical.

As an example of the first method, we cite the empirical formula of Prof. N. V. Ornatskiy[26] for prediction of seasonal moisture accumulation: the increase by weight ΔW of soil moisture during the time t for which it is frozen is determined by the expression

$$\Delta W = \alpha_0 I_\theta^n t \tag{II.2}$$

where I_θ^n is the average temperature gradient over the winter to the nth power, t is time in months, and α_0 is a coefficient that characterizes the hydrogeological conditions of the area subject to heaving.

Working from observations made by Ministry of Railroads frost-heave stations, Ornatskiy recommends that an average value $n \approx 3/2$ be taken for the exponent and, for the hydrogeological coefficient, $\alpha_0 = 0.5$ when the water table is deeper than 1.5 meters and $\alpha_0 = 1$ when it is higher (less than 1.5 m from ground level).

The frost-heave characteristic is then expressed as follows:

$$W_0 + \alpha_0 I_\theta^{3/2} t - W_p > 0 \tag{II.3}$$

where W_0 is the average initial moisture content of the freezing soil layer and W_p is the amount of moisture theoretically needed to fill all of the pores in the soil with water (without consideration of swelling of the soil).

According to Ornatskiy, any soil will heave at a certain critical moisture content equal to

$$W_{cr} \geqslant W_p - \alpha_0 I_\theta^{3/2} t \tag{II.4}$$

Needless to say, the above relationships can reflect reality only if the numerical values of their coefficients have been determined experimentally (at frost-heave stations) for certain natural soil, climatic, and hydrogeological conditions.

In the second method of determining the cryogenic accumulation of moisture in freezing soils, we have a whole series of theoretical proposals based to a certain degree on initial experimental data and supported in one degree or another by field observations.

Let us take note of the most important of the proposed analytical relationships, and dwell in greater detail on those of the recent solutions that are more rigorous in the theoretical sense.

On the basis of the migration principle in its most general form, we can write the following schematic expression:

$$W_{\text{mig}} = f[I_\theta, w, \varsigma, p \ldots] \tag{II.5}$$

i.e., the amount of water due to cryogenic migration is a certain function of the motive gradients.

The following notation has been used in (II.5):

W_{mig} is the amount of migrating water, $I_{\theta, w, \varsigma, p}$ is the motive gradient (θ is the temperature in degrees, W is the moisture content, ς is the chemical potential, and p is the external pressure).

If the moisture migration rate v_{mig} is known, the amount of heaving on freezing of the soil (the increase in the height of a layer of the frozen soil) h_{heave}, will be determined by the expression

$$h_{\text{heave}} = \alpha W_d \gamma_{\text{sk}} \xi i_r + (1 + \alpha) \int v_{\text{mig}} \, dt \qquad \text{(II.6)}$$

where α is the coefficient of volume expansion of the water on freezing, W_d is the total gravimetric moisture content of the frozen soil layer, γ_{sk} is the specific gravity of the soil skeleton, ξ is the freezing depth, i_r is the relative ice content of the frozen soil layer, determined from the content of unfrozen water, and t is the migration time.

Since the migration flow Q_t across a unit area equals

$$Q_t = \int_0^t v_{\text{mig}} \, dt \qquad \text{(II.7)}$$

expression (II.6) can be written in the form

$$h_{\text{heave}} = \alpha W_d \gamma_{\text{sk}} \xi i_r + (1 + \alpha) Q_t \qquad \text{(II.6′)}$$

Taking the "absorption pressure" u_s as the generalized migration force and assuming Darcy's law to be valid for the migration process, M. N. Gol'd-shteyn[27,28] obtained the following value for the unit migration flow Q_t:

$$Q_t = k \frac{u_s - [p + \gamma \xi + \gamma_W (H - \xi)]}{\gamma_W (H - \xi)} t \qquad \text{(II.7′)}$$

where u_s is the "absorption pressure," determined experimentally as the pressure that completely absorbes the excess ice formation in the soil, k is a filtration coefficient, p is an external uniformly distributed pressure, γ is the specific gravity of the frozen soil layer, γ_W is the specific gravity of the water, H is the distance to the water table (Fig. 30), and t is the migration time.

We note that the use of relation (II.7′) to calculate the migration flow and calculation of soil heaving on freezing from this flow are complicated by the need for data on the absorption pressure u_s, which is a complex function of the adsorption-crystallization forces, depends on a number of factors (soil composition, rate and depth of freezing, viscosity of pore liquid, etc.) and cannot now be determined with sufficient accuracy. In addition, the quantity $(H - \xi)$ in (II.7′) is a variable and depends on several factors. All of the above shows that solution of the formulated problem on the basis of the theory of absorption forces and the Darcy equation is not rigorous.

Taking as the motive forces of migration "absorption forces" u_s that depend on the adsorption properties of the soils and the crystallization forces of the ice (which depend, in turn, on the negative-temperature gradient), and assuming further that the course of the soil-freezing process varies parabolically in time under natural conditions and that the external pressure influences only the critical freezing depth ξ_{cr} at which excess deposition of ice ceases, N. A. Puzakov[29] obtained the following expression for the water-migration flow rate Q_t in freezing soils:

$$Q_t = \frac{ku_s}{\beta\gamma_w} \; 2,3H \log \frac{H}{H-\xi} - \xi \qquad (II.8)$$

where $\xi = \sqrt{2\beta t}$ is the variable depth of seasonal freezing over a time t that can vary from zero to the time t_{cr} corresponding to the critical freezing depth ξ_{cr}, i.e., to $t_{cr} = \xi_{cr}^2/2\beta$ ($\beta = \max\xi^2/2t_{win}$ is a climate factor, with t_{win} the duration of winter in the region under consideration).

The approximate expressions (II.7') and (II.8) can be used only for rough calculations.

It is sounder, in our opinion, to take the potential heat and moisture conduction as the motive force of migration, using A. V. Lykov's theory of heat and mass transport in capillary-porous colloidal media.

In general form, Lykov's equation of heat and mass transport takes the form

$$\frac{\partial W}{\partial t} = \nabla \left(a'\nabla W + a'a\nabla\theta \right) + \chi \, \frac{\partial W}{\partial t} \qquad (II.9)$$

where a' is the coefficient of potential conductivity (the film moisture conductivity), a is a temperature-gradient coefficient, and χ is a phase-transition criterion.

Neglecting from now on the moisture-flow components that arise under the influence of temperature gradients (since they are small compared to other components) and neglecting the phase-transition criterion [the last term in the right-hand side of (II.9)], which results in an error no larger than 3–4 percent, G. M. Fel'dman[30] arrives at the following simpler potential conductivity (moisture conductivity) equation, which he solves to determine the migrational water flux in freezing soils:

$$\frac{\partial W}{\partial t} \approx \nabla \left(a'\nabla W \right) \qquad (II.10)$$

With a vertically constant initial soil moisture content W_0 and, consequently, a constant moisture conductivity (potential conductivity) coefficient a', this equation assumes the form

$$\frac{\partial W}{\partial t} = a' \, \frac{\partial^2 W}{\partial x^2} \qquad (II.11)$$

Then, considering a stepwise rhythmic motion of the ice-segregation front during freezing of the soils (which, according to observations, is consistent with natural conditions) at a constant initial moisture content of the soils underlying the freezing layer and taking the moisture content at the plastic limit as the critical moisture content at which migration of water does not occur (which leaves a certain margin for error), Fel'dman solved the differential equation (II.11) numerically and presented the solution for the case of an open system (in the zone of constant migration flow) in the form

$$q_{mig} = A + B\frac{d\xi}{dt}   \text{(II.12)}$$

q_{mig} is the migration flux of water in $kg/(m^2)(hr)$ or mm H_2O/hr and ξ is the variable freezing depth.

The migration flux was calculated (the coefficients A and B determined) with the boundary conditions:

$W(x, 0) = W_0$, the initial soil moisture content; $W(\xi, t) = W_\xi$, the amount of unfrozen water at the freezing boundary; $W(H, t) = W_0$, the initial soil humidity; $\xi = f(t)$, the freezing depth ξ as a rhythmical function of t.

With the above prerequisites and assumptions, three basic cases were computed on a Luk'yanov hydraulic integrator by the method of hydraulic analogies: (1) open system, zone of constant migration flux; (2) zone of ground-water effect, and (3) closed system.

In the first case, the migration flux will be proportional to

$$\Delta W = W_0 - W_\xi \text{ and } \gamma_{sk}$$

where W_ξ is the moisture content corresponding to the amount of unfrozen water at the ice-precipitation front (since $W_\xi \gamma_{sk}$ equals the volumetric soil moisture content) and, recognizing that not all of the water participates in the migration process, but only the excess over the soil moisture content at the plasticity limit W_r, the following numerical expressions were obtained with the hydraulic integrator for the coefficients A and B in (II.12):

$$A = 5.85 \times 10^{-4}\gamma_{sk} (W_0 - W_\xi)(0.34 + 670a')$$

$$B = 1.83\gamma_{sk} (W_0 - W_\xi)(0.34 + 670a') - (W_s - W_\xi)\gamma_{sk}$$

According to the results of migration modelling on the hydraulic integrator, the limits of validity of the values of A and B in (II.12) are as follows:

$$0.12 \times 10^{-4} \leqslant a' \leqslant 3 \times 10^{-4} \text{ m}^2/\text{hr},$$

$$0.35 \times 10^{-3} \leqslant d\xi/dt \leqslant 3 \times 10^{-3} \text{ m/hr}$$

These limits are rather wide, and include the ranges of variation of potential conductivity and the freezing rates usually observed in silty soils under natural conditions.

Numerical values of the migration fluxes were obtained in exactly the same way for the other two cases.

Taking the cryogenic heave of the soil layer to be equal to the total thickness of the ice interlayers in it, we obtain

$$h_h = 1.09 \int_0^t q_{mig}dt   \text{(II.13)}$$

Then, according to Fel'dman, the heave figures will be as follows:

1) for the open system (in the zone of constant migration flux)

$$h_{h,1} = 1.09\{5.85 \times 10^{-4}\gamma_{sk}(W_0 - W_\xi) \times (0.34 + 670a')$$
$$+ [(0.34 + 670a') \times 1.83\gamma_{sk}(W_0 - W_\xi) - (W_r - W_\xi)\gamma_{sk}] \times \nu_\xi\}t \qquad (II.14')$$

2) for the zone of ground-water influence

$$h_{h,2} = \left\{\frac{1.23\sqrt{a'}\gamma_{sk}(W_0 - W_\xi)}{(R_{cr} - R_c)(108/a' \times 10^4 + 70)} \times D - (W_r - W_\xi)\gamma_{sk}\frac{d\xi}{dt}\right\}t \qquad (II.14'')$$

where

$$D = \sqrt{(108/a' \times 10^4 + 70)R_{cr} - 6} - \sqrt{(108/a' \times 10^4 + 70)R_c - 6}$$

R_c is the distance between the ice-segregation front and the water table at the end of the freezing period and R_{cr} is the thickness of the ground-water influence zone, which depends on the freezing rate of the soil ν_ξ in m/hr and on the soil's potential conductivity a' in m²/hr, determined from the diagram plotted by Fel'd-man (Fig. 31) from the results of simulation of the migration process on the hydraulic integrator;

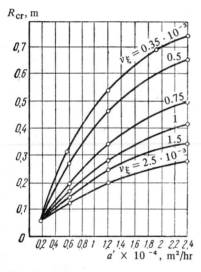

Fig. 31. Thickness of ground water influence zone R_{cr} vs. potential conductivity coefficient $a' \times 10^{-4}$ m²/hr for various soil freezing rates ν_ξ.

3) for the case of the closed system

$$h_{h,3} = 1.09\{5.85$$
$$\times 10^{-4}\gamma_{sk}(W_0 - W_\xi)(0.34 + 670a')$$
$$+ [(0.34 + 670a') \times 1.83\gamma_{sk}(W_0 - W_\xi)$$
$$- (W_r - W_\xi)\gamma_{sk}]\nu_\xi\}t \qquad (II.14''')$$

Although the above expressions appear very complex, calculation of the soil-heave values $h_{h,1,2,3}$ from them is not difficult when the values of γ_{sk}, W_0, W, W_p, ν_ξ, and t are known.

Comparison of the results calculated from (II.14) with observations made under natural conditions shows satisfactory agreement.

Thus, a migration-flux value $q_{mig} = 0.036$ kgf/m² $\times$ hr or mm H_2O/hr was determined for the conditions of the test field of the Vorkuta Scientific Research Station[30] in accordance with (II.12). According to (II.13), the heave is

$$h_{h,1} = 1.09 \times 0.036 \times 4800 = 190 \text{ mm}.$$

The actual amount of heave measured under the same soil-freezing conditions was $h_{h,1} = 201$ mm.

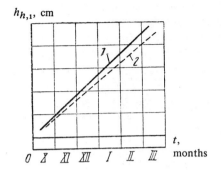

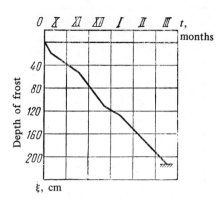

Fig. 32. Data from field observations (1)
and calculated results (2) according to
equation (II.14) for frost heave of soils in
an open field in Vorkuta.

Fig. 33. Frost penetration graph on an experiment field at Vorkuta.

The results of the soil-heave calculations and measurements for the test field
at Vorkuta are compared graphically in Fig. 32; the calculations were made using
formula (II.14), and the line corresponding to the actual amount of heave was
plotted from the two extreme measured points. Figure 33 shows a time curve of
ground frost penetration depths.

Comparison of the calculated predictions and actual data on frost heaving for
other, totaling different climatic and geocryological conditions (including areas
in which frost heaving is most severe, such as the areas with silty soils around
Igarka) showed that the disagreement is only 14.3 percent in any of the cases.

These comparisons show that the theory of heat and mass exchange can be
used successfully to predict cryogenic migration of moisture and frost heaving
for freezing soils.

6. Frost-heaving Forces of Soils in Their Action on the Foundations of Structures

If the increase in soil volume on freezing (cryogenic heaving) takes place
uniformly over a large area in a layer with homogeneous composition and
uniformly distributed moisture, no internal forces will be generated in the
ground—there will be only a simple expansion of the freezing soil layer. But
when the expansion is restricted or made impossible, frost-heaving forces of
considerable magnitude may result and act upon structures.

Simple uniform expansion of freezing soils is almost never observed under
natural conditions because of the inhomogeneity of the soils (vertically and over
area), the more so in the presence of structures that act upon the soils of the
annual freezing layer (changing their temperature conditions, increasing the
pressure on underlying layers, etc.). Moreover, the soils themselves behave in

different ways on heaving, depending on their clay component, the mineralogical composition of the particles, the composition of absorbed cations, etc.

Characteristic heaving graphs measured during freezing for three typical soils (fine sand, fine clay, and silty loam) in the author's experiments appear in Fig. 34. Curve 1, which corresponds to the frost heaving of water-saturated sand when it is frozen from all sides, indicates that sand increases in volume very quickly, though by an insignificant amount, and then (on further cooling) decreases in volume like any other solid. But when water-saturated sand is frozen from one side (as described earlier in Sec. 2) with free drainage of water, its volume remains practically unchanged, i.e., no frost heaving is observed.

The fine clay (Fig. 34, curve 2) behaves completely differently. At first there is a slight contraction of the clay specimen (because of the increase in the absorption forces of the film water and its viscosity on cooling and also possibly because of compression of trapped air bubbles); then we observe heaving of the soil, which occurs not only during the initial phase of freezing, but persists for a considerable time afterward as the soil is cooled further and new portions of loosely bound film water are frozen (similar experiments indicate that the frost heaving of fine clays is still continuing at temperatures of $-10°C$ and lower).

The results of a more detailed experiment carried out to investigate the frost heave of clay (specimen height 95 mm, initial moisture content $W_d = 48$ percent with freezing from the top and seepage of water into the specimen at the bottom) appear in Fig. 35.[31] According to curve a of Fig. 35, heave of the clay is observed continuously (even eight hours after the start of freezing, at soil sample temperatures from -0 to $-10°C$), and only after 25 hours, when the specimen's temperature was down to 26°C, could no further heaving be detected. In clays, therefore, an increase in the heave forces (under confinement or when it is impossible for their volume to increase on freezing) may occur (as is also

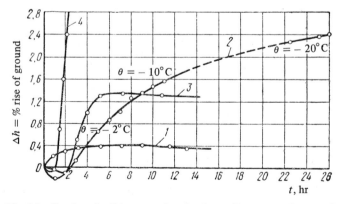

Fig. 34. Graphs of soil heave on freezing (according to experiments of N. A. Tsytovich): (1) sand; (2) fine clay; (3) silty-mud loam; (4) same, with water influx.

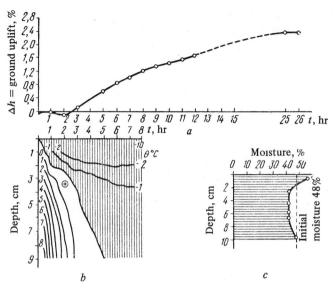

Fig. 35. Heaving curve of Cambrian clay (freezing from the top). (*a*) Frost-heaving curve; (*b*) thermoisopleths of freezing soil sample; (*c*) distribution of soil moisture after freezing.

confirmed by field observations) throughout the time of freezing and subsequent cooling, which causes freezing of more and more of the film water.

The frost heave curve obtained for the silty soil with inflow of water (Fig. 34, curve 4) is highly interesting; at first, it indicates (as in the case of the clays) a certain contraction, followed by very strong heave (which may reach 10 percent and more) in almost direct proportion to time, and then (after the soil sample has been completely frozen), as in the case of the sand (curve 3), a decrease in volume as a result of thermal contraction. The observations indicate that silty loams are the soils most susceptible to heaving, since they are quite permeable to water, their particles are small, and they contain colloids.

The frost heave forces that develop during freezing of soils as a result of the operation of various external and internal factors may assume various values and directions. Two basic types of these forces are now distinguished in accordance with their effects on structures: normal and tangential frost-heave forces.

Normal frost-heave forces act perpendicular to surfaces of structures that limit, resist, or prevent the increase in soil volume during freezing. They are most typically encountered at the footings of structural foundations, where vertical normal forces may arise under certain conditions, and at the side faces of foundation walls, where horizontal perpendicular frost heave forces may operate.

The magnitude of normal frost heave forces is influenced by a number of factors: the properties of the freezing soils (their particle sizes, capacity for adsorption, the level of particle surface free energy, etc.), the compressibility of the underlying soil layers, the external pressure on the soil, and the rigidity (deformability) of the structural elements taking the heaving forces.

The order of magnitude of the maximum normal frost heave forces can be estimated from the pressures developed by the ice crystals during confined freezing of water. As we know, the highest pressures will be developed only when it is totally impossible for the water to expand when it freezes. According to physical data, this pressure reaches an enormous value on the order of 2,115 kgf/cm^2 at a temperature of $\theta = -22°C$. However, the pressures will be much smaller at temperatures above $-22°C$.

To estimate the pressures that can arise when water freezes with no possibility of volume expansion at temperatures above $-22°C$, let us follow the example of Kh. R. Khakimov[32] and the interesting proposal of Yu. G. Kulikov and N. A. Peretrukhin[33] and use the Bridgeman-Tammann empirical relation[34] for the thawing (melting) temperature of ice as a function of external pressure:

$$\Delta p = 1 + 127\theta - 1.519\,\theta^2 \qquad (II.15)$$

where Δp is the pressure in kgf/cm^2 and θ is the absolute negative tempterature in °C.

At normal atmospheric pressure (1 atm or 1 kgf/cm^2), ice melts, of course, at $\theta = 0°C$, as indeed follows from (II.15). At lower temperatures, a higher pressure is required, and ice will melt at the given negative temperature when this pressure is attained. If the pressure is higher than that indicated by the Bridgeman-Tammann formula, the water will not freeze at the given negative temperature $(-\theta°C)$.

A simple calculation by formula (II.15) gives

$$
\begin{aligned}
\text{At } \theta &= -0.01°C & \Delta p &= 2.27 \text{ kgf/cm}^2 \\
\theta &= -0.1°C & \Delta p &= 13.7 \text{ kgf/cm}^2 \\
\theta &= -0.5°C & \Delta p &= 64.1 \text{ kgf/cm}^2 \\
\theta &= -2.0°C & \Delta p &= 249 \text{ kgf/cm}^2 \\
\theta &= -5.0°C & \Delta p &= 598 \text{ kgf/cm}^2, \text{ etc.}
\end{aligned}
$$

These will be the maximum pressures that the ice can develop without melting under the action of an external pressure at negative temperatures. However, these pressures can be developed only when the water is frozen in a rigid container that is sealed on all sides. When water freezes in soils, the actual pressures developed in them will be lower than the magnitudes given.

Calculations indicate (even under conditions favoring the development of normal heave forces) that if soils freeze at temperatures near $-0.01°C$, as occurs in sandy soils and coarse-granular soils in general, then, with consideration of atmospheric pressure, no normal frost heave forces should appear theoretically, even when the soil is placed under an added external pressure of about 1.27 kgf/cm^2, since the soil will not freeze. But for other soils (fine clays) with lower pore-water freezing temperatures, the pressure of the growing ice crystals may (under favorable conditions) reach substantial values, on the order of several tens of kgf/cm^2.

It should be noted that the premises for further direct use of the Bridge-mann–Tammann formula were not in conformity to the problem at hand,[37, 33] namely the compressibilities of the soils were determined from the compression curve, i.e., with vertical constancy of the compressive stresses (conditions under which lateral expansion of the soil is impossible), and the distribution of pressures under the external load took account of the decrease in these pressures with depth (three-dimensional problems). Beside that, the thickness of the compressible zone was determined (for nonconfluent layers of seasonal frozen ground and permafrost) arbitrarily, in accordance with the approximate recom-mendations of the SNiP (Construction Norms and Rules) for the case of operation of a local load, according to which the compressible zone thickness corresponds to the depth at which the external compressing pressure equals 0.2 of the natural (habitual) value. These assumptions are arbitrary and the resulting values for the normal frost-heave forces are clearly low and in contradiction to the results of direct experiments and observations.

However, numerous observations in the field have shown that an external pressure actually reduces soil frost heave forces, offseting them partially or completely, depending on the properties of the freezing soils and the rigidity (compliance) of the structure resisting the soil volume increase during freezing.

Since the above problem of determining the normal frost heave forces has no complete analytical solution at the present time, let us turn to an analysis of direct experiments performed to determine the normal frost-heaving forces of soils using special machines.

Experimental determination of normal frost heaving forces is an extremely complex problem from the methodological standpoint.

Attempts to determine frost-heave forces were made long ago. Thus, M. Ya. Chernyshev[35] drew attention back in 1928 to the upheaving of bridge supports and made the first approximate estimates of the total heave force on the basis of field observations. N. I. Bykov (1940)[36] performed a special field experiment to determine the heave forces in wooden posts (by equalizing the upheave forces with an external load) and, in the absence of detailed data on depth of frost, the force distribution over the adfreezing surface, etc., referred the resulting heave force to a unit foundation perimeter. These and other experiments (through the 1940's) were concerned for the most part with determination of the total heave force when soils froze around foundation posts, without resolution of the force into normal and tangential components. The experiments were also isolated and performed on a small scale.

It appears that the first laboratory studies of normal frost-heaving forces were performed at the Leningrad Construction Engineering Institute (LISI) by N. N. Morareskul[37] under the supervision of the present author; they showed on the basis of experiments performed on a special stand with a beam dynamometer that the normal frost heave forces of soils may reach considerable values (on the order of 5–8 kgf/cm^2 and higher) and depend not only on the rigidity of the structure resisting the heaving, and on the external load, but also on the compliance of the underlying layers.[38]

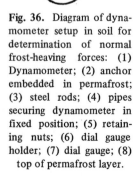

Fig. 36. Diagram of dynamometer setup in soil for determination of normal frost-heaving forces: (1) Dynamometer; (2) anchor embedded in permafrost; (3) steel rods; (4) pipes securing dynamometer in fixed position; (5) retaining nuts; (6) dial gauge holder; (7) dial gauge; (8) top of permafrost layer.

Vertical normal frost-heave forces acting on foundation-column footings when the soil under them freezes were apparently first determined under field conditions in 1958 by V. O. Orlov,[39] and we shall discuss them in some detail.

The normal heave forces were measured with the aid of a special dynamometer plunger with an area of 200 cm^2, which was in contact with the footing of an experimental column simulator (Fig. 36). The total magnitude of the normal frost-heave forces acting on the dynamometer footing was determined from the deflection of the steel dynamometer plate. Simultaneously with measurement of the heaving force, the thickness of the frozen soil layer under the dynamometer footing, its minimum temperature, and the rate of soil heaving under the dynamometer were also determined.

Figures 37 and 38 present time curves of the normal frost-heaving forces of clayey soils that were in a plastic state before freezing (see Ref. 39, Table 3).

The dynamometer force variations in which are shown in Fig. 37 was placed at a depth of 1.65 m (the top of the permafrost layer on this test area was 4.5 m); the dynamometer corresponding to Fig. 38 was placed at a depth of 0.62 m, and in this case the annual freezing layer had been steamed out during the summer down to the permafrost to a depth of 1.70 m.

In addition to these test posts, two more posts were placed with the dynamometers at depths of 1.09 and 1.91 m.

It was established on the unit with the dynamometer buried at 1.09 m that the normal frost-heave force was 1.6 kgf/cm^2 even at a soil-freezing depth of 0.18 m under the dynamometer bottom, and that it continued to increase with further thickening of the frozen layer under the bottom, reaching 47 kgf/cm^2 at a freezing depth of 0.67 m.

The normal heaving forces appeared to increase even further, but the dynamometer had been calibrated only to 50 kgf/cm^2.

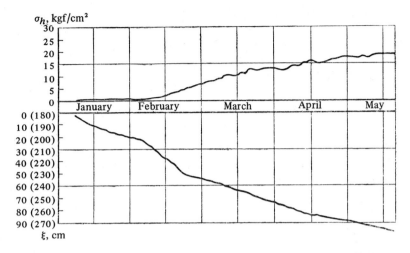

Fig. 37. Time variation of normal frost-heaving forces σ_h and soil-freezing depth ξ (heavy silty sandy loams) under posts (according to V. O. Orlov).

The results of the experiments in which Orlov measured normal frost-heaving forces of soils under foundation footings appear in Table 7 in somewhat condensed form (for three test posts).

Needless to say, these data must be regarded as approximate, reflecting only the order of magnitude of the normal frost heave forces that arise on freezing of particulate soils under foundation footings with severely limited opportunity for volumetric expansion.

According to the data given in Table 7, the normal frost heave forces (according to the results of field experiments) can, under certain conditions, reach enormous values—on the order of 50 kgf/cm² and more. As observations of the volume of the block of soil that supports development of normal frost-heave forces show, it is substantially larger in area than the footing, due to the strength of the structural bonds in the frozen soil.[39] However, the volume of soil that resists compression under the

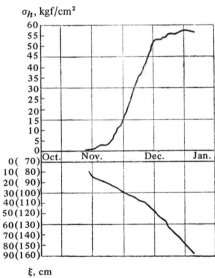

Fig. 38. Time variation of normal frost-heave forces σ_h and soil freezing depth ξ (silty heavy loam) under column footing (according to V. O. Orlov).

Table 7. Normal heaving forces according to field experiments

Location of test foundation	Depth of dynamometer footing, m	Thickness of frost layer under dynamometer, cm	Negative temperature of frozen layer, $\theta°C$	Rate of soil heave under dynamometer, mm/day	Normal forces under dynamometer footing, kgf/cm^2
Annual freezing layer confluent with permafrost layer	0.62	16	−0.4	−	1.7
		20	−0.3	1.38	2.8
		24	−0.7	1.38	6.0
		30	−1.4	1.07	20.5
		35	−3.0	0.92	35.2
		61	−4.8	0.63	56.0
Annual freezing layer not confluent with permafrost layer, the latter at depth of 4.5 m	1.09	18	−0.9	0.95	1.6
		33	−1.4	1.36	5.1
		41	−2.4	1.35	9.5
		51	−4.4	1.22	31.0
		60	−4.7	1.12	38.0
		67	−4.9	0.96	47.0
Same	1.65	21	−0.8	0.6	0.4
		34	−1.8	0.57	1.5
		41	−2.0	0.50	2.6
		51	−2.3	0.47	5.2
		60	−2.5	0.42	10.3
		70	−2.4	0.39	13.0
		82	−2.7	0.35	15.6
		92	−3.2	0.31	19.1

actions to the normal frost-heave froces remains undetermined, as do many other circumstances, not to mention problems of measurement methods, consideration of the effects of weather factors on the instrument readings, etc.—all of which requires further study and consideration.

As a general conclusion from the results of the above field and laboratory experiments on normal frost-heave forces in soils studies, we have a proposition familiar in foundation engineering: *foundations of structures should in no case be laid above the depth of seasonal frost penetration of soils*, since otherwise the appearance of normal frost-heaving forces (which are, in addition, different in different places) will inevitably cause totally unacceptable deformations of the structures.

The statement that normal frost-heave forces can be offset by the light pressure from the structure (on the order of 1–2 kgf/cm^2, and sometimes even slightly larger) is, in our opinion, a misconception: everything depends on what kind of soil is freezing, the conditions of soil freezing under the footings, and the resistance offered by the foundations and superstructures to the volume expansion of the soils on freezing. An exception is the case in which the structure is built on sandy and other coarse-grained soils with full provision for free drainage (pressing-out) of water from the soils; in this case, it is acceptable to lay foundations above the depth of seasonal soil freezing.

The problem of normal frost-heave forces requires further specialized research.

Several experiments were performed (by B. I. Dalmatov[41]) to determine horizontal frost-heave forces exerted normal to the surface of a foundation surrounded by a frost susceptible soil that freezes against its side faces.

The method used to measure the heaving forces was the same as that used in investigating the vertical normal frost-heave forces, i.e., dynamometers set flush against the side faces of a column were used.

The experiments showed that in the process of symmetrical (with respect to the vertical axis of the test pile foundation) freezing of soils at the lateral faces of the foundations, horizontal heave forces on the order of 0.3-0.35 kgf/cm^2 made their appearance at a frozen layer thickness of 0.9-1.0 m, increasing to 0.64-0.66 kgf/cm^2 at freezing depths of 1.45-1.72 m.

Tangential frost-heave forces result from the action of soils that freeze against the side faces of foundations, and are directed along tangents to the foundation surface in contact with the freezing soil. Knowledge of the tangential frost-heave forces is absolutely necessary for the design of foundations for structures to be erected on permafrost and deep winter frost penetration, since designs for heave of foundations are based on tangential frost heave forces, which were previously identified with the so-called forces of adfreezing of the soils to the foundation material. But the latter, according to M. Ya. Chernyshev,[35] connote the total force that must be applied to the foundation to break its bonds with the soil frozen around it.

Many experiments have been designed for determination of adfreezing forces, or, more precisely, the strength of ground adfreezing to foundation materials as a function of the composition of the soils, their moisture (ice) contents, the magnitude of negative temperature, and the rate of application and duration of the shearing load. (N. A. Tsytovich, 1932; I. S. Vologdina, 1936; M. N. Gol'dshteyn, 1940; S. S. Vyalov, 1956, and others).[42]

The results of these experiments indicated high strength of ground adfreezing to foundation materials (on the order of 1 to 35 kgf/cm^2) and established a number of important relationships, namely:

1) water-saturated (icy) frozen sands have the highest adfreezing strengths;

2) gravelly and pebbly soils (thoroughly soaked with free run-off of the water) show the lowest adfreezing forces, not greater than 1 kgf/cm^2;

3) adfreezing strength decreases substantially with decreasing soil moisture content;

4) the adfreezing forces of moist soils and ice increase with decreasing temperature (approximately linearly down to $\theta = -20°C$);

5) the adfreezing strength of many grounds, especially clayey soils, depends both on the rate of load increase (increasing with load) and on the time of operation of the load, decreasing considerably with increasing time of load action.

These tenets, most of which were established back in the early 1930's,[42,43,28,20] are fully in use today in the development of preventive measures against the heaving of foundations, as will be discussed in Part Two of the present book.

In its time, the All-Union Standard (OST 9032-39) recommended that the maximum adfreezing forces of the soils be used to design for frost heave stability of foundations, but this however, introduced heave forces several times higher than those observed in nature.

The adfreeze force problem was developed further in the work of N. A. Tsytovich, M. N. Gol'dshteyn, S. S. Vyalov, and others. It was established (N. A. Tsytovich, 1936) that the average adfreezing force is at least twice the heave forces on foundation models, that adfreezing strength decreases considerably with increasing time under load (M. N. Gol'dshteyn, 1948), and, finally, that a long-term adhesion-strength limit (long-term strength) that depends nonlinearly on the negative temperature is reached under prolonged action of a load (S. S. Vyalov, 1956, and others).

The two-year field experiment of N. I. Bykov (1940) at the Skovorodino Scientific Station of the People's Commissariat of Railroads (NKPS), which was described briefly above, had a strong influence on the investigation of frost-heave forces; according to the results, the frost-heave forces acting on a wooden foundation pile embedded in the active layer was only about 0.6 kgf/cm² on the side surface of the pile (120 kgf per one linear centimeter), so that later investigators were moved to concentrate their attention on field experiments for the determination of the heave forces of freezing soils. We note that much later experiments (1958-1963) carried out under the same conditions (at the Skovorodino Station) indicated heave forces approximately 2.5 to 3 times larger than those obtained by Bykov.

Foundation frost-heave forces are now determined under natural field conditions with special machines of two types: a beam type (Fig. 39a) proposed by V. F. Zhukov et al. and a balanced-force scheme (Fig. 39b) suggested by N. I. Bykov et al.

In the former, the dynamometer is a rigid metallic beam whose deflection under the heaving force is transmitted from a foundation column surrounded by freezing soil and measured with a sensitive dial gauge. The effective heave forces are calculated from the measured beam deflections.

In the second scheme, any motion of the test foundation column is compensated by an additional load, and the load that stops the heaving of the column completely during freezing of the soil is taken as the maximum total heave force.

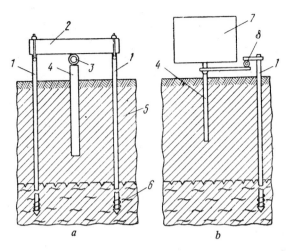

Fig. 39. Setup for field determination of soil frost heaving forces. (*a*) Beam type; (*b*) force-balance type. (1)
Anchor posts; (2) beam for pressure measurement; (3)
deflection gauge; (4) foundation column; (5) active layer;
(6) permafrost layer; (7) load box; (8) dial gauge.

Detailed observations made over a number of years on a beam-type heave-gauging device (designed by the Central Scientific Research Institute of Transportation Construction (TsNIIS))[44] frost susceptible silty loams at the Skovorodino Station established the jumpwise nature of the heave process, in which the frost-heave forces increased with increasing depth of frost (Fig. 40), and that the relative heave forces (per linear cm of column) measured during 1958-1963 varied in the range from 200 to 300 kgf/cm, while the average tangential heaving forces (referred to a unit lateral area of a test column surrounded by freezing soil) was found to be approximately 1-2 kgf/cm^2.

While the results of the above field experiments merit close scrutiny, the design of field tests to determine the frost-heaving forces of soils also requires that observations be made over a number of winter seasons, and only special experimental stations are capable of this. In addition, the results of experiments with beam-type heave-gauging devices, while they give us the order of magnitude of the ground frost heaving forces acting on foundations, require the application of corrections for the rigidity of the beam system, temperature effects, etc., and this somewhat complicates their use in practice.

All of the above described, called for continued improvement of the laboratory method of frost-heaving forces determination simultaneously with the field experiments; here the results of the laboratory tests were improved by reference to field observations.

Important results were obtained here by B. I. Dalmatov in his doctoral dissertation, which was prepared in consultation with the present author.

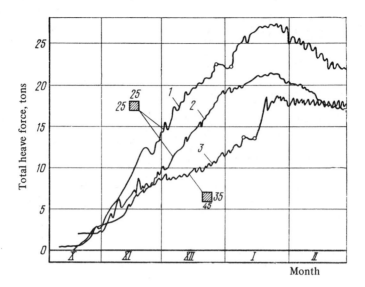

Fig. 40. Time variation of total heaving force for test columns 1, 2, 3 during the 1958–1959 season.

In laboratory exerpiments, Dalmatov investigated the forces of adfreezing of frozen soils to foundation materials on a special Dalmatov-Minin beam press[45] by freezing posts into frozen soil and pressing them through (in the scheme of Tsytovich's 1932 experiments) at different rates; he established that the so-called stable (maximum) adfreeze forces are actually very close to the tangential soil frost heave forces. Experiments performed by Dalmatov's method enable us to obtain curves of the adfreeze forces (more precisely, the resistance of the frozen soil to shear along the lateral surface of a post frozen into it) as functions of application time of the shear load as the post is moved through at various rates.

The beam press (Fig. 41) can be used to determine the shear strength of soil during continuous motion of a post frozen into it; here the speed of the post can be set at values from 2.2 to 20 mm/day, and hence placed close to the soil-heave rate under natural conditions. However, since no abrupt changes in the rate of displacement of ground against the foundation are observed under natural conditions, it is acceptable laboratory practice[46] to use a certain rather high average speed on the order of 10–20 mm/day in determining the shear strength of the soil against the post, velocities that are quite achievable with the Dalmatov-Minin press.

Dalmatov's detailed experiments showed that the curve of soil shear strength against the column (Fig. 42) has three segments: (1) a rapid increase in resistance to shear without any change in the force of adfreezing of the soil to the foundation column (at very small shears of the column with respect to the soil);

Fig. 41. Dalmatov-Minin beam press for determination of "stable adfreezing forces."

(2) a drop in the resistance and (3) a gradual approach of the shear resistance to a certain constant, stable value τ_{st}, when the strain has reached a certain maximum value for the soil in question.

Appropriate experiments[47] showed that this limiting value of strain in continuous shear will determine the "stable adfreezing" for icy clayey soils (silty sandy loams and loams) is approximately 10 mm.

Yu. D. Dubnov[48] showed that the maximum adfreezing force (the highest point on the curve of Fig. 42) can be taken as a working character-

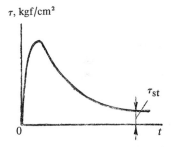

Fig. 42. Time curve of soil resistance to shear at surface of post model frozen into soil.

istic of the tangential heaving forces only for the initial phase of operation of the heave forces on the foundation, before the strength of adfreezing of the soil to the foundation has been broken and while the height of the frozen-ground layer is no greater than 0.3 m (under the Skovorodino conditions).

On the other hand, the average tangential heaving forces, as was recommended by B. I. Dalmatov[49] on the basis of laboratory and field tests, is practically equal to the stable adfreezing strength (more precisely, to the steady-state resistance of a post frozen into the soil to motion with respect to the latter), namely:

$$\tau_{heave} \approx \tau_{st}. \tag{II.16}$$

Relation (II.16) can now be recommended for use in practice.

Later detailed research by Dalmatov and others (e.g., the All Union Scientific Research Institute of Transportation Construction (VNII Transstroy)) have shown that the steady-state resistance of a frozen soil to shear against a foundation post depends primarily on the negative temperature ($-\theta°C$), increasing as it becomes lower, and that this relation can be assumed linear if the temperature is not very low (down to approximately $-15°C$):

$$\tau_{st} = c + b(\theta), \tag{II.17}$$

where c and b are the parameters of the linear relationship $\tau_{st} = f(\theta)$ and (θ) is the absolute value of the negative temperature in $°C$.

The following values of the parameters c and b have been obtained for various soils:

> B. I. Dalmatov
> For morane loam, $c = 0.5, b = 0.12$
> For silty loam, $c = 0.4, b = 0.10$
> For heavy silty loam with sand, $c = 0.4, b = 0.16$
> Yu. D. Dubnov
> For Skovorodino silty loam, $c = 0.356, b = 0.147$
> V. I. Puskov
> For loess type sandy loam, $c = 0.7, b = 0.22$
> For loess type silty loam, $c = 0.5, b = 0.18$.

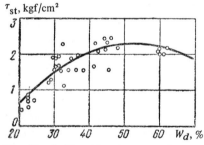

Fig. 43. Stable adfreezing strength τ_{st} of silty loam to wood vs. total moisture content W_d of frozen soil according to Yu. D. Dubnov's experiments.

Thus, the average values of the parameters of icy silty clayey soils (loams and sandy loams) can be put equal to $c = 0.4$–0.7 kgf/cm² and $b = 0.1$–0.2 kgf/cm² × deg.

The value of the stable adfreezing strength or, more precisely, the steady-state resistance of soils to shear against foundation materials, depends nonlinearly on the moisture (ice) contents of the soils (Fig. 43), and has a maximum as these contents change; this was first established by M. L. Sheykov[50] back in 1935–1936 for the shear resistance of frozen soils.

The values given for the stable forces of adfreezing of soils to foundation materials (and, consequently, for the tangential forces of frost heaving) pertain to a homogeneous soil layer for the case of a constant negative temperature throughout its entire depth. Under natural conditions, however, the negative temperatures in freezing soils are not constant, but vary from a maximum at the exposed surface of the soil to the freezing point of the pore water at the freezing front, which is very slightly below zero. The tangential heave forces (stable adfreezing forces) will also vary as functions of temperature, and this must be taken into account[51] in foundation design for heave resistance.

Interest attaches to the actual vertical distribution of the tangential frost-heaving forces acting on posts surrounded by freezing soils and to their variations in time, which were first determined by Dalmatov at Igarka using special dynamometers.[49] Further improvements of the method were proposed by K. Yu. Yegerev for laboratory conditions and by Yegerev and A. A. Zhigul'skiy for field work. Yegerev developed a special "pile instrument" for study of the magnitude and distribution of tangential frost-heave forces in foundation posts (piles); this device took the form of a sheet-duralimin tube whose elastic strains resulting from frost heaving of the soil surrounding it were determined by measuring the variations of the ohmic resistance of electrical strain gauges bonded with carbomide adhesive to the inner surface of the tube (or to special transmitting brackets).[52] The total tangential heave force was calculated from the measured results and used to determine the relative heave force τ_z expressed in kg/linear cm of column perimeter.

By way of example, Fig. 44 (which has been taken from Orlov's book) shows the curves obtained by Yegerev for the distribution of the relative tangential frost-heave forces acting on a pile surrounded by a freezing silty loam (in the Igarka area).

The following conclusions can be drawn on inspection of the curves in Fig. 44: (1) after appearing near the soil surface, the maximum of the relative heaving forces τ_z [kgf/cm] shifts downward with increasing depth of frost in the direction toward the freezing boundary, to approximately 2/3 of the freezing depth; (2) with decreasing temperature of the frozen soil layer, we observe an increase in the average specific value of the tangential frost-heave forces τ_{heave} [kgf/cm^2], and relaxation of the frost-heave forces with increasing displacement of the frozen soil against the column; (3) as the value for the specific (per unit adfreezing area) tangential frost heave forces (according to S. S. Vyalov[20] and V. O. Orlov) one may accept the long-term shear resistance τ_{lt} [kgf/cm^2] of the frozen soil against the foundation material (more precisely, the long-term shear limit.—N.T.).

We present certain values of the specific tangential frost-heaving force τ_{heave} (kgf/cm^2) obtained by generalizing field and laboratory investigations performed over many years at the Igarka Scientific Research Station[39] for sandy-loam soils of the Igarka region with total moisture contents $W \geqslant 30$ % (Table 8).

As a general conclusion from study of data on the tangential frost-heave forces in soils, we should note that it is preferable to determine these tangential forces experimentally to establish design characteristics, preferably using field heave gauges, but that their values can also be estimated from the results of laboratory tests; here the so-called stable adfreezing forces (according to

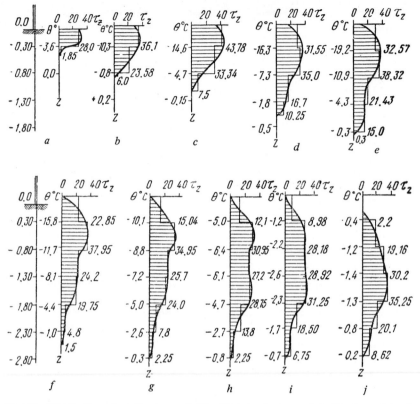

Fig. 44. Distribution of relative tangential frost-heaving forces, τ_z in kgf/cm, over lateral surface of pile instrument (according to experiments of K. Ye. Yegerev): (*a*) 4 November 1957; (*b*) 14 November 1957; (*c*) 25 November 1957; (*d*) 28 December 1957; (*e*) 11 January 1958; (*f*) 11 March 1958; (*g*) 16 April 1958; (*h*) 5 May 1958; (*i*) 26 May 1958; (*j*) 11 June 1958; *z* is depth measured from the surface and $\theta°$ is the temperature.

Dalmatov) or the ultimate long-term resistance of the frozen soil to shear against the foundation surface (according to Vyalov) are, in our opinion, very closely similar and can serve with accuracy sufficient for practical purposes to estimate the tangential soil frost heaving forces effective under natural conditions.

7. Frost Heave Countermeasures*

Countermeasures against the frost heaving of soils first came into use in our country in highway construction during the first half of the last century,[53] but in foundation work only during recent decades.

*Materials for this section were prepared by Ya. A. Kronik, a senior scientific staff member of the Moscow Construction Engineering Institute (MISI) (on the basis of his dissertation, 1970) and were reviewed by the present author.

Table 8. Specific tangential soil-heaving forces, averaged over depth: τ_{heave} in kgf/cm²

Geocryological conditions	Foundation material	Depth of frost penetration, m				
		0–0.5	0–1.0	0–1.5	0–2.5	0–3.5 and more
Freezing layer confluent with permafrost	Wood	2.7	1.4	1.0	–	–
	Reinforced concrete	3.0	1.6	1.1	–	–
Freezing layer not confluent with permafrost stratum (or permafrost absent)	Wood	2.7	1.4	1.2	1.0	0.8
	Reinforced concrete	3.0	1.6	1.35	1.0	0.8

Two basic trends can now be distinguished in the development and use of antiheaving measures.

1. Antiheaving melioration of the soils in the earthworks elements of structures and bases for them, the object of which is to reduce or totally eliminate frost heaving.

2. Antiheaving stabilization of foundations and structures, which ensures their stability to frost heaving of both natural and specially treated soils.

This trend will be discussed in Part Two of the book.

We shall dwell in greater detail on ground antiheaving melioration methods, which include four groups of measures:

1) mechanical measures, consisting in extremely simple treatment that changes the composition (completely or in part) and density of the soils or increases the loads on them;

2) thermophysical measures, whose purpose is to change the temperature-humidity conditions of the soils to reduce or completely eliminate the migration of moisture and frost heaving;

3) physicochemical measures, whose purpose is qualitative modification of the soils (to lower their free surface energy) in order to obtain nonheaving optimum soil mixtures or stabilized frost-resistant earth bases;

4) combination measures, which incorporate almost all of those described above and primarily artificial salinization and layer-by-layer dynamic recompacting of the soils.

The combined measures are based on studies made at the Laboratory of Engineering Permafrostology in Hydraulic Engineering of the Mechanics of Soils, Bases, and Foundations Department of the MISI (S. B. Ukhov, Y. A. Kronik, and others) in collaboration with the Construction Offices of the Vilyuy and

Khantaysk Hydroelectric Power Stations (GES) (G. F. Biyanov, Yu. N. Myznikov, and others).[54-55]

At the present time, the most effective and more highly developed methods of antiheaving soil melioration are the physicochemical and combination methods, and we shall examine them here in somewhat greater detail.

The physicochemical method embraces the following precedures:[56]

1) artificial salinization of the soils, which lowers their freezing temperatures and prevents freezing (sources: S. B. Ukhov, 1959; B. I. Dalmatov and V. S. Lastochkin, 1960; Ya. A. Kronik, 1968, and others);

2) artificial modification of the absorbing complex, by introducing inorganic compounds that sharply change the specific surface areas of the soils and their filtration and capillary properties (sources: T. W. Lambe, 1962; I. A. Tyutyunov and Z. A. Nersesova, 1957 and 1967, and others);

3) treatment of soils with waterproofing compounds that change the surface properties of the soil particles and render them less wettable (sources: M. T. Kostriko, 1957; H. Winterkorn, 1955; I. I. Cherkasov, 1957; I. V. Boyko, 1968, and others);

4) electrochemical treatment of the soils (sources: K. N. D'yakov, 1968; G. N. Zhinkin, 1959, and others).

In this author's opinion, a major contribution to the development of artificial soil salinization as a countermeasure against frost heaving came from research done by S. B. Ukhov[57] and Ya. A. Kronik et al. at the MISI under the author's supervision,[58] and by B. I. Dalmatov and V. S. Lastochkin (at the LISI), which made it possible to develop a basis for artificial antiheaving salinization of soils and to adopt it successfully to construction.

It is most advantageous to use technical sodium chloride in antiheaving salinization of soils; the soil can be kept from freezing down to $-21.2°C$ and the salinization operation can be performed either with crystalline salt or with a concentrated salt solution, the amount of either determined with the aid of the following formulas, which were derived by S. B. Ukhov.[59]

The amount z (kg) of the crystalline salt NaCl to be introduced into V m^3 of soil with a specific gravity γ [tons/m^3] and a moisture content of $W\%$ to obtain a soil-solution concentration (in g/cm^3) that prevents freezing of the soil at a temperature $-\theta°$ will be

$$z = k \frac{10\gamma W_{so} C_\theta V}{(1 + 0.01W)(\gamma_s - C_\theta)}$$ (II.18)

where k is a coefficient equal to the ratio of the weight of salt to its content of chemically pure NaCl, W_{so} is the moisture content of the soil solution ($W_{so} = W - W_{s.b}$) (%), where $W_{s.b}$ is the content of strongly bound water, which can be assumed equal to about 0.7 of the soil's maximum hygroscopicity; γ_s is the specific gravity of a solution of concentration C_θ [g/cm^3], and $\gamma_s = 1 + 0.65C_\theta$; $C_\theta = aC^n$ is the concentration of the soil solution that does not freeze at $-\theta°C$ in g/cm^3, and a and n are dimensionless parameters that have the following values for NaCl:

At θ = from 0 to $-4°C$ $a = -0.0167$; $n = 1.00$
At θ = from -4 to $-14°C$ $a = -0.021$; $n = 0.86$
At θ = from -14 to $21.2°C$ $a = -0.045$; $n = 0.59$

When a concentrated NaCl solution is introduced into the soil, the amount of solution in liters, V_{os} with a salt concentration C_0 (g/cm^3) to be introduced into a volume V m^3 of soil with a specific gravity γ and a moisture content W_0 to produce a soil-solution concentration C_θ, g/cm^3 that prevents freezing at $-\theta°C$ will be determined by the expression

$$V_{os} = \frac{10\gamma W_{so} V}{1 + 0.01W} \times \frac{C_\theta}{C_0\gamma_{s\theta} - C_\theta\gamma_{os}} \qquad (II.19)$$

where $\gamma_{s\theta}$ is the specific gravity of the solution of concentration C_θ in g/cm^3 and γ_{os} is the specific gravity of a solution of concentration C_0 in g/cm^3.

The cited papers [54-58] also discuss questions as to the redistribution of the salt with the passage of time in a salinized block of soil as a result of its diffusion, desalinization of the soil block by filtering waters, and the influence of the degree of salinization of the soil on frost heaving and variations of the soil's physicomechanical property indicators (compressibility and resistance to shear).

In practical antiheaving salinization of soils, it is recommended that the soil be treated no less than 2-3 months before the onset of frost with consideration of the rate of redistribution of the salt in ground mass (as determined by the appropriate calculations).

In recent years, artificial soil salinization has been used successfully in a number of localities to prevent soils from freezing up and to make it possible to work them in winter and use them as fill in earthworks.

Thus, successful examples of the use of this method are found in the experimental salinization of ground at the base for the lightly loaded foundations of the Vorkuta Telecommunications Center (under the supervision of I. V. Boyko) and in a number of successful experiments by American highway engineers (H. Smith, R. Hardy, and others).

Later research and practical experience has shown that by itself, artificial salinization of soils as a countermeasure against freezing and heaving is not effective for a sufficiently long time, since the soil is desalinized after two to four winter seasons and heaving resumes. On the other hand, salinization of soils for one-season prevention of freezeup or as an antiheaving treatment for two to three seasons can be recommended for extensive practical use.

Artificial modification of the absorbing complex in soils as a method against heaving is based for the most part on the studies of Z. A. Nersesova[19] and I. A. Tyutyunov,[15] who showed that, by changing the composition of the exchange cations in the soils, it is possible to reduce frost heaving substantially, and that the exchange cations array themselves in the following series on the basis of heave-control effectiveness:

$$Na^+, K^+ > Ca^{++}; Mg^{++} > Fe^{+++}, Al^{+++}$$

As we noted previously (see Sec. 2 of this chapter and Figs. 27, 28, and 29), salinization of soils with chlorides having univalent cations (KCl and NaCl) completely eliminates migration of water toward the freezing front and frost heaving, even when free water influx is possible.

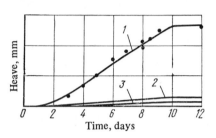

Fig. 45. Frost heaving of specimens of Skovorodino loam: (1) natural; (2) treated with KCl; (3) after removal of excess KCl.

Thus, Fig. 45 presents the results of Nersesova's experiments[56] with specimens of a natural silty loam 1 and loam treated with KCl (curves 2 and 3) in amounts corresponding to their absorption capacity, with the KCl treatment followed by prolonged washing with water to eliminate water-soluble salts. These experiments showed that the antiheaving effect of the potassium ion persists at full strength, since there was practically no migration of water or heaving of the soil samples that had been treated with KCl and washed free of the salt (Fig. 45, curves 3 and 2).

However, field experiments at Skovorodino[60] showed that the antiheaving effectiveness of KCl persists only for a few (about three) years.

Thus, physicochemical countermeasures against the frost heaving of soils, which are, in pure form, highly efficient and promising, nevertheless require the development of measures to prolong the duration of the antiheaving effect.

Treatment of soils for water repellancy completely eliminates their heaving properties, and the use of dispersing agents ($Na_2P_2O_7$) and coagulators ($FeCl_3$) lowers them substantially; however, these countermeasures against frost heaving are again not effective for long enough times, so that further research is necessary.

Electrochemical treatment of soils supplemented by the use of chemical additives (e.g., $CaCl_2$ solution) not only reduces frost heaving of excessively moist clayey soils, but also strengthens them considerably.[61] However, the long-term effectiveness of this procedure has not been adequately verified.

As we noted above, *the combination method of antiheaving soil melioration* incorporates a number of antiheaving measures, which form a soil structure that is stable over a rather long term: salinization to the equilibrium concentration of the pore salt solution, dynamic compacting of the soil to a density higher than the standard for earthworks consisting of unfrozen soils, and (if necessary) thermophysical measures (insulation) and surcharging.

This method was developed by Ya. I. Kronik in the MISI Department of the Mechanics of Soils, Bases, and Foundations[54-55] and is now being introduced with success in dam building in the north.

Salinization, which substitutes cations of the injected salt (usually univalent) for absorbed soil cations is followed by dynamic compacting of the soil to produce a water-stable soil structure that persists for a long time.

At the particular optimum density, chemical forces of cohesion become the predominant interaction forces between soil particles; they result from the conversion of coagulation structures to condensation structures (in P. A. Rebinder's terminology), which are distinguished by higher strength and do not develop residual deformations. Ya. A. Kronik established experimentally that the formation of a coagulation-condensation structure in clayey soils begins at a certain density, e.g., at a soil skeleton specific gravity $\gamma_{sk} = (0.90-1.03)\gamma_{sk\,opt}$ (according to VSN 97-63) for Vilyuy loams.

But to obtain a stable coagulation-condensation structure in clayey soils that persists over the long term, antiheaving salinization must be followed by dynamic compaction of the salinized soil to a density somewhat higher than the optimum for unsalinized soils—for example, to $\gamma_{sk} = (1.05-1.08)\gamma_{sk\,opt}$ (VSN 97-63) for Vilyuy loams.

The stability obtained by combining antiheaving measures for soil stabilization was checked on a special apparatus[62] by repeatedly freezing and thawing soil samples under open-system conditions.

Figure 46 shows the results of one of the MISI experiments, in which the samples were frozen and thawed 16 times.

It has also been established experimentally that the decrease in the relative heave $K_{h.s}$ of a salinized soil as compared with the same coefficient K_h for an unsalinized soil is exponential in nature (Fig. 47) and can be described quite accurately by an equation of the form

$$K_{h.s} = K_h e^{-\alpha(C_{p.s}/C_\theta)} - \beta \frac{C_{p.s}}{C_\theta} \qquad (II.20)$$

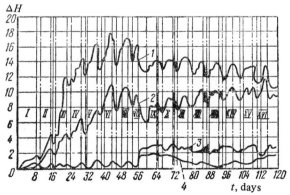

Fig. 46. Heaving of samples of Vilyuy loam on repeated freezing and thawing: (1) unsalinized; (2) salinized with NaCl; (3) salinized with KCl; (4) salinized with CaCl$_2$ (average $\gamma_{sk} = 2$ gf/cm^3; average $W \sim 15\%$).

where α and β are empirical coefficients that have the approximate values $\alpha = 7$ and $\beta = 9.12 \times 10^{-4}$ for the loams studied, C_θ is the equilibrium solution concentration that prevents freezing at the particular temperature $-\theta°C$, $C_{p.s}$ is the actual concentration of the pore solution in g/cm^3, determined from the formula[62]

$$C_{p.s} = \frac{s\gamma_{p.s}}{W_s - W_{n.v} + (1 + 0.01W_{n.v})s} \qquad (II.21)$$

s is the salinity of the soil in percent, W_s is the moisture content of the salinized soil in percent, $W_{n.v}$ is the moisture content of the "nondissolving volume," $W_{n.v} \approx W_{h.s}$, i.e., it may be put equal to the amount of strongly bound water (after A. F. Lebedev), and $\gamma_{p.s}$ is the specific gravity of the pore solution in g/cm^3.

In construction practice, it can be assumed with sufficient accuracy that $\beta = 0$; this simplifies (II.20) to the form

$$K_{h.s} \approx K_h \exp\left(-\alpha \frac{C_{p.s}}{C_\theta}\right) \qquad (II.22)$$

The resulting exponential relationship indicates that even on salinization to a pore-solution concentration $C_{p.s}$ equal to 0.3-0.5 of the equilibrium concentration C_θ, the relative heave is reduced by up to 15 times, while salinization to $C_{p.s} = (0.8-0.9)C_\theta$ suppresses heaving completely; here the minimum amount of salt that ensures the necessary replacement of absorbed cations (according to the experiments of F. I. Tyutyunova) should be about three quarters of the absorption capacity of the particular soil.

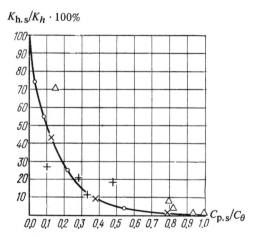

Fig. 47. Relative heave of soils as a function of the degree of their salinization. ○—Moscow-area loam; △—Vilyuy loam; ×—subsoil loam; +—Far Eastern loam.

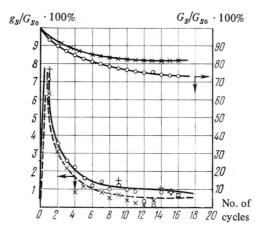

Fig. 48. Experimental plots of amount of salt removed (g_s) and retained by specimen of salinized soil (G_s) against number of freezing-thawing cycles (in %) and initial salt content (G_{so}): +–salinization with $CaCl_2$; ○–salinization with NaCl; ×–salinization with KCl.

MISI experiments indicate that desalinization of soils under repeated freeze and thaw occurs almost exclusively during the thawing phases, and that at least one to two days are needed to initiate this process. Only 20–27 percent of the salt was removed from salinized loam specimens as a result of 16 freezing-thawing cycles, most of the desalinization occuring during the first 4 or 5 cycles (Fig. 48). These data agree with the results of Motla's field observations, according to which up to half of the injected salt was still present in a salinized soil after 10 years.

We mention that in construction of the impermeable core of the Vilyuy Hydroelectric Power Station dam, where more than 300,000 cubic meters of loam fill were placed at negative air temperatures down to −40°C, it was found more economical to place the fill with layer-by-layer antiheaving salinization with NaCl and $CaCl_2$ solutions and appropriate compacting than to salinize the entire volume of winter fill in advance.[63]

The above research results and the practical experience gained in construction of the Vilyuy and Khantaysk Hydroelectric Power Station dams from locally available materials indicate that the combined soil antiheaving melioration method can be recommended for wider use in the construction practice of hydraulic-engineering facilities and bases for lightweight structures, and also to make possible the placement of cohesive-soil fill in winter.

References

1. Bozhenova, A. P.: Importance of Osmotic Forces in the Migration of Moisture. Materialy po laboratornym issledovaniyam (Laboratory Research Materials), Collection no. 3, USSR Academy of Sciences Press, 1957.

2. Puzakov, N. A.: "Vodno-teplovoy rezhim zemlyanogo polotna avtomobil'nykh dorog" (Water and Temperature Regime in Dirt Roadbeds), Avtotransizdat, 1960.

3. Fel'dman, G. M.: Migration of Moisture in Soils on Freezing, in "Teplofizika promerzayushchikh i protaivayushchikh gruntov" (Thermophysics of Freezing and Thawing Soils), Nauka Press, 1964.

4. Bliznin, G. Ya.: "Vlazhnost' pochvy po nablyudeniyam Elisavetgradskoy meteorologicheskoy stantsii 1887–1889 gg" (Soil Moisture Content According to Observations of the Yelisavetgrad Meteorological Station from 1887 to 1889), St. Petersburg, 1890.

5. Chirikov, F. V. and A. Malyugin: Humidity Variation in Podzolic Soil During Freezing and Thawing, *Nauch-Agron Zh.*, No. 1, 1926.

6. Kachinskiy, N. A.: Freezing, Thawing, and Moisture Content of Soil, *Trudy Instituta Pochvovedeniya MGU*, 1927.

7. Sumgin, M. I.: "Fiziko-mekhanicheskiye protsessy vo vlazhnykh i merzlykh gruntakh v svyazi s obrazovaniyem puchin na dorogakh" (Physicomechanical Processes in Moist and Frozen Soils as Related to Heaving on Roads), Transizdat Press, 1929.

8. Taber, S.: The Mechanics of Frost Heaving, *J. Geology*, vol. 38, no. 4, 1930.

9. Fedosov, A. Ye: Mechanical Processes in Soils During Freezing in Their Liquid Phase, *Trudy IGN AN SSSR*, no. 4, 1940.

10. Bozhenova, A. P.: A Development of Taber's Experiments on the Heaving of Soils, *Trudy GIN AN SSSR*, no. 22, 1940.

11. Tsytovich, N. A.: Influence of Freezing Conditions on the Porosity of Water-Saturated Sands, *Vop Geol Azii*, vol. 2, USSR Academy of Sciences Press, 1955.

12. Bozhenova, A. P., and F. G. Bakulin: Experimental Studies of the Mechanisms of Moisture Migration in Freezing Soils, Collection No. 3, *"Materialy po laboratornym issledovaniyam merzlykh gruntov," (Materials on Laboratory Studies of Frozen Soils)*, USSR Academy of Sciences Press, 1957.

13. Bastamov, S. L.: On the Freezing of Soils, Collection 12, *Izd. Nauchno-Issledovatel'-skogo Instituta NKPS*, 1933.

13a. Hoextra, P. and E. Chamberlain: Electro-Osmosis in Frozen Soil, *Nature*, 203 (4952): 1406–1407, Sept, 1964.

14. Tyutyunov, I. A.: "Migratsiya vody v torfyanogelevoy pochve v periody zamerzaniya i zamerzshego yeye sostoyaniya v usloviyakh neglubokogo zaleganiya vechnoy merzloty" (Migration of Water in a Peat-Gel Soil During Periods of Freezing and in the Frozen State Under Conditions of a High Permafrost Table), USSR Academy of Sciences Press, 1951.

15. Tyutyunov, I. A., and Z. A. Nersesova: "Priroda migratsii vody v gruntakh pri promerzanii i osnovy fiziko-khimicheskikh priyemov bor'by s pucheniem" (The Nature of Water Migration in Soils During Freezing, and the Fundamentals of Physicochemical Heave Prevention Measures), USSR Academy of Sciences Press, 1963.

16. (Eliminated).

17. Ananyan, A. A.: *Dokl. vyss. shkoly*, 1959, No. 2.

18. Bokiy, G. B.: Crystal-Chemical Considerations Pertaining to the Behavior of Water in Frozen Clayey Soils, Vestnik MGU, *Geologiya*, no. 1., 1961.

19. Nersesova, Z. A.: The Influence of Exchange Cations, "Laboratorniye Issledovaniya po fizike i mekhanike merzlykh gruntov" (Laboratory Studies in the Physics and Mechanics of Frozen Ground), Collection 4, USSR Academy of Sciences Press, 1961.

20. Vyalov, S. S.: "Reologicheskiye svoystva i nesushchaya sposobnost' merzlykh gruntov" (Rheological Properties and Bearing Strength of Frozen Ground), USSR Academy of Sciences Press, 1959.

21. Lebedev, A. F.: Migration of Water in Soils and Subsoils, *Izvestiya Donskogo Sel'skokhozyaystvennogo Instituta*, vol. 3, 1919.

22. "Teplofizika promerzayushchikh i protaivaiyshcikh gruntov" (Thermophysics of Freezing and Thawing Soils), panel of authors, G. V. Porkhayev (ed.), Nauka Press, 1964.

23. Gol'dshteyn, M. N.: The Phenomenon of Absorption as a Cause of Soil Heaving, *Vestnik Inzhenera i Tekhnika*, no. 4, 1949.

24. Puzakov, N. A.: *Teoreticheskiye osnovy nakopleniya vlagi v dorozhnom polotnye i ikh prakticheskoye primeneniye (Theoretical Bases for Accumulation of Moisture in Roadbeds, and their Practical Application)*, Dorizdat, 1948.

25. See Ref. 17.

26. Ornatskiy, N. V.: *Design of Antiheaving Measures, Regulirovaniye Vodyanogo rezhima* DORNII Collection, (Regulation of Soil Water), Dorizdat, 1946.

27. Gol'dshteyn, M. N.: Migration of Moisture in Soils, NII NKPS Collection, *Issledovaniye raboty grunta v zh-d. sooruzheniyakh (Investigation of Ground Performance in Railroad Beds)*, NKPS Press, 1940.

28. Gol'dshteyn, M. N.: "Deformatsii zemlyanogo polotna v osnovanii sooruzheniy pri promerzanii i ottaivanii" (Earth-Bed Deformations in Base Fills in Earthwork Structures on Freezing and Thawing), Transzheldorizdat, 1948.

29. See Ref. 24.

30. Porkhayev, G. V., G. M. Fel'dman, et al.: "teplofizika promerzayushchikh i protaivayu-shchikh gruntov (Thermophysics of Freezing and Thawing Soils)," chap. II. Nauka Press, 1964.

31. Tsytovich, N. A., and M. I. Sumgin.: "Osnovaniya mekhaniki merzlykh gruntov" (Fundamentals of Frozen Ground Mechanics, p. 99, USSR Academy of Sciences Press, 1937.

32. Khakimov, Kh. R.: "Voprosy teorii i praktiki iskusstvennogo zamorazhivaniya grun-tov" (Problems in the Theory and Practice of Artificial Freezing of Ground), USSR Academy of Sciences Press, 1957.

33. Kulikov, Yu. G., and N. A. Peretrukhin: Determination of Normal Heaving Forces, *Trudy VNII Transportnogo Stroitel'stva*, "Transport" Press, no. 62, 1967.

34. Veynberg, B. P.: "Led" (Ice), Gostekhteorizdat, 1940.

35. Chernyshev, M. Ya.: Deformation of Wooden Bridges due to Ground Frost Heave. *Zheleznodorozhnoye delo*, nos. 1, and 2, 1928.

36. Bykov, N. I., and P. N. Kapterev: "Vechnaya merzlota i stroitel'stvo na ney" (Permafrost and Construction on It), Transzheldorizdat, 1940.

37. Morareskul, N. N.: "An Investigation of Normal Soil Heave Forces," dissertation, LISI, 1950.

38. Dalmatov, B. I., and N. N. Morareskul: An Investigation of Normal Heave Forces of Varved Clay on Freezing, Trudy LISI, no. II, 1951.

39. Orlov, V. O.: "Kriogennoye pucheniye tonkodispersnykh gruntov" (Cryogenic Heaving of Fine-Particle Soils), USSR Academy of Sciences Press, 1962.

40. (Eliminated.)

41. Dalmatov, B. I.: Issledovaniya kasatel'nykh sil pucheniya i vliyaniya ikh na fundamenty sooruzheniy (Study of Tangential Heave Forces and their Influence on the Foundations of Structures), Institut merzlotovedeniya AN SSSR, 1954.

42. Tsytovich, N. A.: Nekotoryye opyty po opredelennyu sil smerzaniya (Certain Experi-ments to Determine Adfreeze Forces), Bulletin No. 25, LO VIS, "Kubuch" Press, 1932.

43. Vologdina, I. S.: Sily smerzaniya merzlykh gruntov s derevom i betonom, Sb. 1, Laboratorniyye issledovaniya merzlykh gruntov (Forces of Adfreezing of Frozen Soils to Wood and Concrete, Collection 1, Laboratory Investigations of Frozen Ground), USSR Academy of Sciences Press, 1936.

44. Peretrukhin, N. A.: Sila moroznogo vypuchivaniya fundamentov (Frost-Heave Forces on Foundations), Trudy VNII Transportnogo Stroitel'stva, no. 62, "Transport" Press, 1967.

45. Dalmatov, B. I.: The Dalmatov-Minin Mechanized Beam Press, "Pribory i Stendy," vol. 2, PS 55–496, 1955.

46. Dubnov, Yu. D.: Frost Heave of Soils and Methods of Protecting Structures Against its Effects, Trudy VNII Transstroya, no. 62, N. A. Peretrukhin (ed.), "Transport" Press, 1967.

47. Tsytovich, N. A., *et al.*: "Materialy po laboratornym issledovaniyam merzlykh gruntov" (Materials on Laboratory Studies of Frozen Soils), Collection 1, USSR Academy of Sciences Press, 1953.

48. Dubnov, Yu. D.: Laboratornyye issledovaniya kasatel'nykh sil pucheniya (Laboratory Studies of Tangential Heave Forces), Trudy VNII Transstroya, no. 62, 1967.

49. Dalmatov, B. I.: "Vozdeystviye moroznogo pucheniya gruntov na fundamenty sooruzheniy" (Effects of Soil Frost Heave on the Foundations of Structures), Gosstroyizdat, 1957.

50. Sheykov, M. L.: Resistance of Frozen Soils to Shear, Collections I and II "Laboratornyye issledovaniya mekhanicheskikh svoystv merzlykh gruntov pod rukovodstvom N. A. Tsytovich" (Laboratory Investigations of the Mechanical Properties of Frozen Soils under the Supervision of N. A. Tsytovich, USSR Academy of Sciences Press, 1936.

51. Tsytovich, N. A.: Design of Foundations for Structures to be Built on Permafrost, *Trudy Gipromeza*, no. 2, 1928.

52. Yegerev, K. Ye.: An Electrical Method for Determination of Tangential Reactions Distributed Over the Lateral Surface of a Loaded Pile Frozen into Soil, Trudy Instituta Merzlotovedeniya AN SSSR, vol. XIV, 1958.

53. Tulayev, A. Ya.: "A Survey of Published Papers Devoted to the Study of Heaving and Countermeasures Against It," No. II, Dorizdat, 1941.

54. Tsytovich, N. A., S. V. Ukhov, and Ya. A. Kronik: Combined Physicochemical Countermeasures Against Frost Heaving of Soil Fill, Papers at Danube European Conference on Soil Mechanics in Highway Construction (May, 1968, Vienna), NIIosnovanii Press, 1968.

55. Kronik, Ya. A.: "Protivopuchinnaya melioratsiya glinistykh gruntov Kraynego Severa v plotinostroyenii" (Antiheaving Melioration of Clayey Soils of the Far North in Dam Construction), dissertation, MISI, 1970.

56. Nersesova, Z. A.: Heave of Silty Loams and Physicochemical Countermeasures Against It, *Trudy NIITransstroy*, no. 62, 1967.

57. Ukhov, S. B.: "Iskusstvennoye zasoleniye svyaznykh gruntov dlya vozvedeniya nasypey v zimneye vremya" (Artificial Salinization of Cohesive Soils to Build Embankments in Winter), dissertation, MISI, 1960.

58. Dalmatov, B. I., and V. S. Lastochkin: "Iskusstvennoye zasoleniye gruntov v stroitel'stve" (Artificial Salinization of Soils in Construction), Gosstroyizdat, 1966.

59. Ukhov, S. B.: Physicochemical Prevention of the Frost Heaving of Soils, Materials of Eighth All-Union Conference on Geocryology, No. 8, Yakutsk, 1966.

60. Dubnov, Yu. D.: An Experiment in the Use of KCl as a Measure Against Heaving of Structure Foundations, *Sb. Trudov VNII Transstroya*, no. 62, 1967.

61. Zhinkin, G. N., *et al.*: Electrochemical Treatment of Heaving Soils, Collection no. 62, VNII Transportnogo Stroitel'stva, 1967.

62. Kronik, Ya. A., S. B. Ukhov, and N. A. Tsytovich: Artificial Salinization of Soils as a Countermeasure Against Frost Heaving, "Osnovaniya, Fundamenty i Mekhanika Gruntov," 1969, No. 1.

63. Kronik, Ya. A.: Antiheaving Salinization of Loams for Winter Construction of the Core of the Vilyuy Hydroelectric Power Station Dam, Ekspress-Informatsiya OES, Seriya "Stroitel'stvo Gidroelektrostantsii", No. 8 (236), 1968. Batenchuk, Ye. N., G. F. Biyanov, *et al.* "Zimnyaya ukladka svyaznykh gruntov na kraynem severe" (Winter Placement of Cohesive Soils in the Far North), "Energiya" Press, 1968.

Chapter III

RHEOLOGICAL PROCESSES IN FROZEN GROUND AND THEIR SIGNIFICANCE

1. On Internal Bonds in Frozen Soils

Frozen and perenially frozen ground are bodies in which the stresses and strains that arise under the influence of an external load are not constant, but vary with time, giving rise to relaxation of stresses and creep (an increase in the strains with the passage of time).

These processes are called rheological (from the word rheology: Greek $\rho\epsilon\omega$, flow; $\lambda\omega\gamma\omega$ science, i.e., the science of the time variations (flow) of the stresses and strains in bodies).

The strong development of rheological processes in frozen soils is due to a peculiarity of their internal bonds, in which ice, which is an ideally flowing solid, plays a role of the first importance.

Three basic types of internal bonds must be distinguished in frozen soils.

1. Purely molecular bonds (van der Waals-London forces) at the points of contact between the solid mineral particles of the soils, the magnitude of which depends on the area of direct contact, the distance between the mineral particles, their compactness, and the physicochemical nature of the particles. These bonds increase with increasing external pressure, but the stability of the mineral particles may be disturbed at some of the contacts.

2. Ice-cement bonds, the most important bonds which are almost entirely responsible for the strength and deformation properties of frozen soils, depend on very many factors: degrees of negative temperature, the total content of ice in the frozen soils (their iciness), the structure and coarseness of ice inclusions and their positions with respect to the direction of the forces in operation, the contents of unfrozen water, gas inclusions, and cavities in the ice, etc.

3. Structural-textural bonds, which depend on the conditions of formation, shaping, and the subsequent existence of the frozen and permanently frozen soils. Different structural elements of frozen soils will be deformed differently, depending on their composition and structure. Here, structural inhomogeneity (the presence of aggregates, open porosity, etc.) will become an important factor; the greater the inhomogeneity of a frozen soil, the greater will be the numbers of structural and constitutional defects in it and the weaker will its structural elements and the frozen soil as a whole become.

The extremely complex structure of frozen soils must be kept in mind: together with the direct contacts between mineral particles, there will also be contacts and cohesion between mineral particles and ice and cohesion of the ice with film water; the ice-cement and ice-interlayer interaction forces will evidently be different (if only because they will have different areas of contact with mineral particles), and they will have different effects on the deformabilities of frozen soils.

Ice-cement bonds will also be of particular importance in a general evaluation of the state of stress and strain in frozen soils, and we shall discuss them in somewhat greater detail.

As we know, the mechanical properties of bodies are chiefly governed by the shear resistance of the particles with respect to one another. For homogeneous solids, this is a complex process, and it is all the more complex for frozen soils as four-component particle systems that owe their basic cohesiveness to ice, which exhibits a distinct anisotropy that becomes especially conspicuous in shear. Thus, the process of pure shear occurs in ice only when the direction of the shear coincides with the basis plane of the ice crystal (Chap. I[20]); in other directions, ice crystals break up and are reoriented (Fig. 49).

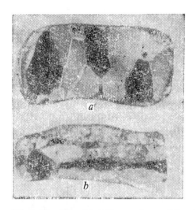

Fig. 49. Reorientation of crystals and recrystallization of ice interlayer in shear. (a) Before testing; (b) after long-term shear.

The cohesion of ice is much weaker in the direction of the principal optical axis (which is perpendicular to the freezing plane) than in the direction parallel to it, as can be judged (according to Shumskiy) even from the coefficient of linear expansion of ice, which is $29 \times 10^{-6} /°C^{-1}$ in the direction of the principal axis and $17 \times 10^{-6}/°C^{-1}$ in the perpendicular direction (at temperatures from 0 to $-66°$); this is explained by the "high intensity of motion of the atoms along the principal axis."

According to B. P. Veynberg (Chap. II[34]), the resistance of ice to crushing (which also characterizes its internal bonding) is 31-32 kgf/cm^2 parallel to the principal optical axis (the average of 1,246 determinations at a temperature $\theta = -3°C$) and 20-25 kgf/cm^2 perpendicular to the principal axis; according to B. A. Savel'yev[1] (average of 2,256 measurements) the ratio of the strength perpendicular to the principal axis to the strength parallel to it is 0.8.

The viscosity of ice for a force perpendicular to the principal axis is $\eta_{\perp} = 10^{10}-10^{11}$ P, and for a force parallel to the principal axis $\eta_{\parallel} = 10^{14}-10^{15}$.

This anisotropy of ice corresponds to an internal structure "like that of a deck of playing cards that have been smeared with a tacky adhesive" (D. McKennel). "The gaps between these platelets (the planes most densely occupied by atoms) are weakness planes and weakly cemented together (Chap. I[20]). This weakening has a considerable influence on the strength properties of ice, and it is along these planes that internal thawing propagates.

The magnitude of the critical shear stress at which ice goes into plastic flow is very low, to judge from the elastic limit in shear, which is below 0.1 kgf/cm^2.

The anisotropy of ice is manifested most strikingly in its plastic properties (which are governed by internal shears in the ice crystals), influences essentially the strength properties, and has almost no influence on elastic properties.

Apart from their distinct anisotropy, the internal bonds of ice are highly sensitive to changes in negative temperature, becoming stronger as the temperature is lowered. Thus, according to the experiments of N. K. Pekarskaya,[2] the instantaneous cohesion of ice (in a ball-indentor test) increased from 22 to 45 kgf/cm^2 as its temperature was lowered from -1.5 to $-3.5°$, i.e., it was approximately doubled. This can be explained by a decrease in the mobility of the hydrogen atoms in the structural lattice of the ice, with the result that the molecules of the ice become more stable and the ice stronger as the temperature is lowered.

The forces of cohesion between ice and mineral particles have not been determined, although experiments to determine the strength of adfreezing of ice to foundation materials (wood and concrete) were made back in the early 1930's (Chap. II[35]). These experiments, in which columns frozen into the soil were pushed through it, yielded (for a standardized load-increase rate of 20 kgf/cm^2 $\times$ min) the following results: at $\theta = -1°C$, $\tau_{adf} = 5$ kgf/cm^2; at $\theta = -5°C$, $\tau_{adf} = 6$ kgf/cm^2; at $\theta = -7°C$, $\tau_{adf} = 12$ kgf/cm^2, and at $\theta = -20°C$, $\tau_{adf} = 22$ kgf/cm^2, i.e., the ice-adfreezing forces increased substantially with increasing negative temperature.

In later experiments (S. S. Vyalov, Chap. II[20]), tests were performed by pulling wooden posts out of ice at various load-increase rates; brittle adfreezing failure was observed when the load was applied rapidly, but the column slid relative to the ice at a steadily increasing rate under long-term load application. These experiments showed that the long-term strength limit of ice in shear is very low, not exceeding 0.2 kgf/cm^2 at $\theta = -0.4°C$; this can be explained by the "instability of the ionic lattice of the ice, since the hydrogen ions, which are 100,000 times smaller than the oxygen ions, have high mobility and easily penetrate into the lattice interstices" (cited from S. S. Vyalov).

As we noted above, ice is present in frozen soils either in the form of ice cement (which occupies the pores between mineral particles and their aggregates) or in the form of ice interlayers—lenses, veins, and other forms of excess ice segregation.

The ice cement may take the following forms (according to P. A. Shumskiy): basal (when individual mineral grains are embedded in the ice), pore (when the ice occupies only pores), and film (with pores free of ice and water). A. M. Pchelintsev[3] adds three more types: contact (when films of water and ice are cut through by mineral grains), cellular (of complex structure with aggregates of clay particles with unfrozen water), and overgrowth ice (when film water grows into a layer of ice around a mineral particle). The above list of forms taken by

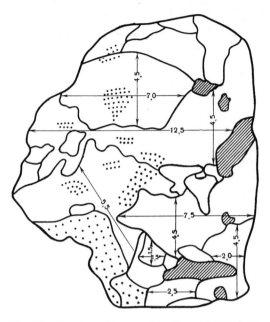

Fig. 50. Drawing of ice crystals (from ice interlayer at depth of 2 m in Igarka region): shaded areas are aggregates of mineral particles in the ice; circles are gas inclusions, and the solid lines represent grain boundaries; crystal dimensions are indicated in cm (drawing by A. M. Pchelintsev).

ice cement shows the complexity of the structure of frozen ground and its cementing by ice.

Although little study has been devoted to the structure of ice cement, crystal-optical investigation of the structure of excess ice (ice interlayers) has indicated extreme complexity of its structure, with different sizes and orientation of the crystals.

Studying the texture of ice interlayers from permafrost of the Igarka region (according to A. M. Pchelintsev, 1964 and) and considering results from studies of the deformation mechanisms of ice in glaciers (according to P. A. Shumskiy, 1961), we may arrive at the following conclusions:

1) the sizes of the crystals in permafrost ice interlayers are larger than those in seasonally frozen soils by factors of 14 to 25, and have areas up to 500 times larger;

2) the shapes of the ice crystals are irregular and often freakish, with orientation of the principal optical axes parallel to the ice interlayer;

3) at least three basic mechanisms of the deformation of ice should be distinguished: (1) flow of ice in slow shear parallel to the basal planes of the

crystals (without a change in the structure of the ice); (2) destruction of the space lattice of the ice with molecular breakdown, recrystallization, intergranular shifting, and breakup into fragments with randomized structure; (3) melting of the ice under very high shear stresses due to the heat of friction along cleavage planes.

At shear stresses larger than 1 kgf/cm^2, ice recrystallizes and the crystals are reoriented with the basal planes parallel to the shear forces; the new crystals are many times smaller in size.

Thus, it follows from the above that the presence of ice (ice cement and ice interlayers) in frozen soils, its plasticity observable under very small shear stresses, and the presence of viscous unfrozen film water (another important constituent of frozen soils) cause rheological processes to develop in a particulate frozen soil; the course taken by these processes also depends on the magnitude of the contact forces of cohesion and on the cohesiveness of the structural aggregates of the frozen soils.

2. Initiation and Development of Rheological Processes in Frozen Soils

The presence, in frozen and permanently frozen soils, of ice inclusions (ice cement and ice interlayers) in which, as we noted in the preceding section, a load of practically any magnitude causes plastic flow and crystal reorientation, together with the presence of viscous films of unfrozen water, initiates and furthers rheological processes when any additional load is applied.

We should note that the strength and deformation properties of frozen and permanently frozen ground differ from those of other solids and those of unfrozen soils primarily in that the application of an external load to a frozen soil always gives rise to irreversible restructuring, which causes stress relaxation and creep deformation even under very small loads, i.e., changes in the strength and deformation properties of the frozen soils with the passage of time. However, as has been shown by the author's experiments (1938-1940)[4] and as will be set forth in detail in subsequent chapters, frozen soils always retain elasticity even in the plastic range (in all stages of creep).

The influence of the time factor on the mechanical properties of frozen soils was established as early as 1935,[5] when the importance of the rate of load increase for the resistance of frozen soils to shear, compression, tension, and adfreezing was pointed out (Table 9), and rheological curves were first obtained experimentally and investigated in 1939 for frozen sand (W_d = 17-18%) and frozen clay (W_d = 54%) (Fig. 51).

The following conclusion can be drawn from examination of the experimental results given in Table 9 and Fig. 51:

1) the load increase rate has its strongest influence on the shear resistance of frozen clayey soils;

Table 9. Influence of load increase rate of resistance of frozen soils to external forces

Designation of soil	Moisture content W_d, %	Temperature θ, °C	Rate of load increase, kgf/cm²/min	Shear strength, kgf/cm²	Moisture content W_d, %	Temperature θ, °C	Rate of load increase, kgf/cm²/min	Adfreezing strength, kgf/cm²
Frozen sandy loam (68% of particles larger than 0.5 mm, 8% smaller than 0.005 mm)	19	−4	156.0	31.2	−	−	22.2	9.3
			46.8	30.4	14	−0.2	7.8	7.3
			23.2	21.8	−	−	1.8	5.7
			20.2	17.1	−	−	1.0	2.8
Frozen clay (36% particles smaller than 0.005 mm)	−	−	−	−	31	−6.6	⎰26.4	13.1
							⎱ 0.1	4.1
Frozen sand (100% particles smaller than 1 mm)	15–18	−1	15	30	−	−	31.9	32
			4	24	−	−	14.6	29
			−	−	18	−4.5	87.0	26
			−	−			4.7	23

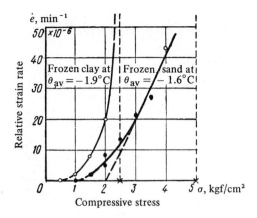

Fig. 51. Rheological curves of frozen clay and sand under uniaxial compression (according to experiments of N. A. Tsytovich, 1939): $\dot{e}$ is the relative strain rate.

2) the lower the rate of load increase, the lower are the strengths of the frozen soils (of all types);

3) the rheological curves for both frozen clay and frozen sand under uniaxial compression are curvilinear, but they become straight lines for sand at pressures $\sigma > 3$ kgf/cm^2.

It has been established on the basis of the data in Fig. 51 and other materials published in the previously cited paper that the stress corresponding to the onset of plasticity for frozen sand (at $\theta = -1.6°C$) is about 2 kgf/cm^2, while that for frozen clay (at $\theta = -1.9°C$) is approximately 1 kgf/cm^2.

Let us consider the creep curves of frozen soils and the forms that they take in the light of the most recent research on the structural changes in frozen soils under load.

To establish clarity for the subsequent analysis (physical and analytical) of the creep of frozen soils, the author considers it advantageous to scrutinize two separate classes of the creep of soils: I) decaying creep (Fig. 52) and II) sustained creep (Fig. 53).

Class I decaying creep occurs only when the stresses in the frozen soil are less than a certain limit, which is well-defined for a given physical state of the soil and a given negative soil temperature. But if the stress on the soil (for example, the external load) is increased beyond this limit, then irreversible structural strains that are sustained in time (class II sustained creep (see Fig. 53)) will appear at a certain value of the stress, and a further stress increase will only accelerate the creep deformation and cause brittle or plastic (shape change with stability loss) failure of the soil at an earlier time. On the basis of extended specially designed experiments performed at the underground (in permafrost

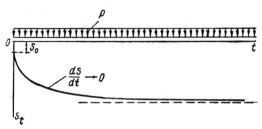

Fig. 52.   Curve of decaying creep.

soils) laboratory at Igarka, S. S. Vyalov arrived at the conclusion "that (sustained—N.T.) creep deformation inevitably goes over into a stage of progressive flow, which terminates in failure" (Chap. II[20]).

We can therefore discuss these two classes (I and II) of the creep of frozen soils, the more so since their analytical descriptions will differ.

It follows from the preceding discussion and from other studies of the physical nature of frozen soils that these soils are extremely inhomogeneous in structure as a result of the complexities of the interactions among the soil skeleton mineral particles, the various forms of ground ice, unfrozen water, vapors, gases, etc., and the complexity of the process in which the frozen soils are formed; this gives rise to conditions favoring the appearance of various local structural defects. In frozen soils, as in other solids but to a substantially greater degree, structural inhomogeneity results in the formation of cavities, unstable contacts between particles and particle aggregates, microscopic cracks, and other weak spots, which, according to the dislocation theory of Academician Yu. N. Rabotnov, can be regarded as points of development of local irreversible shears and local failures of continuity capable of developing and forming new defects under the action of an external load. As was apparently first demonstrated for frozen soils by Ye. P. Shusherina[6] and for unfrozen soils by M. N. Gol'dshteyn, A. Ya. Turovskaya, et al.,[7-8] with recent investigation by S. S. Vyalov, N. K. Pekarskaya, and R. V. Maksimyak,[9] the initiation and development of creep is governed by the development of microcracks, the breakdown of particle

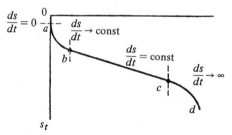

Fig. 53.   Curve of sustained creep. oa—Instantaneous; ab—nonsteady; bc—steady; cd—progressive stages of creep.

aggregates, and the growth of other structural defects in the frozen soil under load.

In the case of decaying creep of frozen soils, closure of microscopic cracks, size reduction of open cavities, and irreversible shears of particles relative to one another predominate. There is also "healing" of microscopic and macroscopic cracks as a result of thawing of ice at the points of contact between mineral particles (an effect to which the author first drew attention in 1937 and for which he supplied certain calculations)[5] and subsequent freezing of the ice in less heavily stressed zones of the frozen soil. Healing of microcracks has been observed in deformed clayey soils in petrographic experiments on thin sections of frozen soils, and secondary freezing of ice in frozen soils was observed in S. S. Vyalov's experiments with compression of the frozen soils (Fig. 54a) and in test-load experiments made on permafrost (Fig. 54b).

Decaying creep is characterized (on strain-time curves) by a progressive decrease in the rate of the irreversible strains, tending to zero as a limit. Damping creep is accompanied by reorientation and recrystallization of the ice, with a decrease in the sizes of the crystals (Fig. 55), so that the density of the ice is increased.

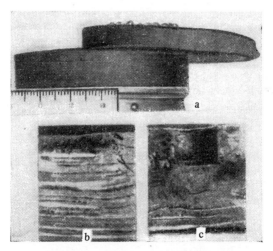

Fig. 54. Melting of ice and displacement of moisture in frozen ground (according to experiments by S. S. Vyalov). (a) Ice crystals formed from moisture pressed out under compression (at $\theta = -1.2°C$) in frozen-soil specimen after shear; (b) melting of ice and transfer of moisture under pressure ($\theta = -0.4°C$) in a block of frozen soil after impression of ram (ice inclusions formed at the boundary of the stressed zone are visible); (c) same, with puncture of ice interlayer in varved clay.

The equations of the rheological stress-strain state of frozen soils in decaying creep will be different from the corresponding equations for sustained creep.

As we noted above, sustained creep of frozen soils occurs only when the stresses on the soil are greater than a certain limit (the ultimate long-term strength, which will be discussed in the chapters that follow) and has (in addition to the instantaneous deformation, which is so small that it is often neglected in frozen soils) three stages, as established by the present author.[10]

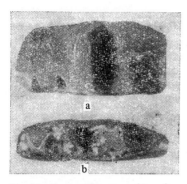

Fig. 55. Recrystallization of ice after prolonged compression. (*a*) Shape of crystals before deformation; (*b*) after deformation.

The first stage is that of nonsteady creep (Fig. 53, curve segment *ab*), in which the strain rate tends not to zero as it does in decaying creep, but to a certain constant value.

The second stage is one of steady creep, or plastoviscous flow at a practically constant rate of deformation; it may continue for various spans of time (which are sometimes very short under high stresses but then sometimes very long), but when the strains reach a certain value and a certain time has elapsed as necessary for restructuring of the soil, it invariably goes over into a third stage—a stage of progressive flow (the term coined by the author back in 1939), which is characterized by steadily increasing strain rates and terminates in brittle failure or plastic loss of stability of the soil.

In the first stage, according to the crystal-optical and microscopic studies that we mentioned earlier, the dominant process is one of closing of microcracks, healing of structural defects by moisture that has been pressed out of overstressed zones and refrozen, a decrease in free porosity due to particle dislocation, and partial closure or size reduction of macrocracks; all of these factors cause rheological compacting of the frozen soils reaching perceptible magnitude (in the experiments of Ye. P. Shusherina, by up to 2-2.5% by volume by the start of the steady-creep stage in a clayey soil at a temperature $\theta = -5$ to $-10°C$ and pressures up to 60 kgf/cm^2).

Closing of microcracks predominates in the second stage, and the decrease in the free porosity of the frozen soil is offset by the formation of new cracks, for the most part microscopic (although no macrocracks visible to the unaided eye were found in the experiments of Ye. P. Shusherina, S. S. Vyalov, et al.); after a certain time (sometimes a very long time), equilibrium is established between healing of existing structural defects and the generation of new defects, so that the strain rate becomes constant and the volume of frozen soil remains practically constant during this creep stage.

Among the constituents of frozen soils, the ice is the first to enter creep, since this requires the smallest shear stress; However, its viscoplastic flow is much slower than that of the mineral interlayers, since the viscosity of ice is higher than the viscosity of the frozen soils.

In the third, progressive stage of creep we observe the development of microscopic cracks, the appearance of new microscopic cracks at a steadily increasing rate with their transition to macroscopic cracks and the resulting weakening of the frozen soil; in addition, the ice inclusions have by now recrystallized and reoriented themselves with their basal planes parallel to the shear direction, causing a substantial decrease in the shear strength of the ice inclusions and, consequently, of the frozen soil as a whole.

Of greatest practical importance among all of the stages of sustained creep of frozen soils is the stage of steady-state creep, in which the frozen soil is in viscous flow without failure of continuity; here, if the time of the transition to progressive flow exceeds the design life of the structure (on the basis of the appropriate forecast), the erection of structures can be approved even on frozen soils that will come under load from the structures in the stage of viscoplastic flow.

However, the conditions for the transition from the stage of steady creep to the unacceptable stage of progressive flow, which terminates in failure of continuity and stability loss of the frozen soils, requires a special analysis.

3. Rheological Equations of the Stress Deformed State and Strain in Frozen Ground

The rheological equations of the stress-strain state of a soil establish relationships between stresses, strains, and time. Consideration of the time factor is the basic difference between rheological and ordinary equations.

The fundamental relationships to be scrutinized in the present section are the time variations of the deformations in frozen soils (their flow) under various types of loading. Rheological flows of frozen soils under load constitute the most important property of frozen soils and are of greater practical importance than, for example, their short-term compressive strength.

The physical aspect of the problem, namely the generation and development of rheological processes in frozen ground, was described in the preceding section, and the development of other types of strains in frozen soils, including those governed by changes in the proportions of the individual phases (components) in a given volume of soil, will be discussed in Chap. 4, which is devoted to general deformations of frozen soils. This section, however, will be confined to the equations of the rheological state of frozen soils.

If a frozen soil is placed under load in individual steps and the load at each step is maintained until deformation has stopped or until the strain rate has reached a practically constant value, the time variation of the strains can be represented by a curve of the kind in Fig. 56.

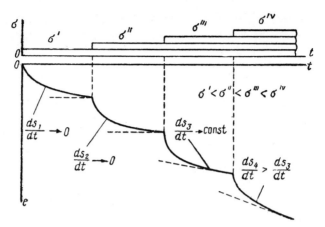

Fig. 56. Time variation of strains in frozen soil under stepped load.

The corresponding experiments indicate that the larger the effective load, the sooner will the soil go into viscoplastic flow at a steady strain rate whose value increases with increasing load[10] (Chap. II[42]).

The largest stress that can be imposed without sustained deformation of frozen soils (viscoplastic flows) determines the so-called long-term resistance or long-term strength of frozen soils. The accuracy of long-term strength determinations depends both on the size of the load steps and on the accuracy with which the strains are measured.

However, as will be shown in the next section, methods have now been developed for direct and quite accurate determination of the ultimate long-term strengths of frozen soils irrespective of the method used to measure creep deformations.

The stepped-load strain curve can be replotted as a family of curves by laying off the relative-strain values for various load steps from the same coordinate origin (Fig. 57). Then the upper curves (1 and 2 in Fig. 57a) will describe the variation of the strains in decaying creep, and the others will describe it for sustained creep.

The curves of Fig. 57a can be used to plot the strain e against the effective stress σ, as indicated in Fig. 57b, by measuring the strains corresponding to a given time interval (t_1 or t_2, etc.) under various stresses σ.

Experiments show that the $e_t = f(\sigma)$ plot is curvilinear for all values of t (except $t \approx 0$, when it may be a straight line), and that all curves (for the various t_i) for icy frozen soils can be regarded as similar[11] to one another and described by a single equation (but with various values of the parameters), e.g., by a power-law equation (as is conventional in soil mechanics):

$$\sigma = A(t)e^m \qquad \text{(III.1)}$$

where $m < 1$ is a strain-hardening coefficient that depends (according to S. E.

Gorodetskiy's experiments)[11,12] neither on temperature nor on the time of load action, $A(t)$ is a variable strain modulus (kgf/cm^2), which varies (as shown by the same experiments) as a function of the time t of load action and the temperature $-\theta$ according to a power law:

$$A(t) = \zeta t^{-\lambda} \tag{a$_1$}$$

where

$$\zeta = \omega(1 + \theta)^k \tag{a$_2$}$$

the parameters λ, ω, and k are determined experimentally.

Considering expressions (a$_1$) and (a$_2$) and determining the relative strain e from (III.1), we obtain

$$e = \left(\frac{\sigma}{\zeta}\right)^{1/m} t^{\lambda/m} \tag{III.2}$$

where the parameters m and λ and also k [formula (a$_2$)] are smaller than unity.

Relation (III.2) was derived for uniaxial compression and does not take account of the initial instantaneous deformation; however, experiments performed at the USSR Academy of Sciences Institute of Permafrostology and Moscow State University in connection with the problem of sinking deep mine shafts (down to 500 meters) at a deposit in the Kursk magnetic anomaly (KMA) with artificial freezing of the soils have shown that it gives a fully satisfactory description of the rheological deformation of frozen soils.

In the case of a complex state of stress, equation (III.2) becomes more complicated, since the forces acting on the soil must, in the general case, be regarded as consisting both of isostatic pressures p, which cause only volume deformation, and of the shear-stress intensity T, which determines the deformation of shape.

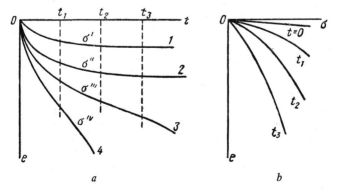

Fig. 57. Rheological curves for frozen soils under stepped loading. (a) Creep curves (1, 2 decaying; 3, 4 sustained); (b) relation between stress σ and relative strain e.

If the compressive and tensile strengths of the body undergoing deformation are the same, it can be assumed that the shear stress intensity T is a certain function (e.g.,[13]) of the shear strain intensity γ, i.e.,

$$T = \overline{A}(t)\Gamma^m \tag{III.3}$$

where

$$T = \sqrt{\frac{2}{3}(\tau_1^2 + \tau_2^2 + \tau_3^2)} \quad \text{and} \quad \Gamma = \sqrt{\frac{2}{3}(\gamma_1^2 + \gamma_2^2 + \gamma_3^2)}$$

whereby τ_1, τ_2, and τ_3 are the highest shearing stresses and γ_1, γ_2, and γ_3 are the largest relative shear strains.

In the case, for example, of triaxial compression, taking account only of the shape change of the frozen-soil specimen without volume change, we have

$$T = \frac{1}{\sqrt{3}} \times \frac{N}{F} \qquad \Gamma = \sqrt{3}e_z$$

where $e_z = \Delta h/h$, h is the height of the soil specimen, N is the effective axial load, and F is the cross-sectional area of the specimen.

In expression (III.3), the coefficient $\overline{A}(t)$ is the strain modulus in the complex state of stress. In the case of a triaxial stressed state, this modulus equals

$$\overline{A}(t) = 3^{-(m+1)/2}A$$

where A is the strain modulus for uniaxial compression (kgf/cm^2).

Equation (III.3) does not take account of the influence of the average principal stress $\sigma_{av} = \sigma_1 + \sigma_2 + \sigma_3/3$, whose value in the case of a triaxial compression test on a frozen soil (in a stabilometer) will be $\sigma_{av} = 1/3 \times N/F + p$ [where N is the axial compressing force, F is the cross-sectional area of the soil sample, and p is the isostatic (lateral) pressure]. However, S. S. Vyalov[14] has shown that the influence of the average principal stress may be quite substantial for frozen soils that do not resist compression and tension equally.

Considering the average pressure σ_{av}, the rheological equation of the stress-strain state of frozen soils assumes the form

$$T = \overline{A}(t)\Gamma^m + \sigma_{av}B(t)\Gamma^n \tag{III.4}$$

where $\overline{A}(t)$, $\overline{B}(t)$, m, and n are parameters determined from the experimentally established curves of $\phi_1(\Gamma) = T_0$ and $\phi_2(\Gamma) = \tan\psi$ shown in Fig. 58.*

We note that Yu. K. Zaretskiy used expressions (III.3) and (III.4) in creep calculations for frozen ground cylinders and in design for sinking the KMA deep shafts by the frost stabilization method.[11]

---

*The method of determining the parameters of (III.4) is set forth in detail by Ye. P. Shusherina in the paper by S. S. Vyalov and Ye. P. Shusherina entitled "Resistance of Frozen Ground to Triaxial Compression," Sec. 4 in the collection "Merzlotnyye issledovaniya" (Frost Research), Moscow State University Press, 1964.

A more general rheological equation of the stress-strain state of frozen soils, and one that is now often used in engineering creep and relaxation calculations for frozen soils and yields simpler solutions, is the equation of the Boltzmann-Volterra hereditary creep theory, according to which the deformation at any point in time depends not only on the stress at that time, but also on the history of the prior deformation.

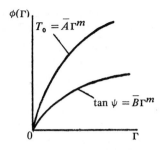

Fig. 58. Graph for determination of parameters of rheological equation of stress-strain state of frozen soils (with consideration of σ_{av}), plotted from experimental data.

Under continuous loading (or at a constant load), the total relative strain e will be composed of the instantaneous strain e_{inst} and the creep strain e_t, i.e., $e = e_{inst} + e_t$, a situation that is described in general form in the theory of hereditary creep by the equation

$$e = \frac{\sigma(t)}{E_{inst}} + \int_0^t K(t - t_0)\sigma(t_0)dt_0 \qquad (III.5)$$

The first term on the right-hand side describes the instantaneous strain, and the second the creep strain, which varies in time and is assumed proportional to the effective stress $\sigma(t_0)$, the time of load action dt_0, and a certain decreasing function $K(t - t_0)$, which is known as the creep coefficient.

The creep coefficient characterizes the effect of a load applied at an earlier time t_0 on the strain at time t, and equals the time-variable creep rate under constant stress referred to unit effective stress.

We note that the power-law equation (III.2) given previously for the total strain is a particular case of Eq. (III.5) with $m = 1$.

Depending on the analytical expression for the creep coefficient, Eq. (3.5) can describe either decaying or sustained creep.

We present expressions for decaying-creep coefficients that have come into practical use:

1) the exponential-type coefficient

$$K_1(t) = \delta e^{-\delta_1 t_1} \qquad (III.6)$$

2) the Rzhanitsyn coefficient

$$K_2(t) = \delta e^{-\delta_1 t_1}/t_1^{\delta_2} \qquad (III.6')$$

3) the Zaretskiy coefficient

$$K_3(t) = T/(T + t_1)^2 \qquad (III.6'')$$

where $\delta, \delta_1, \delta_2$, and T are experimentally determined creep parameters.*

*See Sec. 5 of this chapter.

We note that according to extensive experimental research done by Prof. S. R. Meschan at Yerevan, by this author at the MISI, and others, the exponential coefficient K_1 gives a good description of the decaying creep of unfrozen and thawed particulate clayey soils, while K_3, the hyperbolic coefficient, is most suitable according to studies by Yu. K. Zaretskiy[15] for description of the decaying creep of plastic frozen (icy) soils.

The process of sustained creep can be described by the following types of creep kernels:

4) the Duffing (power-law) coefficient

$$K_4(t) = \delta t^{-\delta_2} \qquad \text{(III.6''')}$$

5) an exponential coefficient, but one that is added to a constant rate ($\dot{e}_0$), i.e.,

$$K_5(t) = \delta e^{-\delta_1 t} + \dot{e}_0 \qquad \text{(III.6}^{\text{IV}}\text{)}$$

We note that the hereditary creep theory permits the use of any equation for the creep coefficient that describes the time variation of the creep rate referred to a unit effective stress if it is justified by experiments for the particular type of soil. Description of sustained creep is more complex and it is necessary in many cases to use the sum of several coefficients instead of a single one (for example, a sum of exponential coefficients), or to describe the stages of sustained creep (nonsteady and steady, with viscoplastic flow) with different equations.

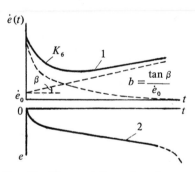

Fig. 59. Time variation of strain rate corresponding to coefficient K_6 (curve 1), and sustained creep (curve 2).

Thus, the following creep coefficient can be used to describe sustained creep with a gradual approach of the flow rate (in a uniaxial stressed state) to a nearly constant value:

$$K_6(t) = \delta e^{-\delta_1 t} + \dot{e}_0(1 + bt) \qquad \text{(III.6}^{\text{V}}\text{)}$$

where b is a proportionality coefficient.

Figure 59 illustrates the nature of the time variation of deformation rate that corresponds to coefficient K_6.

When creep of a frozen soil is dominated by the stage of viscoplastic flow with a practically established constant strain (flow) rate, the nonsteady-creep stage (segment ab, Fig. 53) can be disregarded if the time intervals involved are long enough; then the dependence of the steady flow rate on the effective stress can be used as the rheological equation for the state of stress and strain of icy frozen soils. In this case, we can put

$$\dot{e}_t = \frac{1}{\eta_{t,\theta}}(\sigma - \sigma_0)^n \qquad \text{(III.7)}$$

where $\eta_{t,\theta}$ is a time-variable viscosity coefficient that depends on the temperature of the frozen soil; σ_0 is the initial stress below which viscoplastic steady flow has not yet started sometimes known as the creep threshold, and n is a dimensionless coefficient greater than or equal to unity.

Experiments indicate that the creep threshold σ_0 is somewhat higher than the ultimate long-term strength, i.e., $\sigma_0 \geqslant \sigma_{lt}$.

If we take $\sigma_0 = \sigma_{lt}, n = 1$, and a constant coefficient of viscosity, i.e., $\eta_{t,\theta} = \eta_0$, then equation (III.7) assumes the form of the familiar Bingham-Schwedoff equation for the flow of plastoviscous bodies, namely:

$$\dot{e}_t = \frac{1}{\eta_0} (\sigma - \sigma_{lt}) \qquad (III.7')$$

where η_0 is the true value of the viscosity coefficient.

Finally, we may assume for ice that $\sigma_{lt} = 0$ and $n \neq 1$; then equation (III.7) assumes a simpler form that has been successfully used in a number of forecasts of glacier flow:

$$\dot{e}_t = \frac{1}{\eta_0} \sigma^n \qquad (III.7'')$$

The applicability of the above equations of the rheological stress-strain state, (III.1)–(III.7), for frozen soils has been thoroughly confirmed by the results of direct experiments.[16-17]

A method of testing frozen and permanently frozen soils for creep in simple[18] and complex[19] stressed states has now been elaborated in detail (chiefly at the NIIOSP); in the subsequent sections of this work, therefore, we shall describe only the determination of the parameters of the rheological state equations most frequently used in practice for frozen soils.

4. Stress Relaxation in Frozen Soils, and Ultimate Long-Term Strength

The need to consider the time factor in estimating the strength of frozen soils was demonstrated by the author's experiments back in the early 1930's, in which it was established that the resistance of frozen soils to external forces depends to a great degree on the time of action of the load and the rate at which it increases—the more slowly the load on a specimen of frozen soil is increased (for example, in a compression or shear test, etc.), the smaller is the load required for its failure. The strength of frozen soils decreases enormously when they are subjected to constant loads—according to our experiments, by a factor of 3 to 15, depending on composition, ice content, and temperature. This decrease is due to a rheological process of stress relaxation (slackening) in the frozen soil under the action of the constant load.

As we noted earlier (see Sec. 2 of this chapter), the action of a constant load on frozen ground results in recrystallization of ice inclusions, the development of microscopic cracks, and, at a certain stress value, growth of these cracks to macroscopic dimensions (up to loss of continuity of frozen ground or plastic loss of stability), together with a restructuring that lowers the shear strength and results in progressive flow.

The stress decrease $\sigma(t)$ at a constant strain is accurately described by the relaxation equation, which derives from the equation of hereditary-creep theory (III.5) when it is solved for $\sigma(t)$ at $e = $ const:

$$\sigma(t) = e\left[E_{inst} - \int_0^t R(t)dt\right] \qquad \text{(III.8)}$$

where $R(t)$ is the resolvent of the creep coefficient.

Differentiating Eq. (III.8) and solving it for $R(t)$, we obtain

$$R(t) = \frac{1}{e} \times \frac{d\sigma}{dt} \qquad \text{(III.9)}$$

i.e., the resolvent of the creep coefficient is a time-varying stress referred to the unit constant strain and determined from the relaxation-rate curve $(d\sigma/dt)$ by dividing its ordinates by the constant value of e.

Equation (III.8) is the one used for mathematical description of the relaxation (strength-loss) curves of frozen soils.

The strength decrease of a frozen soil can be determined in approximation as a function of the time of load action t by the use of a somewhat arbitrary formula recommended by the NIIOSP:[18]

$$\sigma_{lt} = \frac{\beta}{\ln \dfrac{t_\infty}{B}} \qquad \text{(III.10)}$$

where β and B are experimentally determined parameters, β having the dimensions kgf/cm^2, and B being equal to the number of years corresponding to the theoretical strength value at a sufficiently large value of t; t is usually taken equal to 100 years (with a 5% accuracy).

The strength-loss curves of frozen soils are known as long-term strength curves, since the ultimate long-term strengths of frozen soils (ult σ_{lt}) are determined from them.

Creep curves (for example, those shown in Fig. 57a) are used to plot curves of long-term strength.

The times corresponding to the initial increase in the rates, i.e., the times at which progressive flow begins (the inflection points on the creep curves corresponding to various stresses) are determined from sustained-creep curves (Fig. 60a). These data are used to plot the long-term-strength curves (Fig. 60b). The initial ordinate of this curve corresponds to the instantaneous strength σ_{inst} of the frozen soil, the ordinate at any time t' to the strength of frozen soil at that time, i.e., $\sigma(t')$, and, finally, the ordinate after a sufficiently long time interval,

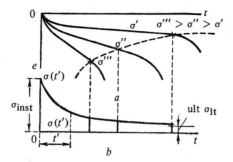

Fig. 60. Plotting of long-term strength graph for frozen soil from sustained-creep curves.

when the variations of $\sigma(t')$ can be neglected, determines the ultimate long-term strength ult σ_{lt} of the frozen ground.

This quantity is of overriding practical importance, since when $\sigma >$ ult σ_{lt} the creep will be sustained, but when $\sigma <$ ult σ_{lt} it will decay. Thus, ult σ_{lt} is the largest stress at which progressive flow has not yet started.

We note that the relaxation of shear stresses τ (kgf/cm^2) will be described by a long-term-strength curve quite similar to that of Fig. 60b.

The ultimate long-term strengths of frozen soil can also be determined directly from the results of a single test performed on the soil.

Two methods exist: (1) the well-known ball-plunger (ball-test) method proposed by the author back in 1947[20] for study of the cohesive forces and determination of the ultimate long-term strength of frozen soil, and (2) the dynamometer method developed by S. S. Vyalov (1964).[21]

The simplest method is to test the frozen soil with a ball plunger in a special machine (Fig. 61) with the frozen soil held at a strictly constant subzero temperature: in a cryostat, in a frost chamber, or in an underground cryological laboratory (the simplest expedient may be an appropriately equipped pit dug into permafrost).

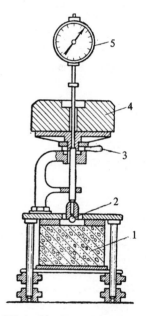

Fig. 61. Tsytovich ball plunger for testing cohesion and long-term strengths of cohesive soils. (1) Soil specimen; (2) ball plunger; (3) locking screw; (4) weight; (5) dial.

The ball plunger of the instrument is brought to rest against the surface of the frozen-soil sample or on a smoothed area of the frozen soil at the bottom of the pit (ball-plunger devices for field tests of soils in pits have now been developed and put to use) and then placed under a constant load P (kgf), which is transmitted to it via a rod.* The sinking of the ball is measured from the beginning of loading until the deformation of the soil has practically ceased, so that the complete long-term strength curve of the soil can be plotted later.

Knowing the diameter D of the ball plunger and having measurements of the sinking distances at various time intervals s_t under the given load P, we determine the average resistance of the soil (referred to a unit area of the spherical plunger impression) corresponding to a given sinking distance s_t from the formula

$$\text{Av } \sigma \approx \frac{P}{\pi D s_t} \tag{III.11}$$

With time, the plunger will sink farther into the frozen soil, but the distance increments will become smaller and smaller until, quite automatically and unequivocally, the resistance of the soil is balanced and the sinking distance reaches a stabilized state s_{lt}. Then the ultimate long-term strength ult σ_{lt} will be determined by the same very simple expression as before (III.11) with s_{lt} substituted for s_t:

$$\text{Ult } \sigma_{lt} = \frac{P}{\pi D s_{lt}} \tag{III.11'}$$

The author's studies have shown that in order to obtain invariant experiments, the following relation must exist between the stabilized sinking distance and the ball-plunger diameter:

$$0.01 < \frac{s_{lt}}{D} < 0.1$$

Under very small loads and at values of $s_{lt}/D < 0.01$, elastic (and not only plastic) deformations of the frozen soil will strongly influence the sinking distance, and at large loads and values of $s_{lt}/D > 0.1$ the formula for the average normal stresses will become highly approximate. We note that the NIIOSP "Instructions" (1970) recommend loads of 2 to 3 kgf on the plunger rod, for example, with a ball-plunger diameter of 22 mm on plastic frozen clayey soils and 4 to 5 kgf on frozen fine-sand soils.

Ball-plunger tests can be used to plot the complete long-term strength curve of a frozen soil for any time interval t (Fig. 62); here the stress that causes the ball to sink 10 seconds after the start of loading is usually taken as a conventional instantaneous strength.

As for determination of the steady-state sinking distance s_{lt}, the sinking of the ball plunger is usually observed for several hours, until deformation of the frozen soil ceases entirely. However, research at the Siberian Division of the USSR Academy of Sciences Institute of Geocryology has shown that the

---

*See, e.g., Fig. 32 in N. A. Tsytovich's book "Mekhanika gruntov" (Kratkiy kurs) [Soil Mechanics (A Short Course)], Izd-vo "Vysshaya Shkola," 1968.

long-term settling of the ball plunger into permafrost can be assumed approximately to be equal twice the sinking distance of the ball 30 minutes after the start of loading, i.e., $s_{lt} \approx 2.5's_{30°}$.

Thus, the ball-plunger method provides an extremely simple way of determining the ultimate long-term strength of frozen soils with accuracy sufficient for practical purposes and in unequivocal and automatic fashion.

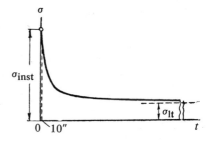

Fig. 62. Long-term strength curve of frozen soil (from ball-plunger test data).

The second method of determining the long-term strengths of frozen soils, the dynamometer method, was, as we noted above, proposed by S. S. Vyalov for use of a standard dynamometer as a spring press.

The use of a spring press to investigate the mechanical properties of soils was proposed and used successfully in tests of permafrost by the present author back in 1937.[22] Figure 63 shows a universal spring press of the model used for the permafrost field studies: determination of compressive strength, elastic modulus, and settling in the frozen and thawing states. The loading mechanism and dynamometer was a stiff steel spring whose deformations were determined by two dial gauges, while the deformations of the test specimen of frozen soil were measured with two sensitive (accurate to 0.001 mm) strain gauges of the type developed by Prof. N. N. Aistov. The press was used to test samples of cubical, prismatic, and cylindrical shape under conditions such that the soil could expand freely and conditions under which expansion was impossible (in odometers). This portable spring press could be used for field tests in any convenient excavation and yielded highly accurate measurements of the strains and the load transmitted to the soil specimen.

According to S. S. Vyalov, determination of the long-term strength limit ult σ_{lt} is almost automatic with the dynamometer (spring) press designed by V. F. Yermakov (Fig. 64).

In a dynamometer test of a frozen soil (in the form of a cubical or prismatic specimen), a load (somewhat

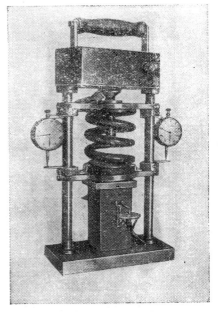

Fig. 63. Field-type spring press for testing frozen soils (1937 design).

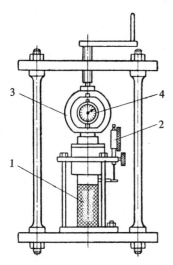

Fig. 64. Dynamometer press for determination of long-term compressive strength of frozen soils. (1) Soil specimen; (2) dial gauge for measurement of specimen deformation; (3) dynamometer; (4) dynamometer gauge.

smaller than the short-term compressive strength) is applied by pressing the dynamometer against the soil specimen, and the initial deformation λ_0' of the spring of standard dynamometer and the initial deformation λ_0'' of the frozen-soil specimen are recorded.

With time, stress relaxation will decompress the spring dynamometer, its deformation will decrease, and the deformation of the soil test specimen will increase until it reaches a final value (in a stabilized state) λ_f' for the dynamometer and λ_f'' for the soil specimen.

For a rigid dynamometer, the load P_f that corresponds to the long-term strength of the test soil will be

$$\text{lt } P_f = E\lambda_f' \qquad \text{(III.12)}$$

where E is the calibration coefficient of the dynamometer.

On the other hand, the long-term strength (per unit specimen area) is

$$\text{ult } \sigma_{\text{lt}} \approx \frac{\text{lt } P_f}{F} \qquad \text{(III.13)}$$

where F is the area of the frozen-ground specimen.

Adequate rigidity of the dynamometer, when its deformability can be disregarded, is determined, according to the 1970 NIIOSP Instructions, by the condition

$$\left(\frac{\lambda_0''}{\lambda_f''}\right)^{0.5} = 1 \pm \Delta \qquad \text{(III.14)}$$

where Δ is the acceptable error (usually put equal to 0.1).

If the dynamometer rigidity condition (III.14) is not satisfied, a correction is introduced into (III.13):

$$q = \left(\frac{\lambda_0''}{\lambda_f''}\right)^m \qquad \text{(III.15)}$$

where m is a parameter (work-hardening coefficient) determined from a graph (Fig. 65) plotted in logarithmic coordinates: pressure $\ln P_i$ vs. soil deformation $\ln \lambda_i''$, as the arc tangent of the slope angle of the experimental line to the strain axis. Then the expression for determination of the ultimate long-term strength of a frozen soil from dynamometer tests assumes the form

$$\text{ult } \sigma_{\text{lt}} = \frac{\text{lt } P_f}{F} q \qquad \text{(III.13')}$$

The results of direct experiments to determine the long-term strengths of frozen and permanently frozen soils will be examined in detail in the next chapter. Here we make only the most general statement that, according to the experiments, the strength loss of frozen soils due to relaxation of their internal stresses under long-term load application is quite considerable and reaches values from 1/3 to 1/15 of their instantaneous strength.

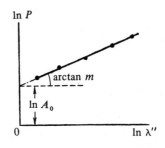

Fig. 65. Plot of experimental data (in logarithmic coordinates) for determination of dynamometer rigidity correction parameter for press shown in Fig. 64.

Thus, relaxation of the resistance of frozen soils, which is due to reorientation of mineral and ice-inclusion particles and to conversion of elastic to plastic strains (chiefly as a result of a decrease in the cohesive forces), is the most important factor determining the strength properties of frozen and permanently frozen soils, and this must be recognized in specifying design strength values for frozen ground, an operation that should proceed from the ultimate long-term rather than the instantaneous strengths.

5. Determination of Creep Parameters of Frozen Soils

The power-law equation (III.1), the equations of hereditary creep theory (III.5), and the equation of viscoplastic flow (III.7) are at present those most widely used to describe decaying creep of frozen soils. We shall now discuss determination of the parameters of these equations.

The relation between the stresses and strains that correspond to a given elapsed time t_i from the beginning of loading is established from creep curves (Fig. 57a) and represented by a family of $e(t) = f(\sigma)$ curves (see Fig. 57b), for which the time t is the independent variable.

According to equation (III.1), the relation between the stress σ and the total relative strain e of a frozen soil is given by the expression

$$\sigma = A(t)e^m$$

A graph is plotted in logarithmic coordinates, with $\ln \sigma_i$ on the vertical axis and $\ln e_i$ on the horizontal axis with their signs (Fig. 66) to check the validity of this equation and to determine the parameters of the variable modulus $A(t)$ [kgf/cm^2] and the strain-hardening coefficient m (an abstract number) from experimental values of e_i corresponding to various time intervals t_i and various stresses σ_i. Then, in accordance with the curves of Fig. 57b, we have as many graphs as there were recorded time intervals t_i.

If equation (III.1) is legitimate, the plots will be straight lines inclined to the e axis whose intersections with the σ axis will yield the value of $\ln A(t)$ directly, while the slopes of the lines will give the dimensionless strain-hardening parameters m, i.e.,

$$m = \frac{\Delta (\ln \sigma)}{\Delta (\ln e)} \qquad (III.16)$$

Experimental results indicate that in most cases the parameter m is a constant for various time intervals except the initial one, i.e., most of the $e = f(\sigma)$ lines are parallel to one another.

To determine the parameters of the more complex equation (III.2), the same type of logarithmic plot is constructed as in Fig. 66, but in the coordinates $\ln e^m/\sigma$ and $\ln t$. Then the segment cut off by the line on the $\ln e^m/\sigma$ axis will give the value of the parameter $1/\zeta$, and the slope of the line the parameter λ.

As we noted above, the exponential coefficient (III.6) and the hyperbolic coefficient (III.6″) have come into extensive use when the creep of frozen soils is described with the aid of the most general theory—the rheological theory of heredity (of Boltzmann-Volterra).

According to equation $K_1(t) = \delta e^{-\delta_1 t}$, the parameters of the decaying-creep coefficient will be the creep-decay coefficient δ_1 (min^{-1}) and the coefficient δ of the creep coefficient (min^{-1}).

To determine the parameter δ_1 from strain curves, the relative creep strain rates are determined for various time intervals t and a plot (Fig. 67) of the logarithm of the relative strain rate per unit effective stress $\ln \dot{e}(t)/\sigma$ against the time t is constructed (Fig. 67). Then, as follows from the form taken for the creep kernel, the slope of the resulting semilogarithmic line will be numerically equal to the creep decay coefficient, i.e.,

$$\delta_1 = \tan \alpha \qquad (III.17)$$

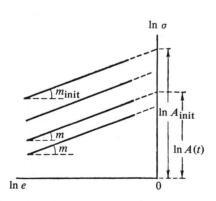

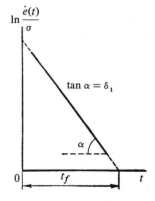

Fig. 66. Determination of parameters of the power-law equation $\sigma = f(e)^m$ from a logarithmic plot.

Fig. 67. Determination of the parameters of the exponential coefficient K_1 for decaying creep of frozen soils.

Knowing the creep decay coefficient δ_1, we determine the coefficient δ of the creep coefficient from the formula*

$$\delta = \delta_1 \frac{e_f - e_{init}}{e_{init} \left(1 - e^{-\delta_1 t_f}\right)} \tag{III.18}$$

where e_f is the final relative strain, e_{init} is the initial relative strain, which corresponds to the onset of creep (not including the instantaneous strain), e is the base of natural logarithms, and t_f if the time for practically complete stabilization of the settling rate at the particular load level.

The above expressions (III.17) and (III.18) can be used for unequivocal determination of decaying-creep parameters needed to describe the creep process according to the rheological theory of heredity.

Determination of e_f and t_f requires a long observation time, but this time can be shortened substantially for thawing soils (as is shown elsewhere[23]) by using measurements of the water pore pressure made during the soil test.

As we noted earlier, the decaying creep of plastic frozen soils is accurately described by the parabolic coefficient (III.6″)

$$K_3(t) = \frac{T}{(T + t)^2}$$

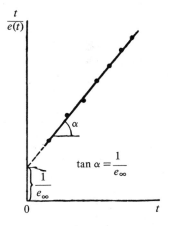

Fig. 68. Determining the parameters of decaying-creep coefficient K_3 for plastically frozen soils.

If the experimental data from determination of the relative creep settling $e(t)$ lie along a straight line when plotted in the coordinates $t/e(t)$ and t, the use of (III.6″) to describe the decaying creep is quite legitimate. Then the segment cut on the axis of ordinates (the $t/e(t)$ axis) will give the value of the parameter T/e_∞, and the slope of the line will give the reciprocal of the steady-state (final) relative creep strain, i.e., $1/e_\infty$ (Fig. 68).

Further analysis of the coefficient K_3 shows that the parameter T corresponds to half of the time elapsed from the start of the creep deformation to its attainment of the steady value e_∞, and that e_∞ (determined from the slope of the line in Fig. 68) gives us an opportunity to determine the final value of the total stabilized creep deformation without continuing the experiment until the deformations have stabilized—a very important advantage.

Remembering that $1/\tan \alpha = e_\infty$, and $e_\infty = s_\infty/h$, we have

$$s_\infty = \frac{1}{\tan \alpha} h \tag{III.19}$$

*See footnote to p. 120 (pp. 236, 237).

Since the relative compressibility of the soil is $a_{0\infty} = e_{\infty}/p$ (where p is the external pressure in compactive compression of the soil), the stabilized (final) value of this coefficient for the present case with the value of e_{∞} known from the curve of Fig. 68 will be

$$a_{0\infty} = \frac{1}{p \tan \alpha} \qquad \text{(III.19')}$$

The values of the decaying-creep parameters described here are used in predicting the creep deformations of frozen soil and permafrost under structures, as will be discussed in the chapters that follows.

As we saw above, the steady-state sustained creep of frozen soils, i.e., their viscoplastic flow at a constant rate, is described by equation (III.7)—a Bingham-Schwedoff relation for the flow of the frozen soils as a function of effective stress (but only power-law):

$$\dot{e}_t = \frac{1}{\eta_{t,\theta}} (\sigma - \sigma_0)^m$$

Let us denote by $\sigma - \sigma_0 = \sigma_e$ the effective stress or the stress excess above the threshold corresponding to the onset of viscoplastic flow; we then obtain

$$\dot{e}_t = \frac{1}{\eta_{t,\theta}} (\sigma_e)^n \qquad \text{(III.7'')}$$

The parameters of (III.7'') are determined in the same way as those of the power-law equation (III.1), from a logarithmic plot of the strain flow rate $\dot{e}_t$ against the effective stress e (similar to the plot in Fig. 66).

But in determining σ_0 it is often possible to confine ourselves to the rheological rate curve of plastoviscous flow plotted in the usual coordinates (flow rate vs. stress σ) (Fig. 69). Then the segment cut on the axis by the estention of the linear part of the curve [i.e., assuming $n = 1$ in (III.7'')] can be taken approximately as the initial parameter of viscoplastic flow, i.e., as σ_0.

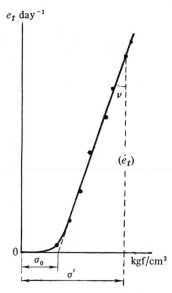

Fig. 69. Rheological viscoplastic flow rate curve for frozen soils.

But if the curve of $\dot{e}_t$ against σ is nearly straight at $\sigma > \sigma_0$, we have directly from the curve of Fig. 69 (with $n = 1$)

$$(\dot{e})'_t = (\sigma - \sigma_0)\cot \nu \qquad \text{(III.7''')}$$

then the coefficient of viscosity is

$$\eta = \tan \nu \qquad \text{(III.20)}$$

Strictly speaking, the viscosity coefficient, which characterizes the rate of viscoplastic flow (steady-state sustained creep) is not a constant for permafrost and frozen soils, but depends on the time of load application and the negative temperature of the soil.

Thus, according to studies made of frozen soils in the USSR Academy of Sciences Institute of Permafrostology, the viscosity coefficients of frozen soils are described as functions of negative temperature by the expression

$$\eta = u(1 + \theta)^q \qquad \text{(III.21)}$$

where θ is the absolute value of the negative temperature in $°C$ and u and q are parameters to be determined from experiments.

An equation by Prof. N. N. Maslov can be used for the dependence of the viscosity coefficient on the time of deformation:

$$\eta_t = \eta_f - (\eta_f - \eta_0)e^{-rt} \qquad \text{(III.22)}$$

where η_0 and η_f are the initial and final viscosity coefficients and r is a parameter that reflects the properties of the soil (and is determined from experimental data).

According to experiments in which the author collaborated at the MISI with A. Sh. Patvardkhan,[24] the viscosity coefficient is sufficiently described as a function of effective pressure for clayey soils by the equation

$$\eta_i = \eta_0(1 + \psi \ln \sigma_i) \qquad \text{(III.23)}$$

where η_0 and ψ are parameters determined from the semilogarithmic plot.

6. Experimental Data on Values of Creep Parameters of Frozen Soils

Let us cite certain experimental data from determinations of the parameters of the rheological stress-strain state equations for frozen soils.

In experiments performed in connection with strength and creep calculations for the frozen ground barriers used in sinking the deep KMA shafts by the artificial soil-freezing method (for the most part, in experiments performed by scientific staff members S. E. Gorodetskiy and Ye. P. Shusherina), the parameters of the power-law equation (III.1) for the stress-strain state of frozen soils were determined by the procedure set forth earlier.

The work-hardening coefficient (the parameter m) was found to be quite constant for the two soils tested (a sandy loam and a clay), but the parameter $A(t)$ (kgf/cm^2)—the strain modulus—depends strongly on the independent variables when measured at various time intervals t and various temperatures $-\theta°C$ (Table 10).

They also determined the parameters of equation (III.2), which takes account of the dependence of the modulus $A(t)$ on a power of the time (i.e.,

Table 10.   Values of the Parameter $A(t)$ of equation (III.1)

Time, t	Kelovey sandy loam, $m = 0.27$, $W = 26\%$, $\gamma = 1.7$–2.1 g/cm³			Bat-Bayos clay, $m = 0.40$, $W = 20$–24%, $\gamma = 2.06$–2.15 g/cm³		
	Values of $A(t)$, kgf/cm², at $\theta°$C					
	-20	-10	-5	-20	-10	-5
1 min	204	116	66.5	520	270	149
30 min	145	81.5	47.5	271	149	82
1 hr	135	76.5	44.0	245	134	74
2 hr	126	71.5	41.0	221	115	63
6 hr	113	64.0	37.0	181	95	52
12 hr	106	59.5	34.5	156	81	44

on $t^{-\lambda}$) and on the negative temperature [according to the expression $\zeta = \omega(1 + \theta)^k$], i.e.,

$$e = \left[\frac{\sigma t^\lambda}{\zeta}\right]^{1/m} \tag{III.2'}$$

The experimental data obtained for the Kelovey sandy loam at $\theta = -10°$C and $\sigma = 30$ kgf/cm² are plotted as points on Fig. 70, where the solid line represents the curve constructed from (III.2'). As we see from the plot of Fig. 70, the deviation of the analytical curve from the experimental data is insignificant.

The parameters of the exponential coefficient of hereditary creep theory were not determined specifically for the frozen soils by the method set forth above, but the manner in which the creep coefficient $K_1(t - t_0)$ varies can be inferred from Table 10, since

$$K_1'(t) = \frac{d}{dt}\left[\frac{1}{A(t)}\right]\sigma(t)$$

$\dot{e} \times 10^2$

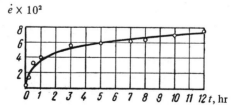

Fig. 70. Analytical creep curve constructed from equation (III.2'), and experimental points for the Kelovey sandy loam. $\theta = -10°$C, $\sigma = 30$ kgf/cm².

The parameters of the exponential creep coefficient $K_1(t-t_0) = \delta e^{-\delta_1 t}$ were determined by direct experiments for dense clayey soils and were found equal to:[25]

$$\delta_1 = 0.005\text{-}0.040 \text{ min}^{-1} \qquad \delta = 0.1\text{-}0.5 \text{ min}^{-1}$$

i.e., the parameter δ is several (sometimes 10 to 20) times larger than the parameter δ_1, and both decrease with increasing compactness of the ground.

The parameter T of the hyperbolic creep coefficient

$$K_3(t) = \frac{T}{(T+t)^2}$$

was calculated for various frozen soils (by the method set forth above) by Yu. K. Zaretskiy,[26] who worked from time curves of the settling of frozen soils under test loading with a plunger, and were found to be as follows:

For frozen sand ($\theta = -0.5°C$)
    At $\sigma = 4\text{-}8$ kgf/cm^2, $T = 151$ hr
    At $\sigma = 8\text{-}12$ kgf/cm^2, $T = 482$ hr
For frozen varved clay ($\theta = -0.5°C$)
    At $\sigma = 13$ kgf/cm^2, $T = 37.5$ hr
For unfrozen fluvioglacial clay ($\theta = +5°C$, $\epsilon_{init} = 0.7$, $\gamma = 1.92$ g/cm^3)
    At $\sigma = 1.5\text{-}2$ kgf/cm^2, $T = 1.0$ hr
    At $\sigma = 3\text{-}3.5$ kgf/cm^2, $T = 1.9$ hr

These data (for example, those for the frozen sand) and others indicate that the parameter T depends on the compacting pressure, increasing with increasing load.

Back in 1935-1937 (Chap. II[31]), the author investigated the viscoplastic flow of frozen soils and determined their viscosity coefficients; the effects of soil composition, effective stress, and negative temperature on the flow of the frozen soils were studied experimentally. As an example, Fig. 71 shows rheological curves obtained at that time for the variation of the relative-strain rate as a function of effective stress for two types of frozen ground: a sandy loam and a silty clay.

Although the temperatures of the frozen-ground specimens rose slightly during the experiments (the temperatures are indicated on the curves), the results clearly indicate curvilinear variation of the relative strain rates as functions of effective pressure.

It was found in experiments performed at the same temperature ($-0.8°C$) that the viscosity coefficient (determined from the steady-state relative strain rate in compression) for the frozen sandy loam ($W = 19.1\%$) has the value $\eta = 1.9 \times 10^{12}$ P, but that $\eta = 0.9 \times 10^{12}$ P for the frozen clay (with a 36% content of the fraction < 0.005 mm and a moisture content of 27.7 percent), or almost a whole order of magnitude smaller than the viscosity of pure ice, which, according to B. P. Veynberg, is $\eta = 1.2 \times 10^{13}$ P at $\theta = 0°C$.

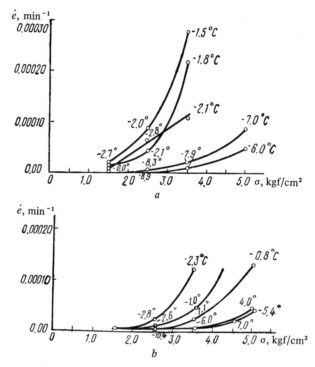

Fig. 71. Variation of relative strain rates of frozen clay (*a*) and frozen sandy loam (*b*) under compression as a function of effective stresses (according to experiments of N. A. Tsytovich, 1936).

We explain this last observation as follows: in frozen soils, and especially at temperatures no lower than the range of substantial phase transitions of water, a certain part of the pore water, and sometimes a rather large part, is in the unfrozen liquid state, causing the frozen soil to flow at higher rates than ice. It is interesting to note that we advanced this proposition (which we later demonstrated experimentally) back in 1937 (Chap. II[31]), when no one had surmised that unfrozen water was present in permafrost soils.

These experiments also indicated that the deformations of viscoplastic flow decay up to a certain limiting value of the compressive stress (pressure) that is well-defined for the particular frozen soil, and that sustained viscoplastic flow is established at stresses above this limit, proceeding at a rate inversely proportional to the viscosity η of the frozen soil.

The results of our initial experiments to determine the viscosities of frozen soils in compression as functions of temperature and effective compressive stress are given in Table 11.

More detailed experiments were carried out in an underground laboratory in permafrost with a practically constant negative temperature at Yakutsk with the

Table 11.   Viscosity coefficients of frozen ground η (P) as functions of temperature $-\theta°$C and effective compressive stress σ (kgf/cm^2)

Designation of frozen soil	Average moisture content by weight W, %	$\sigma = 1.5$		$\sigma = 2.5$		$\sigma = 3.5$		$\sigma = 5$	
		$-\theta$	$\eta \times 10^{12}$	$-\theta$	$\eta \times 10^{12}$	$-\theta$	$\eta \times 10^{12}$	$-\theta$	$\eta \times 10^{12}$
Clay	33.2	-2.8	4.4	-2.0	2.1	-1.7	0.8	$-$	$-$
	30.8	$-$	$-$	$-$	$-$	-7.9	10.3	-6.5	4.2
Silty mud	32.6	-2.8	8.8	-1.0	7.3	-0.8	1.2	$-$	$-$
(clay)	37.6	$-$	$-$	-3.8	14.7	-3.5	13.7	-2.8	4.9
	13.1	$-$	$-$	-2.1	14.7	-1.0	6.8	-0.9	1.1
Sandy loam	12.0	$-$	$-$	$-$	$-$	-6.0	20.6	-4.0	7.3

objective of determining the rheological flow parameters of frozen soils (in the form of equation (III.7)): the long-term strength ($\sigma_0 = \sigma_{lt}$), the nonlinearity exponent n, and the viscosity coefficient η with the frozen-soil specimens (sand and loam) held at a strictly constant negative temperature ($\theta = -3°$C).

The frozen-soil specimens were tested in compression, tension, and shear with torsion; the times of the experiments ranged from a few days to several tens (up to 82) of days.

The results of experiments[27] performed at the Institute of Permafrostology of the Siberian Division of the USSR Academy of Sciences are presented in summarized form in Table 12.

The data in Table 12 indicate that both the long-term strength limits of the frozen soils and their viscosities are different for different types of stressed states, and that (according to the experiments of S. Ye. Grechishchev) the dependence of the viscoplastic flow rate on the effective stress (the stress excess over the long-term strength) is invariably linear for frozen sands (Fig. 72), but

Table 12.   Parameters of viscoplastic flow equation (III.7) of frozen soils at $\theta = -3°$C

Designation of soil	Compression			Tension			Shear		
	σ_{lt}^{comp}, kgf/cm^2	$\eta \times 10^{-4}$, $\dfrac{\text{kgf}^n \times \text{day}}{\text{cm}^{2n}}$	n	σ_{lt}^{ten}, kgf/cm^2	$\eta \times 10^{-4}$, $\dfrac{\text{kgf}^n \times \text{day}}{\text{cm}^{2n}}$	n	τ_{lt}, kgf/cm^2	$\eta \times 10^{-4}$, $\dfrac{\text{kgf}^n \times \text{day}}{\text{cm}^{2n}}$	n
Frozen sand	6.5	0.40	1.00	1.8	0.10	1.00	1.7	0.05	1.00
Frozen loam	3.6	1.20	1.64	2.5	0.67	1.69	$-$	$-$	$-$

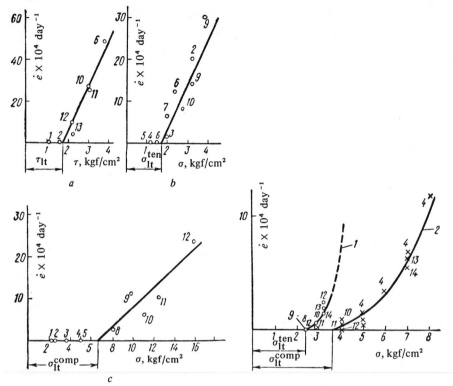

Fig. 72. Rates of viscoplastic flow as functions of stress for frozen sand (according to experiments of S. Ye. Grechishchev, 1958–1961). (*a*) Pure shear; (*b*) tension; (*c*) compression (at $\theta = -3°C$).

Fig. 73. Viscoplastic flow rates as functions of stresses for frozen loam. (1) Tension; (2) compression ($\theta = -3°C$).

nonlinear for clayey soils, although the relative-strain-rate curve levels off at a certain rather large stress value (Fig. 73), i.e., under large enough stresses (approximately one and a half times the long-term strength limit) it can be assumed in approximation that the viscosity coefficient is also a constant for frozen clayey soils.

Finally, we present experimental data characterizing the viscosities of frozen soils as functions of their negative temperature and the effective stress (Table 13) according to experiments designed to establish the dimensions of frozen ground barriers for KMA deep shafts to be sunk with artificial soil freezing at depths from 360 to 455 m below the surface.[16]

In concluding the present section, we note that by themselves, literature data on the viscosities are not sufficient for quantitative soil forecasts; the viscosities must be determined directly for the particular type of soil at the particular temperature and effective-stress values.

Table 13. Viscosity coefficients η (P) of frozen soils as functions of temperature ($-\theta°C$) and compressive stress σ (kgf/cm²)

Designation of soil	$\theta = -5°C$			$\theta = -10°C$			$\theta = -20°C$		
	σ, kgf/cm²								
	40	30	20	60	40	30	60	50	40
Kelovey sandy loam (0.25–0.5-mm fraction 73.8%; < 0.005 mm 4.8%)	4×10^{11}	5.4×10^{12}	7.3×10^{14}	5×10^{11}	0.9×10^{14}	3.5×10^{14}	1×10^{14}	1.4×10^{15}	—
Kelovey silty sandy loam (0.25–0.05-mm fraction 31.1%; 0.05–0.005 mm 64.7%; smaller than 0.005 mm 4.2%)	—	6.3×10^{11}	1.1×10^{13}	—	6.6×10^{12}	5.1×10^{13}	3.2×10^{13}	8.6×10^{13}	1.0×10^{11}
Bat-Bayos clay (< 0.005 mm fraction 54 to 69%)	—	†	1.2×10^{14}	—	4.1×10^{11}	4.3×10^{12}	4.5×10^{11}	1.6×10^{13}	1.1×10^{14}

7. Certain General Conclusions Regarding the Flow of Frozen Soils Under Load

The general methods described in the foregoing sections of this chapter for the investigation of rheological processes that arise in frozen soils under long-term subjection to constant loads permit quantitative evaluation of the flow processes in frozen soils—both those of general (natural) character, as well as local processes due to the weight of structures.

These processes may significantly influence the stability of natural permafrost massifs, and they may, under certain conditions, give rise to totally unacceptable deformations in the bases of structures built on them.

As we showed earlier, viscoplastic flows of frozen ground are governed basically by the fluidity of ice and by the presence of unfrozen film water bound to the mineral particles of the frozen soils.

As we noted earlier, ice flows even at very small constant effective-stress values, i.e., inelastic viscoplastic deformations (flows) occur in it. The conditions under which viscoplastic flows arise in ice inclusions contained in frozen soils will, of course, be somewhat different from those for solid ice (as in glaciers), since in many cases the ice in frozen soils will be in a "block" formed by soil mineral particles, and the development of flows in individual lenses, interlayers, and other similar ice inclusions will be resisted by friction at the constant surfaces between the ice and mineral layers of the soil.

However, comparison of the values given for the viscosities of frozen soils, e.g., in Tables 11 and 13, with the viscosity of solid ice, which is 1.2×10^{13} P at $\theta = -0°C$ according to V. P. Veynberg (Chap. II[33]) and about 3×10^{14} P at $\theta = -1°C$ according to K. F. Voytkovskiy's observations of stored ice,[28] gives reason to believe that, as we established earlier, frozen soils (and especially clayey soils) have viscosities lower than that of ice.

This permits the conclusion that glacier-like viscoplastic flows are possible in massifs of permafrost situated on mountain slopes.

Still, it must be remembered that the flows of frozen soils differ from those of ice, since frozen soils have a long-term strength limit (i.e., a stress limit below which viscoplastic flows do not occur), while ice has no long-term strength.

Rheological processes may also have a very strong influence in building up the total inelastic deformation in the bases of structures built on permafrost when they remain at negative temperatures (especially if the frozen soils are strongly icy, high-temperature types), even at pressures that do not exceed the long-term strength of the frozen soils.

On the other hand, viscoplastic flow of frozen soils will occur at pressures above the long-term strength, and with time this will cause progressive flow of the base of the structure, i.e., its failure, something that can be predicted if the viscoplastic flow parameters and the ultimate disruptive strains of the base are known. Rheological flows become especially hazardous when the temperature of the frozen soil rises to values close to $0°C$, when the viscosities of the soils decrease substantially and there is a danger of the rapid development of progressive flow.

Local viscoplastic flows of ice present in frozen and permanently frozen soils arise at points of stress concentration under load and bring about a change in the structure of the frozen soils due to plastic extrusion of ice from more heavily stressed areas to less heavily stressed areas, with partial thawing of the ice under load and subsequent freezing of the resulting water; this is indicated by the results of laboratory experiments and especially those of special field tests (in the underground laboratory) at Igarka with long-term maintenance of the load, as described previously.

We have also noted that reorientation of ice-inclusion crystals and mineral particles occurs in stressed zones, primarily because of the fluidity of the ice and the presence of unfrozen water, so that a substantial decrease in the strength of frozen soils under external forces is observed.

On the basis of the above, we can arrive at the following conclusions.

The mechanical properties of frozen soils and the mechanical processes that take place in them must be studied with consideration of their flow under load: whether it is the strength properties of the frozen soils or their deformation properties that are being determined, the time factor must always be taken into consideration, since it has an exceptionally strong influence on all mechanical-property indicators and on the mechanical processes. Tests of frozen soils without consideration of the time factor may lead to totally incorrect conclusions as to the behavior of the frozen soils under load.

We called attention to the importance of the fluidity (in the broad sense of the word) of frozen soils back in 1941, when the fluidity of frozen soils under load was regarded as an important factor that always required consideration in studies of frozen soils as a kind of "axiom" of frozen-ground mechanics.[29] The author's later study (1952) (Chap. I[13]), in which he developed this idea and described all of the forms of flow (nonsteady, steady, progressive), and later work done in this area by S. S. Vyalov (1959) (Chap. II[20]), K. F. Voytkovskiy (1959),[29] S. Ye. Grechishchev (1963),[27] Yu. K. Zaretskiy (1965),[26] and others confirmed the prime importance of fluidity for evaluation of the mechanical processes in and mechanical properties of frozen and permanently frozen soils.

Thus, the flow of frozen soils, which always occurs in them under constant loads, is in fact one of the fundamental principles without knowledge of which it is impossible to investigate the mechanical processes and mechanical properties of frozen soils either in study of frozen ground and permafrost as bases for structures or (in our opinion) even in study of various physicogeological phenomena under the various topographic situations in the permafrost region.

Full account will be taken of the fluidity of frozen soils in our subsequent discussion of the results of study of the mechanical properties of frozen ground (Chap. IV). However, it must be noted that experimental data on the quantitative values of the flow characteristics of frozen soils (for example, the parameters in various stages of creep) are now totally inadequate, especially for frozen-soil temperatures near $0°C$, i.e., for the conditions that are most likely to develop undesirable processes. This calls for further research on the deformation and strength properties of frozen ground as functions of time, with determination of their parameters at various temperatures and especially for frozen-ground temperatures near $0°C$, i.e., for high-temperature permafrost.

References

1. Savel'yev, B. A.: "Stroyeniye, sostav i svoystva ledianogo pokrova morskikh i presnykh vodoyemov" (Structure, Composition, and Properties of Ice Cover on Maritime and Fresh-Water Bodies), Moscow State University Press, 1963.
2. Pekarskaya, N. K.: Shear Strength of Perennially Frozen Soils with Various Textures and Ice Contents, in "Issledovania po fizike i mekhanike merzlykh gruntov" (Research in the Physics and Mechanics of Frozen Soils), USSR Academy of Sciences, 1961.
3. Pchelintsev, A. M.: "Stroyeniye i fiziko-mekhanicheskiye svoystva merzlykh gruntov" (Structure and Physicomechanical Properties of Frozen Soils), Nauka Press, 1964.
4. Tsytovich, N. A.: "Problems of Soil Mechanics in the Practical Design and Erection of Structures," Doctoral thesis, Vol. II, "Issledovaniye deformatsiy merzlykh gruntov" (Study of the Deformations of Frozen Soils), Leningrad Institute of Railway Transportation Engineers (LIIZhT), 1940.
5. Tsytovich, N. A. and M. I. Sumgin: "Osnovaniya mekhaniki merzlykh gruntov" (Fundamentals of Frozen Ground Mechanics), p. 149, USSR Academy of Sciences Press, 1937.
6. Shusherina, Ye. P.: "Coefficient of Transverse Deformation and the Volume Deformations of Frozen Soils During Creep," Sb. MGU (Moscow State University Collection), No. V, 1966.
7. Gol'dshteyn, M. N., et al.: "Doklady k V Mezhdunarodnomu kongressu po mekhanike gruntov i fundamentov" (Papers at the Fifth International conference on Soil Mechanics and Foundations), Gosstroyizdat, 1961.
8. Turovskaya, Ya. A.: Influence of Deformation on the Structure of Clayey Soils, LIIZhT Collection No. 4, 1957.
9. Vyalov, S. S., N. K. Pekarskaya, and R. V. Maksimyak. The Physical Nature of the Deformation and Failure of Clayey Soils, "Osnovaniya, fundamenty i mekhanika gruntov," No. 1, 1970.
10. Tsytovich, N. A.: "Printsipy mekhaniki merzlykh gruntov" (Principles of the Mechanics of Frozen Soils), Izd-vo AN SSSR, 1952.
11. Vyalov, S. S., et al.: "Prochnost' i polzuchest' merzlykh gruntov i raschety ledovgruntovykh ograzhdeniy" (Strength and Creep of Frozen Soils, and Calculations for Frozen Ground Barriers), USSR Academy of Sciences Press, 1962.
12. Loc. cit., Chap. V, and also S. S. Vyalov and Ye. P. Shusherina: Resistance of Frozen Soils to Triaxial Compression, Moscow State University Collection "Merzlotnyye issledovaniya" (Frost Research), No. IV, 1964.
13. Tsytovich, N. A.: "Mekhanika gruntov" (Soil Mechanics), 4th edition, Stroyizdat Press, 1963.
14. Vyalov, S. S.: Plasticity and Creep of Cohesive Media, in "Doklady k VI Mezhdunarodnomu kongressu po mekhanike gruntov" (Papers at Sixth International Soil Mechanics Congress), Stroyizdat, 1965.
15. Zaretskiy, Yu. K.: Rheological Properties of Plastic Frozen Ground, "Osnovaniya, fundamenty i mekhanika gruntov," 1971, No. 2.
16. Vyalov, S. S., S. E. Gorodetskiy, Yu. K. Zaretskiy, et al.: "Prochnost' i polzuchest' merzlykh gruntov i raschety l'dogruntovykh ograzhdeniy" (Strength and Creep of Frozen Soils and Calculations for Frozen Ground Barriers), USSR Academy of Sciences Press, 1962.

17. In collection "Prochnost' i polzuchest' merzlykh gruntov" (Strength and Creep of Frozen Soils), Siberian Division, Institute of Permafrostology, USSR Academy of Sciences, USSR Academy of Sciences Press, 1963.

18. Vyalov, S. S., S. E. Gorodetskiy, V. F. Yermakov, et al.: "Metodika opredeleniya kharakteristik polzuchesti, dlitel'noy Prochnosti i szhimayemosti merzlykh gruntov" (Method for Determination of Creep, Long-Term-Strength, and Compressibility Characteristics of Frozen Soils), NIIOSP, Nauka Press, 1966.

19. Vyalov, S. S., S. E. Gorodetskiy, and N. K. Pekarskaya: "Rekomendatsii po opredeleniyu dlitel'noy prochnosti i polzuchesti merzlykh i ottaivayushchikh gruntov" (Recommendations for the Determination of Long-Term Strength and Creep of Frozen and Thawing Soils), NIIOSP, 1970.

20. Tsytovich, N. A.: "Opredeleniye sil stsepleniya merzlykh gruntov po metodu sharikovoy proby" (Determination of Cohesive Forces in Frozen Soils by the Ball-Probe Method), Fondy Instituta merzlotovedeniya AN SSSR (Records of the USSR Academy of Sciences Institute of Permafrostology), 1947.

21. Vyalov, S. S.: Soviet patent No. 161133, January 21, 1964.

22. Tsytovich, N. A.: Certain Mechanical Properties of the Yakutian Permafrost, Trudy Komiteta po vechnoy merzlote (Transactions of the Permafrost Committee), vol. X, USSR Academy of Sciences Press, 1940.

23. Tsytovich, N. A., and Z. G. Ter-Martirosyan: A Method of Determining the Creep Parameters of Partially Water Saturated Clayey Soils in Undrained Tests, Osnovaniya, fundamenty i mekhanika gruntov, No. 3, 1966.

24. Patvardkhan, A. Sh., and N. A. Tsytovich: Influence of Normal Stresses on the Viscosity and Strain-Hardening of Clayey Soils, Trudy VI Mezhdunarodnogo kongressa po mekhanike gruntov (Transactions of Sixth International Soil Mechanics Congress), 1965.

25. See Ref. 23.

26. Zaretskiy, Yu. K.: Osnovaniya, fundamenty i mekhanika gruntov, No. 2, 1971.

27. Grechishchev, S. Ye.: Creep of Frozen Soils in a Complex State of Stress, in "Prochnost' i polzuchest' merzlykh gruntov" (Strength and Creep of Frozen Soils), Siberian Division, USSR Academy of Sciences, USSR Academy of Sciences Press, 1963.

28. Voytkovskiy, K. F.: Raschet sooruzheniy iz l'da i snega (Design of Snow and Ice Structures), Izd-vo AN SSSR, 1959.

29. Tsytovich, N. A.: Nachala mekhaniki merzlykh gruntov. Monografiya (Principles of the Mechanics of Frozen Soils, A Monograph), Archives of the USSR Academy of Sciences Permafrostology Institute, 1941.

30. See Ref. 28.

Chapter IV

STRENGTH PROPERTIES OF FROZEN GROUND; CRITICAL AND DESIGN STRENGTHS

1. Instability of Mechanical Properties of Frozen Ground and Causes of Changes in Its Strength

In using permafrost as a basis or environment for structures, the engineer is confronted with a totally unique natural material with properties different from those of all others—a material that is so sensitive to external disturbances that even insignificant changes in their values or their nature and time of action affect its mechanical properties. It is therefore important to know which changes in the mechanical properties must be taken into account in the design of the structures (e.g., in their bases and foundations, underground pipelines, etc.) on frozen soils and the consequences to which these changes lead, i.e., it is very important to establish the determination methods of the magnitudes of design for the mechanical properties of the frozen soils with consideration of their *instability* and to predict the changes that may take place during the service life of the structure.[1]

It is necessary to evaluate beforehand the practical importance of variations in the strength of the frozen soils and its limiting value, the variations in deformability of the frozen soil bed over its depth, and other mechanical-property indicators, among which the following are fundamental: the strength indicators (compression, shear, and adfreezing strengths) and the deformability moduli of the soils in the frozen and thawed states.

Factors that govern the mechanical-property instability of freezing, frozen, and thawing soils include:

a) the temperature variations of the soils under natural conditions and under the influence of structures built on them;

b) changes in the state of stress in freezing, frozen, and thawing soils under the influence of internal and external factors;

c) time under load, which governs stress relaxation and creep in frozen and thawing soils.

The temperature variation in permafrost is insignificant under natural conditions, but the vertical temperature distribution is, as a rule, nonuniform. This nonuniformity renders permafrost soils inhomogeneous, since it is well known that the lower the temperature of a frozen soil the higher is its resistance to external forces and the lower its deformability. However, the intensity of the temperature influence on the mechanical properties of frozen soils varies according to the region of the water phase-transformation in which the temperature is varying.

In the range in which water has strong (significant) phase transformations (from 0 to $-0.5°C$ for sandy soils and from 0 to $-5°C$ for clayey soils), the factors that determine the strength of frozen and permanently frozen soils are the quantitative contents of ice and unfrozen water and the dependence of these contents on negative-temperature variations.

Thus, as the temperature falls from -1 to $-2°C$, the ultimate strength of frozen sand under simple compression changes from 64 to 75 kg/cm², i.e., by about 15 percent, while the change for frozen clay at the same temperature change is from 10 to 15 kg/cm², i.e., by 50 percent, which is quite natural, since the amount of unfrozen water has been reduced by no more than 0.1 percent in the case of the sand, but by 5 percent in the clay.

Thus, the ice-content variation, or the unfrozen-water content of the frozen soil, is the dominant factor in the range of significant water phase transformations.

Analysis of similar data on the ultimate-strength increase of frozen soils as the temperature falls through the range in which water has only insignificant phase transformations indicates that the strength increase cannot be explained solely by an increase in ice content (or a decrease in the unfrozen-water content). Here the simultaneous influence of a second factor—a qualitative change in the ice (the increase in its strength with lowering negative temperature) becomes significant.[2]

In the range of the practically-frozen soil state, the strength properties of a given frozen soil are determined basically by the strength of the ice cementing it and its strength increase as the temperature falls.

We should note that ice is characterized by very weak hydrogen bonds between the atoms, with a sharp decrease in their mobility with decreasing temperature; this factor governs the strengthening of the ice structure. However, ice becomes stronger with decreasing temperature only up to a certain limit (apparently close to $-70°C$, when the ice strength is nearly at its maximum).

The anisotropy of frozen soils, which is especially pronounced in the ice, is of significant importance for their strength properties. Thus, according to S. S. Vyalov,[3] the strain rate of ice in shear parallel to the basic planes of the crystals is $v_{\parallel} = 0.34$ mm/hr, but $v_{\perp} = 0.01$ mm/hr perpendicular to this plane. In addition, as we pointed out earlier (Chap. III, Sec. 6), ice is more viscous than frozen soils, so that rheological processes take place more slowly in ice than in frozen soils.

The lower the negative temperature of a frozen soil, the higher does its contact shear strength become. Thus, according to N. K. Pekarskaya,[2] $\tau_{ul} = 1$ kgf/cm² at positive temperatures, but $\tau_{ul} = 6.5$ kg/cm² at $\theta = -0.8°C$ and $\tau_{ul} = 9.0$ kg/cm² at $\theta = -2°C$.

All of the above factors are responsible for the instability of mechanical-properties of frozen soils as their temperatures vary.

Negative-temperature variations also have a strong influence on the deformability (compressibility) of frozen soils.

Experiments have shown[4] that the compressibility of a high-temperature frozen soil (in the range of strong water phase transitions) is high and approaches that of dense clay (relative compressibility coefficient $a_r \approx 0.005-0.03$ cm^2/kg), while the compressibility of soils in the range of the practically frozen state is so small that it can be neglected in engineering calculations.

Thus, as a consequence of the vertically variable natural temperature conditions in permafrost, we observe *nonuniformity* of its strength and deformation properties, so that special conditions apply in calculations for and in the design of bases for structures to be built on permafrost soils.

Prediction of ground-temperature variations below the footing of the foundation, and especially the rise of the base temperature to above-zero values, are of cardinal importance for all forms of construction on permafrost.

When the structures are built under preservation of the frozen state at their base, it is very important to know the temperature at which the frozen soil will stabilize after a certain rather long interval of time (for example, after 10-15 years), since the magnitude of negative temperature will, in the final analysis, be the principal factor determining the strength properties of the frozen soil for a particular composition (its ice content). It is also important to know the temperature distribution in the permafrost below the thawing bowl in order to make proper use of its bearing capacity for placing deep supports (columns, piles, etc.). These problems will be examined in Part Two of the book.

Property instability occurs not only when the soils are kept frozen, but also *during thaw under structures.* First, the boundary conditions change during the thaw, since the depth of compression-resistant (permafrost) ground increases steadily; second, the ground undergoes compaction during thaw and, for clayey soils there is consolidation for a long time after thawing, a prime manifestation of which is variation of the strain modulus of the soil over the depth of the thawing layer.

Thus, it was found in experiments with models of thawing clayey bases performed in the laboratory headed by the author, that the change in the porosity coefficient of the thawed soil under load decreases exponentially with depth. This has a strong effect on the vertical pressure distribution in the thawed layer and especially on the pressure value at the contact surface between the thawed and frozen layers. The experiments were connected under two-dimensional conditions as shown schematically in Fig. 74b.

Figure 74a presents graphs of porosity coefficient Δe for a soil during thaw, and Fig. 74c the vertical distribution of the principal-stress sum Θ as obtained on an electrolytic integrator in an inhomogeneous base whose modulus of deformability decreases with increasing depth; here graph 1 corresponds to a homogeneous half-space under the conditions of a two-dimensional problem,

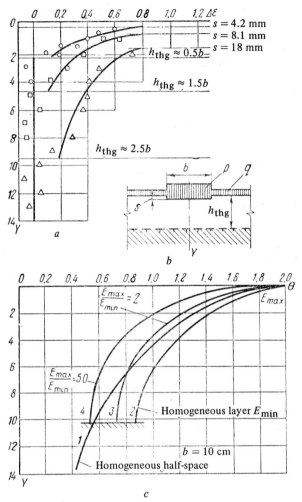

Fig. 74. Nonuniformity of compacting (variations of porosity coefficient $\Delta\epsilon$) and distribution of the sum of principal stresses Θ over depth Y in a thawing clayey bed (according to experiments of V. D. Ponomarev).

graph 2 to an elastic homogeneous layer on an incompressible base, and graphs 3 and 4 to an inhomogeneous layer with a modulus of deformation that decreases linearly with depth; $E_{max}/E_{min} = 2$ for line 3 and $E_{max}/E_{min} = 50$ for graph 4.

The data shown here indicate that, in a layer of nonuniform compressibility on an incompressible base, the area under the curve of the principal-stress sum, which is proportional to the settling due to thaw consolidation, may be much smaller than the area under the same curve for a homogeneous layer or even for a homogeneous half-space.

Thus, the vertical variation of compressibility has a very strong influence on the engineering deformability characteristics of thawing soils.

Changes in the stressed states of freezing, frozen, and thawing soils strongly influence their resistance to external forces and their deformability.

Investigations performed at the Igarka Scientific Research Station of the USSR Academy of Sciences[5] indicate that not only do stresses appear and changes in pore pressure occur in the freezing layer of the ground but pressures also increase in the already frozen layers of the soil.

In confirmation of the above, Fig. 75 shows graphs of pressure variation in a frozen soil that were obtained at Igarka from mechanical-dial gauge measurements and plots of ground-temperature variation with depth at points near the positions of the gauges.

The graphs indicate that the arising pressures generally follow the ground temperature variations. To explain these data, it must be assumed that, in accordance with the principle of dynamic equilibrium between the unfrozen water and the ice in frozen soils, water phase transitions occurring in freezing and frozen soils are the basic factor influencing the intensity of the force field: variation of the temperature in a frozen ground in the subzero range affects not only its ice content, but also its state of stress, and this, in turn, influences its mechanical properties.

The effect of external pressure on the properties of frozen soils is manifested in different ways: first, the unfrozen-water content increases in frozen soils with increasing pressure; second, localized pressures increase substantially at the points of contact between mineral particles.

Thus, according to the author's experiments,[6] a clayey soil contained 42 percent unfrozen water at $-1.7°C$, but on being subjected to an external pressure of 2 kg/cm² at the same temperature, it contained 58 percent unfrozen

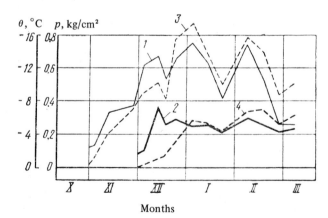

Fig. 75. Variation of pressure in frozen soil during freezing and further cooling: (1 and 2) pressures at depths of 0.4 and 1 m; (3 and 4) temperatures at depths of 0.5 and 1 m.

water. This substantial change in the unfrozen-water content was reflected both in a decrease in the ultimate strength of the frozen ground and in a decrease in its modulus of deformability. For example, the author's experiments showed that the normal modulus of elasticity E of a permanently frozen sandy loam with undisturbed structure with 8 percent clay had the value $E_1 = 100 \times 10^3$ kg/cm^2 at a moisture content $W_d = 40\%$ and a temperature $\theta = -4°C$ under a pressure of 1 kg/cm^2, but that $E_2 = 60 \times 10^3$ kg/cm^2 at a pressure of 2 kg/cm^2, and $E_3 = 47 \times 10^3$ kg/cm^2 at 3 kg/cm^2.

External pressure has less influence on the unfrozen water content of very dense soils and, consequently, on their deformability than on those of nondense soils; which is explained by the stronger binding of pore water in dense soils.

The mechanism by which external pressure affects the mechanical properties of frozen soils consist, as we indicated above, in the transformation of the external pressure at points and areas of contact between mineral particles into enormous local stresses, which cause melting and flow of the ice and transfer of water into less heavily stressed regions.

In addition, recrystallication of ice, a decrease in the size of the ice crystals, and changes in their orientation in response to the direction of effective stresses are observed in ice interlayers under the action of the stress field. Specially designed experiments indicate that these changes occur quite slowly.

Thus, the amount of unfrozen water in a frozen soil and, consequently, the degree to which it is cemented by ice change under the influence of stresses.

The time of load action is one of the prime factors in the instability of mechanical-property of frozen ground.

The mechanical properties of frozen soils change insignificantly on instantaneous application of a load at constant temperature, since the melting of pore ice at the contact points do not occur instantaneously, but require a certain time to develop.

But, as we noted in the preceding chapter, stress relaxation and, under certain conditions, damped and undamped creep occur when an external load acts for a considerable time. These processes naturally also affect the properties of the frozen soils, since contact bonds are subject to progressive breakdown, microscopic and macroscopic cracks are formed and developed, particles are reoriented with recrystallization of the ice, and the mineral particles are packed somewhat more compactly (the porosity of the soil is reduced); here, the longer the time of load action, the stronger the influence of the load on the properties of the frozen soils. However, the restructuring of frozen soils (the adaptation of the structure to the direction of the effective forces) will not continue indefinitely, since both stress relaxation and creep are processes that subside at pressures below a certain limit for a given state of the soil, and in viscoplastic flow (steady-state creep) a certain balance is achieved between the disturbances to the structure and its strengthening.

The instability of mechanical property of frozen ground and its sensitivity to external disturbances result in vertical inhomogeneity of frozen and thawing soils, and this, together with the nonlinearity of the stress-strain relation for these soils and their creep property, gives rise to complications in calculations for frozen soils as bases and surroundings for structures. In certain cases, however, consideration of the relationships noted above permits more economical design of foundations for structures to be built on permafrost and on thawing ground and to make sufficiently accurate prognosis of the behavior of permafrost and thawing soils under natural conditions and at the bases of structures.

Allowance for the nonlinearity of creep and the inconstancy of the strain modulus of frozen soil with depth below the footing of the foundation (its increase with depth) $(E_0^{1/m} = z^n)$ indicates* that with certain combinations of nonlinearity parameters (e.g., with $m = 0.3$ and $n = 0.5$), there is a substantial decrease in the nonuniformity of the contact-pressure distribution over the base of a rigid continuous footing: the pressure varies from $0.318p_{av}$ (for an homogeneous ground) to $0.476p_{av}$ (for an inhomogeneous ground) at the center of the footing and from $1.431p_{av}$ to $0.597p_{av}$ at the extremities (at a distance of 39/40 of the half-width of the foundation).

The above data indicate that consideration of the inhomogeneity, nonlinearity, and creep of a frozen base makes it possible to assume a much more uniform pressure distribution for design purposes than that which is usually obtained from the solution of linear elasticity theory.

With the more uniform reactive pressure distribution, the bending moments are considerably smaller for foundations designed considering their work in combination with the base, so that they can be designed more economically (by as much as 30 to 40 percent) without detriment to strength.

In exactly the same way, as we have indicated, consideration of the vertical decrease in the modulus of deformability in the case of thawing soils permits the use of more economical solutions.

Thus, in evaluating the design mechanical property and deformability characteristics of frozen and thawing soils, account must be taken of their instability in time; the values of the characteristics must either be defined for a specified time interval (during which a given temporary structure is to stand) or conform to the ultimate long-term steady-state value.

A practically useful engineering theory for prediction of the interaction between structures and permafrost must be developed to consider possible subsequent changes in the mechanical properties of the frozen and thawing soils.

2. Compressive and Tensile Strengths of Frozen Soils

The compressive strength of frozen soils (under instantaneous loading and prolonged load action) is of prime practical importance for evaluation of the strengths of frozen soils under short-term loads, especially for specification of

---

*This example is taken from "Mekhanika gruntov" (Soil Mechanics) by N. A. Tsytovich, pp. 293–295, Stroyizdat, 1963, which gives Zaretskiy's more complete solution of this contact problem.

the design pressures in the bases of structures to be built on permafrost and for strength calculations for frozen soil walls in shaft excavation and pits by the method of artificial freezing.

The first experiments to determine the short-term strengths of frozen ground were performed by this author back in 1928-1929 and were published in 1930.[7] Experimental research on the strength properties of frozen soils was developed further both in the author's later studies and in the previously cited work of M. N. Gol'dshteyn, S. S. Vyalov, Ye. P. Shusherina, N. K. Pekarskaya, S. Ye. Grechishchev, S. E. Gorodetskiy, Yu. K. Zaretskiy, and others.

A 1930 paper by the author discussed the compressive strengths of frozen soils as functions of their composition, negative temperature, moisture content, and structure (on the basis of tests on artificially prepared specimens of frozen ground and structurally undisturbed permafrost specimens). The relationships originally obtained for compressive strength as a function of the various factors were later specified more exactly.

Factors of particularly great importance for evaluation of the strength of frozen soils under normal forces (compression and tension) include: (1) their instantaneous (near-maximum) strength, which is usually equated to the so-called ultimate resistance (or, after a proposal of Ye. P. Shusherina, to the short-term resistance, which is more precise) and (2) the ultimate long-term strength, i.e., the resistance at which the strains are always of exponentially increasing nature and still short of the transition to viscoplastic flow terminating in progressive failure of the ground.

We note that the ultimate compressive strength σ_{ul}^{cm} of frozen soils, determined at a standard rate of load increase (15-20 kg/cm^2/min) can be taken in approximation (or conditionally) as the instantaneous resistance. In addition, research done in recent years (by N. K. Pekarskaya, who tested frozen ground at rates of load increase from 1 to 900 kg/cm^2/min) revealed that the ultimate compressive strength is arbitrary, since frozen grounds with different compositions (sands, clays) deform differently on failure (brittle or plastic), and that exact determination of the ultimate load is not possible in plastic deformation;[8] for such soils, the mechanical yield point is determined from the break in the ln σ − ln λ graph, where σ is the resistance to compression and λ is the longitudinal deformation (Fig. 76). Experiments also show that the limit of mechanical fluidity σ_{lt}^{cm} is not a constant magnitude, but depends upon the rate of load application, increasing with its augmentation (Fig. 77). The ultimate long-term compressive strengths of frozen soils can also be established from the plot of Fig. 77 (by extrapolating the curves of $\sigma = f(\nu)$ to the intersection with the σ axis).

However, as we pointed out in the preceding chapter, the ultimate long-term compressive strength σ_{lt}^{cm} can be established much more simply from direct tests of frozen-soil specimens using a ball plunger (in which case σ_{lt}^{cm} is determined automatically) or with dynamometer-type instruments.

The value of the long-term strength in uniaxial compression, i.e., the pressure at which progressive flow has not yet started, can also be determined without

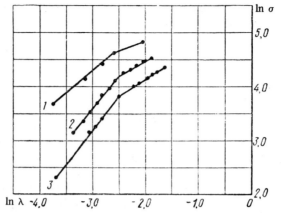

Fig. 76. Compression diagram of frozen sand ($\theta = -20°C$, $W_d = 17\%$) for various rates of load application: 21.7 (1); 6.5 (2); and 1.2 kg/cm²/min (3); logarithmic coordinates.

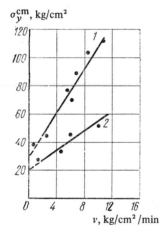

Fig. 77. Yield point σ_y^{cm} of frozen soil in compression as a function of the load-application rate v: (1) frozen sand ($\theta = -20°C$; $W_d = 17\%$); (2) frozen clay ($\theta = -20°C$; $W_d = 32\%$).

waiting for complete damping of the strains in the frozen-soil specimen at the various load steps if the parameter T of the hyperbolic creep coefficient (Eq. III.6″) is known.

According to Yu. K. Zaretskiy[9]

$$T = T_0 \frac{\sigma_{ul}}{\sigma_{ul} - \sigma} \qquad (IV.1)$$

where T_0 is the creep parameter of the frozen soil, which does not depend on the value of the applied load.

Applying Eq. (IV.1) for several load steps (at least two) at a known value of the parameter T (see Chap. III, Sec. 5), we obtain two equations with two unknowns (T_0 and σ_{ul}); solving them simultaneously, we determine the unknown value of σ_{ul}.

We now cite certain numerical values of the ultimate compressive strength σ_{ul}^{cm} of frozen soils and examine their dependence on the negative temperature $-\theta°C$ and the total moisture content W_d of the frozen soil, comparing them with the values of the ultimate long-term compressive strengths.

Table 14 presents certain data on the ultimate compressive strengths σ_{ul}^{cm} of frozen and permanently frozen soils that differ in mechanical composition and temperature. The data were obtained from tests of frozen-soil cubes on hydraulic presses at a standard loading rate of 15–20 kg/cm²/min.

Table 14.   Ultimate strength of frozen soils in uniaxial compression

Designation of soil	Total moisture W_d, %	Tempera-ture θ, °C	Strength σ_{ul}^{cm}, kg/cm²	Investigator
Artificially frozen soils				
Quartz sand (100% content of 1–0.05-mm fraction)	14.7 14.3 14.0 14.1 14.9 14.3	− 1.8 − 3.0 − 6.0 − 9.0 − 12.0 − 20.0	62 78 99 118 134 152	N. A. Tsytovich (1930)
Silty sandy loam (61.2% of 0.05–0.005-mm fraction; 3.2% < 0.005 mm)	21.6 23.1 22.1 21.3	− 0.5 − 1.8 − 5.1 − 10.3	9 36 78 128	Same (1940)
Clay (50% content of < 0.005-mm fraction)	34.6 36.3 35.0 35.3	− 0.5 − 1.6 − 3.4 − 8.2	9 13 23 45	Same
Quartz sand (100% content of 1–0.05-mm fraction)	16.7	− 20.0	150	N. K. Pekarskaya (1966)
Cover clay (44.3% content of < 0.005-mm fraction)	32.0	− 20.0	91	Same
Structurally undisturbed permafrost				
Silty sand (76.4% content of > 0.05-mm fraction; 2.8% of < 0.005 mm)	19.8 19.1 19.8 29.3	− 1.3 − 3.9 − 12.0 − 11.0	105 140 174 97	Permafrost Study Committee (KoVM) Team (compilation by L. S. Khomichevskaya, 1940)
Sandy loam (10% content of < 0.005-mm fraction)	24.8 26.5	− 3.3 − 6.0	58 80	Same
Heavy loam (14.8% content of < 0.005-mm fraction)	24.9 25.0 25.1	− 1.5 − 4.8 − 11.8	29 38 65	Same
Gravelly loam (fractional contents: 43–63% > 2 mm; 19–29% 2–0.05 mm; 14–28% 0.05 mm)	12–17	− 9.8	49–59	V. N. Taybashev and V. G. Gol'dtman (VNII-1, Magadan, 1963)

These values will, of course, be somewhat smaller than the instantaneous strength of the frozen soil under uniaxial compression. As examples, Table 14 includes both some of the results of the author's original 1930 experiments and very recent data that permit more complete characterization of the resistance of the frozen soils to compression.

Examining these and other similar data, we observed first of all that the ultimate compressive strength of frozen soils is very high even at the standard (not the maximum) load-increase rate: it is in the tens and even hundreds of kg/cm². But under faster loading (according to Pekarskaya's experiments, up to 500–900 kg/cm²/min), the compressive strength of frozen sand at $-40°C$ reaches 154 kg/cm² and more, while that of frozen clay rises as high as 750 kg/cm²; the clay specimens deformed plastically with no signs of failure.

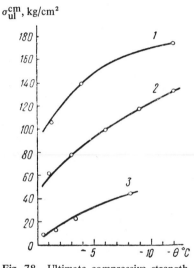

σ_{ul}^{cm}, kg/cm²

Fig. 78. Ultimate compressive strength σ_{ul}^{cm} of frozen soils as a function of negative temperature $- \theta°C$: (1) sand; (2) sandy loam; (3) clay.

It follows from the above data that frozen ground offer, very good resistance to the action of short-term loads.

The ultimate compressive strength of frozen ground depends very strongly on negative temperature. By way of illustration, Fig. 78 shows curves of $\sigma_{ul}^{cm} = f(-\theta°)$ for three frozen soils: (1) sand; (2) sandy loam; (3) clay. Similar curves were obtained earlier by the author as well as later by other investigators.

Analysis of these and similar curves showed that they are described quite accurately by an equation of the form[1]

$$\sigma_{ul}^{cm} = a + b(\theta)^n \qquad (IV.2)$$

where a, b, and n are parameters and θ is the absolute value of the negative temperature.

According to experiments performed at the Igarka Scientific Station of the USSR Academy of Sciences (1959), the parameter n has a value near 1/2, while the studies of Ye. P. Shusherina and S. S. Vyalov,[10] which were concerned with evaluation of the strengths of frozen soils connected with the sinking of shafts by the method of artificial freezing indicate that $n \approx 1$ can be assumed with accuracy sufficient for practical purposes in the case of naturally compacted soils (sandy loams and clays), i.e., a linear relation can be used between compressive strength and the negative temperature $-\theta°C$, while the parameters a and b are variable and depend on the time of load action, decreasing as it increases.

We note that the highest rate of increase of the compressive strength of frozen ground with decreasing negative temperature is observed in the range of significant water phase transitions (approximately from 0 to $-1°C$ for sands and from -0.5 to $-5°C$ for clays), when pore water is freezing at the highest rate; however, compressive strength also increases at lower temperatures, a fact that can no longer be explained solely in terms of increasing ice content of the soils, the strength increasing at a rate that varies in more complex fashion.

Ye. P. Shusherina and Yu. P. Bobkov[11] showed that the latter is due to the fact that with the overall increase in the strength of the ice, the rate of strength increase becomes lower as the temperature is lowered to $-20°$ and then increases on further cooling; this can be explained by the differing effects of reduced translational motion of hydrogen atoms in the structural lattice of the ice (with a simultaneous decrease in the dimensions of the structural lattice itself) at temperatures up to $-20°C$ and at lower temperatures (down to $-55°C$).

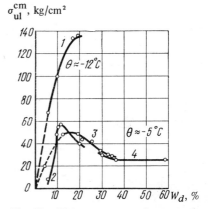

The total moisture content W_d of frozen soils (including its ice and unfrozen-water contents) has a strong influence on their compressive strength. We shall confine ourselves here to an introduction of a graph indicating the variation of the ultimate compressive strengths of frozen ground with the total moisture content (Fig. 79) according to the author's 1937–1940 experiments.

Fig. 79. Ultimate compressive strength σ_{ul}^{cm} of frozen soils as a function of their total moisture content W_d: (1) sand; (2) sandy loams; (3) clay (51% content of 0.005 mm fractions); (4) silty clay (63% content of fraction < 0.005 mm).

We should note that the compressive strength of all frozen soils increases at moisture contents short of complete water saturation and, as a rule, decreases when the soil is fully saturated or oversaturated with moisture. The compressive strength σ_{ul}^{cm} of frozen soils was investigated in detail by Ye. P. Shusherina as a function of total moisture at low temperatures (from -10 to $-55°C$). The general form of the compressive strength relation for frozen ground in the cases of incomplete and complete water saturation, in comparison with that of pure ice, is shown in Fig. 80, according to the above studies, except that we have added the branch of the curve corresponding to incomplete water saturation of the soils.[12]

The nature of the uniaxial compressive strength of frozen soils as a function of total moisture content is basically the same for all types of frozen soils, namely: for incomplete water saturation and loose texture (below W_{min}; segment OA of the graph in Fig. 80), the compressive strength increases; at complete water saturation, it first (segment (AB)) decreases with increasing moisture, reaching (at point B) the compressive strength of ice, and then, at high moisture contents (corresponding to segment CD of the curve), it remains

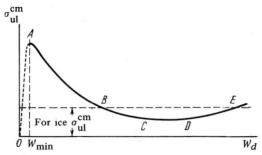

Fig. 80. General nature of compressive strength of frozen soils σ_{ul}^{cm} as a function of moisture content W_d: —— complete water saturation; – – – incomplete water saturation.

practically constant and, on a further increase in moisture content, gradually approaches the compressive strength of pure ice again (point E on the curve of Fig. 80).

The ultimate long-term compressive strength of frozen ground is of special importance for evaluation of the strength of these soils in the bases of structures to be built on permafrost. The corresponding experiments indicate that it is much lower (sometimes by a factor of 5-10) than the ultimate compressive strength, and smaller still compared to the instantaneous strength σ_{inst}^{cm}.

Thus, according to carefully designed experiments performed by S. E. Grechishchev[13] in an underground laboratory with the negative temperature held strictly constant ($\theta = -3°C$) and with stepped loads held until full strain decay (sometimes for as much as 82 days), the following figures apply:

For frozen sand with $W_d = 19.8\%$, $\sigma_{inst}^{cm} = 75$ kg/cm² and $\sigma_{lt}^{cm} = 6.5$ kg/cm² (i.e., smaller by a factor of 11.5).
For frozen loam with $W_d = 31.8\%$, $\sigma_{inst}^{cm} = 35$ kg/cm² and $\sigma_{lt}^{cm} = 3.6$ kg/cm².

Research done at the Permafrost Institute of the Siberian Division of the USSR Academy of Sciences[14] has shown that the long-term compressive strength σ_{lt}^{cm} is described well as a function of the negative temperature $-\theta$ by Eq. (IV.2) with the parameters a_{lt} and b_{lt} taken from Table 15 and the parameter $n = 1/2$.

We note that experiments to determine the compressive strength σ_{lt}^{cm} of frozen soils under prolonged load action have shown that the larger the load step, the slower is the decay of the strain: for example, when $\sigma^{cm} = 2.5$ kg/cm², the strains in loam had damped out after 3 days, but at $\sigma^{cm} = 5$ kg/cm² this occurred only on the 10th day.

The tensile strengths of frozen soils are substantially (by factors of 2 to 6) smaller than their compressive strengths, a fact that was established back in 1937;[15] (as was shown by later experiments), the long-term tensile strength of frozen clayey soils exceeds that of sandy soils. This can be explained by the

Table 15.   Parameter of equation (IV.2) for long-term compressive strength σ_{lt}^{cm}
as a function of temperature

Designation of frozen soil	a_{lt}, kg/km²	b_{lt}, kg/cm² × deg$^{1/2}$
Loam-filled gravel	0.5–1.5	4.5–6.0
Sand-filled gravel	0.0–0.1	5.0–7.0
Sand (W_d = 17–23%)	0.0–0.1	3.5–6.0
Sandy loam (W_d = 20–25%)	0.0–0.2	2.5–4.5
Loam (W_d = 30–35%)	0.2–1.0	2.0–3.0
Clay (W_d = 25–35%)	0.5–1.5	2.0–9.0

increase in the number of contacts between mineral particles under compression, with a decrease in the distance between them, while the distance between particles increases under tension and the number of contacts between them becomes smaller.

The experiments of N. K. Pekarskaya[2] showed that when a clayey soil is brought rapidly to failure, its ultimate (conventional instantaneous) tensile strength is 1.7–1.8 times larger at temperatures from -2 to $-10°C$ than that of frozen sand.

The tensile strength of frozen soil is subject to the same laws as its compressive strength, namely: tensile strength increases with decreasing negative temperature and depends on the composition of the frozen soil, its ice and moisture contents, texture, etc. In absolute value, neither the instantaneous tensile strength of frozen soils nor their ultimate long-term strength is equal to the compressive strength, and this limits the applicability of the equations of the stressed and strained state that are used for isotropic frozen soils.

Table 16 presents a series of values of the instantaneous tensile strengths of frozen and permanently frozen soils and their long-term tensile strengths.

It is important to note that Igarka experiments confirmed that frozen and permanently frozen soils have an ultimate long-term tensile strength: thus, specimen No. 17 (Table 16) did not fail even after 6 years under a tensile stress of 1.8 kg/cm².

3. Shear Strength of Frozen Soils

A condition for strength of any material at a given point is, as we know, a sufficient shear resistance at this point. For this reason, the shear strength of frozen soils is a mechanical characteristic of these materials without knowledge of which it is impossible to calculate either the ultimate strengths of the soils in the bases of structures, or their strength in various types of protective structures (for example, excavation of pits and shafts using artificial freezing of the ground), or the stability of massive blocks of frozen ground subjected to shear loads.

Table 16. Instantaneous and ultimate long-term tensile strengths of frozen soils

Speci-men no.	Designation of soil	W_d, %	$-\theta\,°C$	σ^t_{inst}, kg/cm²	σ^t_{lt}, kg/cm²	Investigator
		Artificially frozen soils				
1	Clay (45% content of fraction < 0.005 mm)	19.4	− 1.2	9.8	−	N. A. Tsytovich, 1952
2	Same	19.4	− 2.5	16.8	−	Same
3	Same	19.4	− 4.0	21.6	−	Same
4	Cover clay	32.00	− 2.0	11.5	−	N. K. Pekarskaya, 1966
5	Same	32.0	− 5.0	13.8	−	
6	Same	32.0	− 10.0	26.5	−	Same
7	Heavy loam (22.5% content of fraction < 0.005 mm)	31.8	− 3.0	12.0	2.6	S. E. Grechishchev, 1963
8	Heavy sandy loam	34.0	− 4.0	17.0	2.0	N. A. Tsytovich, 1952
9	Quartz sand	17.0	− 2.2	6.3	−	N. K. Pekarskaya, 1966
10	Same	17.0	− 5.0	7.9	−	
11	Same	17.0	− 10.0	16.0	−	Same
		Permafrost				
12	Silty heavy loam	39.0	− 4.2	24.0	1.7–1.8	S. S. Vyalov, 1954
13	Light loam	30.0	− 4.0	20.0	1.6–1.7	Same
14	Silty heavy sandy loam	44.0	− 4.2	20.0	1.6–1.7	Same
15	Same	44.0	− 0.2; − 0.4	7–8	0.3	Same
16	Same	30.0	− 4.6	−	1.7–1.8	Same
17	Light silty sandy loam	31.0	− 4.3	20.0	1.8	Same

Experiments performed in recent decades (by M. L. Sheykov, S. S. Vyalov, N. K. Pekarskaya, and the author) have shown that the shear strength of frozen ground depends on a number of factors and is a function of no fewer than three variables:

$$\tau = f(-\theta, p, t) \qquad\qquad (IV.3)$$

where $-\theta$ is the negative temperature of the soil, p is the external pressure, and t is the time of load action.

As carefully designed experiments show, the ultimate (failure) strength of frozen soils under planar shear depends on the normal pressure, i.e., it is governed not only by cohesive forces, but also by internal friction and can be described under not too high pressures (up to 10–15 kg/cm²) by a first degree equation of the normal pressure with variable parameters depending on the negative temperature $-\theta°C$ and the time of load action t, i.e.,

$$\tau_{ul} = c_{\theta,t} + \tan \phi_{\theta,t}p \qquad \qquad (IV.4)$$

The results of determination of the ultimate shear strength of frozen soils at various temperatures[16] indicate that the lower the temperature of the frozen soil, the higher is its shear strength—both total and components: the angle of internal friction $\phi°$ and the cohesion c in kg/cm². Thus, for frozen clay ($W_d =$ 33%) $\phi = 14°$ at $\theta = -1°C$ and $\phi = 22°$ at $\theta = -2°C$. At temperatures near 0°, the angle of internal friction of a frozen soil is practically equal to the same angle for the unfrozen soil, while the cohesion is substantially higher as compared to that of the soil in the unfrozen state.

Cohesion accounts for a substantial part of the total shear strength of frozen soils (Table 17);[16] in many cases, and especially for high-temperature frozen soils (when the angle of friction approaches its value for the unfrozen soil), this makes it possible to consider only the cohesion of the ground in calculations. Further experiments showed* that the angle of internal friction of frozen sands is practically independent of load time, while frozen clays are characterized by changes in cohesion and angle of internal friction with increasing time of failure load, application.

*See Ref. 2, part 2, this chapter.

Table 17. Friction and cohesion in % of total shear strength for soils in the frozen and unfrozen but solid states (cover clay)

Bulk density γ, g/cm³	Moisture content W_d, %	Normal load p, kg/cm²	Shear stress τ, kg/cm²	Cohesion c, %	Friction $\tan \phi$, %
		At $\theta = -1°C$			
—	—	1	5.5	94.7	5.3
—	—	3	6.0	86.7	13.3
1.84	26.5	4	6.2	83.9	16.1
1.88	34.8	8	7.3	71.2	28.8
1.85	29.1	12	8.3	62.6	37.4
		At $\theta = -2°C$			
—	—	1	7.6	94.7	5.3
—	—	3	8.4	85.7	14.3
1.86	32.1	4	8.9	81.0	19.0
1.84	32.3	8	10.5	68.5	31.5
—	—	12	12.2	59.8	40.2
		At $\theta = +20°C$			
2.04	23.2	1	0.84	91.7	8.3
2.07	22.8	3	0.99	77.7	22.3
2.11	22.7	5	1.10	70.0	30.0

Note: The unfrozen specimens were first compacted by a load of 5 kg/cm².

Table 18.   Shear strength of frozen soils under torsion

Designation of soil	W_d, %	$\theta\,°C$	τ_{ul}, kg/cm^2	τ_y, kg/cm^2	Test method
Sand (96% content of fractions from 1 to 0.25 mm)	15	-2.5	17.0	5.0	Twisting of cylindrical specimens
Silty soil (61% content of 0.05–0.005-mm fractions)	22	-0.5	4.8	1.7	Same
	22	-1.8	9.7	3.0	
	22	-6.0	28.0	6.8	
Clay (50% content of $<$ 0.005-mm fractions)	35	-0.5	4.6	1.1	Same
	35	-1.4	6.5	2.3	
	35	-10.0	18.0	6.8	

The time of shear-load action is of substantial importance for evaluation of the shear strengths of frozen soils because of the enormous influence exerted on the shear strength by stress relaxation, which controls the flow of frozen soils under load and the decrease in their strength.

Table 18 presents the results of some of our experiments[17] to determine the ultimate (near-instantaneous) shear strength τ_{ul} in torsion and the yield point τ_y in shear, which we determined with the shear load increasing at a rate of 1.5 kg/cm^2/min in 0.6-kg/cm^2 steps and held until the strains subsided or until viscoplastic flow began.

The data of Table 18 indicate a decrease in the shear strength of frozen soils with the passage of time.

A particularly sharp decrease in the shear strength of frozen soils from τ_{inst} to τ_{lt} occurs under long-term application of load. Thus, for example, in experiments with a frozen clayey soil of reticular structure with a total moisture content $W_d = 33\%$ at a temperature $\theta = -2°C$, it was found that $\tau_{inst} = 13.7$ kg/cm^2, while the long-term strength $\tau_{lt} = 1.1$ kg/cm^2.

The decrease in the shearing resistance of frozen ground under long-term application of a constant load occurs chiefly at the expense of its cohesive forces (Table 19).

Table 19.   Decrease in cohesion of a frozen soil (heavy dusty sandy loam) under prolonged load application (according to experiments of S. S. Vyalov)

Tempera-ture, °C	Cohesion in % of instantaneous value at various times						
	1 min	5 min	30 min	1 hr	2 hr	8 hr	Ultimate
-4.2	72	63	56	52	47	45	37
-1.2	62	43	38	36	31	30	25
-0.4	52	37	26	24	20	19	18

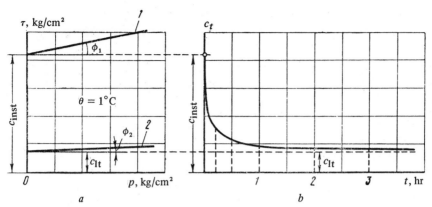

Fig. 81. Shear strength of frozen soils plotted against time of load application. (*a*) Shear diagram, $\tau = f(p)$; (*b*) relaxation of cohesive forces $c = f(t)$.

To illustrate this, Fig. 81*a* shows the results of some of our experiments to determine the strength of the same frozen soil ($W_d = 33\%$), but at a temperature $\theta = -1°C$: line 1 indicates the strength in rapid shear and line 2 the ultimate long-term strength for various values of the normal pressure p, while Fig. 81*b* shows the relaxation of the cohesive forces in time for the same soil. These data indicate that the angle of internal friction of frozen clay dropped from $14°$ (in instantaneous shear) to $4°$ (in ultimate long-term shear) in these experiments, while cohesion decreased from 5.2 to 0.9 kg/cm^2, i.e., several times its value.

The experiments also showed that, as a rule, a particularly substantial decrease occurs with time in the cohesive forces of the frozen soils (approximately from 1/3 to 1/5 of c_{inst}) — this fundamental strength characteristic of these soils; in frozen sands, the angle of internal friction is practically independent of the time of load action, while in frozen clays both the angle of internal friction and especially the cohesion decrease with the passage of time.

We note also that shear strength can no longer be regarded as a linear function of pressure over a broad pressure range for frozen ground, and that it is necessary to consider a shear diagram in the form of a curved envelope of ultimate-stress circles that is different for different time intervals (Fig. 82*a*), approximating the experimental data as a certain nonlinear function or linearizing the envelope piecewise, or else considering the dependence of the ultimate tangential-stress intensity (which corresponds to the appearance of progressive flow) on the average normal stress σ_{av}. Here the $T = f(\sigma_{av})$ curves will also be nonlinear in the general case (Fig. 82*b*).

The cohesion of frozen and permanently frozen soils, which is usually the factor determining their strength because it dominates in their total shear resistance can be determined quite successfully by the ball test (see Fig. 61), as was suggested by the author as early as 1947[18] and described in detail in other papers.[19]

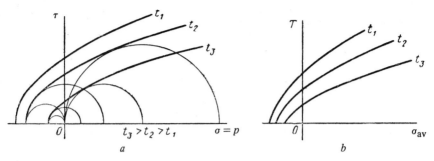

Fig. 82. General case of shear strength of frozen ground as a function of external pressure and load application time. (*a*) Envelopes of ultimate-stress circles in shear, $\tau = f(\sigma, t)$; (*b*) shear-stress intensity T as a function of average pressure σ_{av} and the time t, i.e., $T = f(\sigma_{av})$.

If it is assumed in accordance with a rigorous solution of plasticity theory (A. Yu. Ishlinskiy, 1944)[20] for ideally bound undensifiable solids that the ratio of the hardness to the stress at the yield point is constant at 0.36, the cohesion of bound soils (among which we may include all disperse frozen and nonfrozen clays, disperse frozen sands and ice) is determined, as has been shown by the author, by the expression

$$c_e = 0.18 \frac{P}{\pi D s_t} \qquad (IV.5)$$

where P is the load on the ball plunger, D is its diameter, and s_t is the settling depth of the plunger, which is different at different time intervals.

If the settling (impression) depth s_t of the ball plunger is determined immediately after application of the load (within 5–10 sec), the cohesion thus determined can be taken practically as the instantaneous cohesion; but if s_t corresponds to the steady-state, stabilized sinking depth of the plunger (ultimate long-term), the cohesion determined from Eq. (IV.5) at $s_t = s_{lt}$* will be the ultimate long-term cohesion.

As the author showed earlier,[21] the cohesion determined from Eq. (IV.5) is a certain composite characteristic of cohesive soils that takes account both of their cohesion and, to a certain degree, friction as well. But if the cohesion of frozen soils and their internal friction are taken into account separately, it is necessary to introduce a certain correcting multiplier M into Eq. (IV.5) to determine the cohesion alone; the value of the correction depends on the angle of internal friction ϕ and, as has been shown by V. G. Berezantsev,[22] $M = 1$ (at $\phi < 5°$), $M = 0.615$ (at $\phi = 10°$), $M = 0.285$ (at $\phi = 20°$), and $M = 0.122$ (at $\phi = 30°$).

However, as will be shown below (see Sec. 6 of this chapter), the application of this correction with pure cohesion separated and simultaneous consideration of friction, which is determined in separate experiments, greatly complicates

*The details of ball-plunger tests are set forth in the author's papers (see Ref. 18) and in his book "Mekhanika gruntov" (Kratkiy kurs) [Soil Mechanics (A Short Course)], *Izd-vo* "*Vysshaya shkola*," 1968. For the definition of c_{lt}, see also Chap. III, Sec. 4.

calculations of the bearing strength of frozen ground and permafrost without substantially improving them, especially at internal-friction angles smaller than 20°, and for this reason this refinement is not widely used at the present time.

Based on the above, the cohesion determined by the ball-test method should be regarded as an *equivalent* cohesion[23] c_e which is a composite characteristic of the cohesive forces in plastic soils.

We note that detailed experiments by S. S. Vyalov have shown that the strain-hardening of frozen soils by the ball plunger is quite insignificant and that the cohesion determined can be regarded as an invariant real characteristic of the mechanical properties.[24]

By measuring the sinking depth of a ball plunger set up on a frozen-soil specimen or directly over a leveled area in the excavation at various time intervals t under the action of a constant load and calculating the equivalent cohesion from Eq. (IV.5), we obtain data to build a complete relaxation graph of the cohesive forces with time (Fig. 83).

The cohesive forces in frozen and permanently frozen ground strongly depend on their negative temperatures $-\theta$, as has been shown by appropriate experiments, described quite accurately by an equation of the form of (IV.2) whereby the exponent for θ can be put equal to $1/2$, i.e.,

$$c_\theta = a' + b'(\theta)^{1/2} \qquad \text{(IV.2')}$$

where a' and b' are, as before, parameters of the curve $c = f(-\theta°)$, and θ is the absolute value of the negative temperature of the frozen ground.

Certain values of the instantaneous c_{inst} and long-term c_{lt} cohesion for frozen soils according to Vyalov's experiments appear in Table 20.

These and similar data are used successfully in calculations of the bearing ability of permafrost at the base of structures.

4. Adfreezing Strength of Soils

Data on the adfreezing strength of the soils, evaluated as their shear strength at the surface of adfreeze to the foundation material, are of very great importance in calculating placement of foundations in permafrost, where the

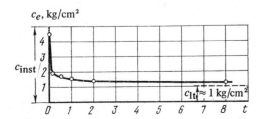

Fig. 83. Relaxation graph of cohesive forces c_e in frozen sandy loam ($W_d = 35\%$, $W_u = 13.5$, $\theta = -0.3°C$).

Table 20.   Instantaneous and ultimate long-term cohesion of permafrost with undisturbed structure, kg/cm^2

Designation of soil	W_d, %	$\theta = -0.3$ to $-0.4°C$		$\theta = -1.0$ to $-1.2°C$		$\theta = -4.0$ to $-4.2°C$	
		c_{inst}	c_{lt}	c_{inst}	c_{lt}	c_{inst}	c_{lt}
Dense varved clay (mineral interlayers)	30–40	5.7	1.8	–	2.6	16.0	4.2
Silty heavy loam	36	4.3	0.6	7.0	1.0	12.0	–
Silty light loam	30	4.1	0.9	7.4	–	11.0	2.0
Heavy silty sandy loam	28–34	4.0–4.5	0.9–1.0	7.3	1.6	8.0–15.0	2.8–3.2
Same, high moisture	43	6.0	0.75	–	–	11.0	2.0
Same, with peat	30	–	–	–	–	9.0	2.0
Silty sand	23	11.0	2.1	14.0	2.7	20.0	3.7–4.5

active layer is subject to frost heave. Adfreezing strength is usually determined by pushing in or pulling out stakes that have been frozen into the ground.

The first (1931–1936) experiments to determine adfreezing strength (or the so-called adfreezing forces) were performed at high load-increase rates and approached the maximum values, while the so-called ultimate adfreezing strength was determined at the standard loading rate of 15–20 kg/cm^2/min. Certain data from determinations of the adfreezing strength (the ultimate value, which approaches the instantaneous shear strength at the adfreezing surface) in experiments of the author and coworkers[25] are condensed in Table 21.

Table 21.   Soil-to-wood adfreezing strengths

Designation of soil	Tempera-ture $-\theta$, °C	Total moisture (on raw weight) W_{tot}, %	Adfreezing strength τ_{af}, kg/cm^2
Sandy loam (68% content of fractions from 1.0 to 0.05 mm; 8% smaller than 0.005 mm)	− 0.2	11.0	1.3
	− 1.2	6.3	2.8
	− 1.2	11.5	7.0
	− 1.2	14.2	8.2
	− 5.6	11.4	20.8
	− 10.5	12.3	24.7
	− 17.4	11.3	27.4
Clayey soil (36% content of fractions smaller than 0.005 mm)	− 0.2	21.3	2.9
	− 1.5	20.9	5.9
	− 5.8	22.1	11.1
	− 10.8	22.1	18.6
Gravel (99% content of fractions from 2 to 20 mm)	− 9.9	1.4	0.9
Same, with 1.8% silt content	− 10.5	2.7	1.6

These data show that the mechanical composition of the soil has a strong influence on the ultimate strength of adfreezing of frozen soils to wood at a given moisture content: the coarser the particles of the soil, the lower is the adfreezing strength. The lowest adfreezing strength was obtained for coarse gravel wetted with water.

To illustrate the above, Fig. 84 shows a graph of adfreeze strength of various soils to wood at the same relative moisture content (about 80 percent of total moisture capacity) and at the same temperature ($\sim -10°C$) as a function of coarseness of the soil mineral particles. It follows from inspection of this graph and the data of Table 21 that the highest adfreeze strength is observed for water-saturated sands, especially when medium-grained, and that this strength is lower for coarse sands and lowest for gravel and pebbles. Frozen clays have a somewhat lower strength of adfreeze to wood (at temperatures above $-20°C$) as compared to water-saturated sands. This is explained by the higher unfrozen-water content of clayey soils as compared to the water content in sandy soils.

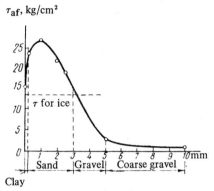

Fig. 84. Ultimate strength of adfreezing of various soils to wood vs. the mechanical properties of the soils.

The relation established here is of significance for practice: columns and foundations will set better in permafrost soils if they are packed with moistened sand (which later adfreezes tightly to the column material), and the segments of the columns and foundations in the active layer should be packed with nonfreezing materials or at least with dry gravelly and pebbly soils to mitigate the effects of heaving; this has been borne out in practice.

Adfreezing strength is largely influenced by the initial total moisture content of the soils, with which it increases (see, e.g., τ_{af} in Table 21 for the sandy-loam soil at $-1.2°C$ for $W_{tot} = 6.3\%$ and $W_{tot} = 14.2\%$.

The magnitude of negative temperature also has a substantial effect on adfreezing strength (Fig. 85).

The above results and numerous others indicate that adfreezing strength increases especially rapidly with decreasing temperature in the range of significant water

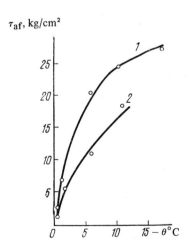

Fig. 85. Adfreeze strength of soils to wood as a function of negative temperature. (1) Sandy-loam and (2) clayey soils.

phase transitions (approximately from 0 to $-1°C$ for sands and sandy loams and from -0.5 to $-5°C$ for clays), i.e., in the range in which most of the pore water turns to ice.

Experiments also indicate that the instantaneous adfreeze strength (or ultimate adfreeze strength) may reach enormous values, on the order of 300–500 tons/m^2, at low temperatures (below the range of intensive water-to-ice phase transitions).

However, the adfreezing strength decreases substantially at below-standard loading speeds. Thus, according to our own data (1936 experiments), the adfreezing strength of a sandy loam was 9.3 kg/cm^2 at a load-increase rate of 22 kg/cm^2/min and only 2.8 kg/cm^2 at 1 kg/cm^2/min.

This is explained by flow of the frozen soils in shear against the surface of the test column.

Table 22 presents certain data on adfreezing strengths obtained with consideration of the flow of frozen soils under load (according to experiments performed in 1939 by the author and E. I. Levin). These data were obtained with stepwise loading, each step of the shear load being held until deformation stopped, and it was assumed that this had happened if the displacement of the column was less than 1 μm in the last 20 minutes (which is, of course, a somewhat arbitrary limit).

The given data show that the adfreeze strength of moist soils to wood, obtained under conditions such that the frozen soils flow at temperatures in the range of rapid water-to-ice phase transitions, is approximately 1/5 to 1/10 of the ultimate adfreeze strength.

In 1940–1943, M. N. Gol'dshteyn[26] made detailed studies of adfreeze strength under consideration of shear-stress relaxation and established that it increased by a factor of 11.2 for a moist medium-grained sand when the time under load was increased from 5 minutes to 2,060 hours, and by a factor of 12.7 for a loam when the time under load was varied from 2 minutes to 521 hours,

Table 22.   Strength of adfreeze of soils to wood corresponding to flow of frozen soils under load

Designation of soil	Adfreeze strength in kg/cm^2 at stepwise shear and temperatures (°C)				Remarks
	-0.5	-1.5	-3	-5	
Sand (W_{tot} = 18%) and silty soil (W_{tot} = 28%)	0.6	1.5	3.0	5.8	Sand (79% content of fractions from 1 to 0.05 mm) Silty soil (72% content of fraction from 0.05 to 0.005 mm)
Clay (W_{tot} = 39%) and ice (W_{tot} = 100%)	0.4	0.7	–	2.4	Clay with 59% content of fraction < 0.005 mm

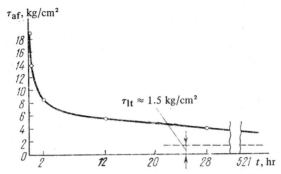

Fig. 86. Adfreeze strength as a function of time of shearing load application (according to experiments by M. N. Gol'dshteyn).

i.e., the long-term adfreezing strength found in the experiment was, on the average, 1/12 of the ultimate (near-instantaneous) adfreeze strength (Fig. 86).

As an example, Table 23 presents certain values obtained for the ultimate long-term strength of adfreeze of soils to wooden columns in pre-steamed ground (according to Vyalov's experiments).

The data of Table 23 have been used in Fig. 87 to plot the long-term strength of ground adfreeze to wood against negative temperature.

We note that Gol'dshteyn's experiments to study the adfreeze strength of ground to wood and concrete during continuous slip of the column relative to the soil and repeated driving of the column indicated that adfreeze strength decreases under these conditions by an average of one-half.

Thus, the above material obviously implies that it is necessary in experimental determinations of the strength of adfreezing of ground to wood and concrete to take full account of the time of action of the shear load and, for constant (long-term) loads, to determine the ultimate long-term adfreeze strength.

If results of direct experiments to determine the strength of adfreeze of ground to foundation materials are not available for the particular type of soil, its temperature, and moisture content, reference may be taken to SNip II-B.6-66 (Sec. 57), which gives standardized resistances of frozen soils to shear relative to

Table 23.   Ultimate long-term strength of adfreezing of soils to wooden columns

Soil	$\theta \degree C$	τ_{lt}, kg/cm²	Site and test method
Silty sandy-loam soils ($W_{tot} = 30$–40%)	− 0.2	0.3	Extraction of piles driven into steamed permafrost (under field conditions in Igarka region)
	− 0.5	0.6	
	− 0.7	0.8	
	− 1.0	1.0	
	− 2.0	1.5	
	− 3.0	2.0	

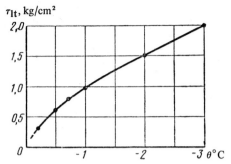

Fig. 87. Long-term strength of adfreeze of ground to wood as a function of temperature $-\theta$.

a lateral surface of adfreezing (R_{sh}^s, kg/cm^2) to wood or concrete for design temperatures at the middle of the ith permafrost layer, with a reducing correction factor of 0.7 applied for determination of the adfreeze strength of soils to metallic surfaces.

5. Resistance of Frozen Ground to Disintegrating by Cutting

The study of disintegration of frozen ground by cutting is especially important in the establishment of methods for effective drifting in frozen ground for various types of tunnelling and earthworks projects under permafrost conditions.

Extensive experiments were carried out in 1951–1955 at the Institute of Mining of the USSR Academy of Sciences under the supervision of Prof. N. A. Zelenin[27] to determine the resistance of frozen ground to cutting.

The following were taken as the basic indicators of the cutting resistance of the frozen soils: cutting effort using a standard cutting tool of elementary profile (a flat wedge) (P in kg) and the specific resistance to cutting

$$k = \frac{P}{hb} \text{ kg/cm}^2 \qquad \text{(IV.6)}$$

where h is the cutting depth and b is the cutting width.

Reduction of data from several thousand experiments indicated that the cutting force P is a linear function of the cutting depth h (Fig. 88) at all negative temperatures and moisture contents of the frozen soil.

The cutting effort P is strongly influenced by the width of the

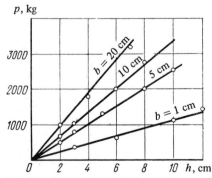

Fig. 88. Cutting force P in kg as a function of cutting depth h for sandy loam ($W_d = 34\%$, $\theta = -1°C$, number of percussions $C = 62$) for various profile widths b.

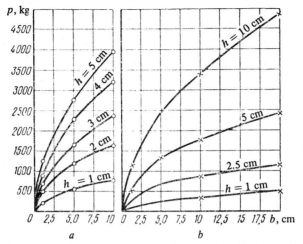

Fig. 89. Cutting force P for sandy loam as a function of width b of elementary profile at various cutting depths h (at a cutting angle $\alpha = 90°$, taper angle $\beta = 180°$). (a) $W_d = 18\%$, number of percussions $C = 143$, $\theta = -3$ to $4°C$; (b) $W_d = 34\%$, $C = 62$, $\theta = -1°C$.

elementary cutting-tool profile b; this influence is qualitatively the same as for unfrozen soils.

The plot of cutting force against the width of the elementary profile is curvilinear (Fig. 89) and can be approximated as a power-law function:

$$P = Ab^m \qquad (IV.7)$$

Direct experiments for various profile widths (from 1 to 10 cm) and various temperatures (from $\theta = -1°C$ to $\theta = -4°C$ and lower) have shown that the exponent m has a value very close to 1/2, i.e., we may assume $m \approx 0.5$.

The value of the coefficient A, which characterizes the physical state of the frozen soil, is determined from a graph plotted in the coordinates $\ln P$ and $\ln b$ (Fig. 90) and corresponds to the force P for a profile of width $b = 1$ cm, reflecting the cutting resistance of various frozen soils to the same profiles.

The experiments also showed that the specific cutting resistance k [kg/cm²] depends on the total moisture content W_d of

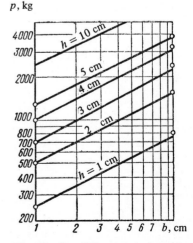

Fig. 90. $P = f(b)$ curves in logarithmic coordinates.

the frozen soils, with the maximum specific cutting resistance (like the resistance to compression) of the frozen ground corresponding to the state in which the pores of the soil are completely filled with ice; a further increase in moisture lowers this resistance slightly.

Figure 91 shows near-maximum specific cutting resistances k for various frozen soils as functions of their negative temperatures $-\theta$.

The specific cutting resistance of frozen soils increases most rapidly (like their other resistances) in the range where the water-to-ice phase transitions are significant (most rapid), i.e., at temperatures in the approximate range 0 to $-5°C$; this should be regarded as due, for the most part, to an increase in the ice content of the frozen ground (or a decrease in the content of unfrozen water) with lowering negative temperature. However, as Zelenin correctly observes, the cutting resistance of frozen soils is determined at temperatures from -5 to $-40°C$ and lower not only by the decrease in unfrozen-water content, but also by the strengthening of the crystalline structure of the ice, since the hardness (and, consequently, the strength) of ice single crystals (which do not contain unfrozen water) increases from 2-3 on the Mohs scale (at $\theta = -3°C$) to 4 (at $\theta = -40°C$) and 6 (at $\theta = -78°C$).

Although this statement, which we have discussed in greater detail in earlier sections of the present book, appears quite reasonable, it requires quantitative

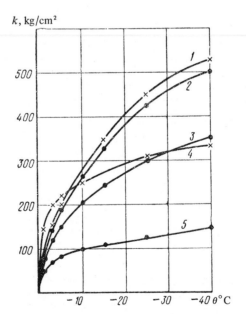

Fig. 91. Curves of $k = f(-\theta)$ for various frozen soils. (1) Sandy loam ($W_d = 19\%$); (2) loam ($W_d = 25\%$); (3) clay ($W_d = 31\%$); (4) sand ($W_d = 18\%$); (5) sand ($W_d = 11\%$).

adjustments, since the properties of the pore ice have not been studied thoroughly enough during the lowering of negative temperatures of frozen ground.

As before, the specific cutting resistance k of frozen soils can be described as a function of the negative temperature $-\theta$ by a power-law equation:

$$k = B\theta^n \qquad \text{(IV.8)}$$

where, according to Zelinin's experiments, the exponent $n = 0.5$ for all types of frozen clayey soils (clays, loams, sandy loams) and $n \approx 0.25$ for frozen sands; B is a parameter determined from a logarithmic plot (similar to that in Fig. 90).

Analysis of numerous experiments to determine the specific cutting resistance k [kg/cm^2] of frozen soils as related to the percussion count C of the standard DorNII (Scientific Research Institute of Roads) percussion machine, i.e., the number of percussions necessary to sink a cylindrical rod with a taper angle $\beta = 180°$ and an area of 1 cm^2 10 cm into the soil with a 2.5-kg weight dropped from a height of 0.4 m, has shown that a proportional relation exists between k and C; which is very important, since it enables us to determine the specific cutting resistance of frozen ground by means of a very simple dynamic percussion-machine test.

We note that a percussion bit with a cylinder area $F = 0.5$ cm^2 and a taper angle $\beta = 30°$ has been found more convenient for tests of frozen ground (giving shorter test times); for greater generality of the conclusions, its readings are converted to the standard readings of the DorNII machine on the basis of an established empirical relationship: $C = 2.85C'$ (where C is the percussion count of the standard machine and C' is the percussion count for frozen ground with the bit described above).

Studies have also shown that frozen ground should be broken by methods in which tensile stresses predominate, and that the energy consumed in excavating frozen ground will be minimized if it is chipped with a narrow wedge bit giving a tearing effect.

Based upon results from several series of experiments with frozen ground, with an adequate number of repetitions (no fewer than 4), Zelenin compiled a cutting resistance scale for frozen soils, which is reproduced in condensed form (for two frozen soils) in Table 24.

This scale was constructed on the basis of two indicators: (1) the number of percussions C of the DorNII machine and (2) the specific cutting resistance k (to a standard cutting bit with width $d = 3$ cm, taper angle $\beta = 180°$, and cutting angle $\alpha = 90°$).

On the basis of his cutting-resistance experiments with frozen ground, Zelenin recommends that the cutting force P be determined (for the case of power shovels and prime movers with engines developing 50 hp or more and the possibility of using them directly for cutting frozen ground with elementary-type cutting bits) from an empirical formula that takes the following simple form for the optimum $\alpha = 30°$ cutting angle for frozen soils:

Table 24. Cutting resistance scale for frozen soils

Designation of soil	Total moisture content W_d, %	Coefficients	At temperature, °C						
			-1	-3	-5	-10	-15	-25	-40
Frozen clay	17	C	35–40	70–80	100–110	150–165	180–200	–	–
		k	35–40	70–85	100–115	150–170	180–200	250–270	290–315
Same	24	C	55–60	90–100	125–135	190–210	220–235	–	–
		k	55–65	90–100	120–130	180–200	215–240	270–285	320–340
Same	31	C	65–70	120–130	140–160	210–220	290–310	–	–
		k	65–70	115–130	140–160	210–230	280–310	290–320	330–360
Same	49	C	40–45	65–70	90–100	135–145	180–190	–	–
		k	40–45	70–75	90–100	135–140	170–190	235–245	280–310
Frozen sand	6	k	12–14	15–18	20–22	25–27	28–30	32–35	40–45
Same	11	k	50–55	65–70	85–90	90–95	100–145	120–130	140–150
Same	18	k	150–160	200–210	220–230	240–250	260–280	285–300	325–340

Note: C is the number of percussions of the dynamic density gauge; k is the specific cutting resistance (kg/cm²) for $b = 3$ cm.

$$P = Ch\Delta \sqrt{b} \qquad \text{(IV.9)}$$

where C is the number of percussions of the dynamic density gauge, which is fitted with a cylindrical tip having an area of 1 cm^2, Δ is a blunting coefficient, which equals 1 for a slightly blunted cutting tool and 0.85 for a sharp one and increases to 2 as wear progresses.

The value of C is taken from Table 24, those of h and b are specified by design criteria (at $h/b \approx 3$ for the shape of the bit has no particular influence on the cutting force), and Δ must be adjusted by trial and error at the site.

The above data form a basis for calculation of cutting forces for frozen soils under various conditions.

6. Magnitudes of Critical and Design Strengths of Frozen Ground

The critical strengths of frozen soils are determined analytically from expressions that follow from the theory of the ultimate stressed state of soils.

As discussed in Sec. 3 of this chapter, a linear dependence of the ultimate (failure) shear strength of the normal pressure (Eq. IV.4) can be used as the equation of the ultimate stressed state of frozen soils at pressures that are not too high (up to 10–20 kg/cm^2), i.e.,

$$\tau_{ul} = c_{\theta,t} + \tan \phi_{\theta,t} p$$

where $c_{\theta t}$ is the cohesion of the frozen ground, which depends (in contrast to the case of unfrozen ground) both on the negative temperature $-\theta\,^\circ$C and on the time of load application t, and it is necessary in the case of a constant load to consider relaxation of the resistance of the frozen ground and use the ultimate long-term cohesion c_{lt}; $\phi_{\theta,t}$ is the coefficient of internal friction, which also depends on the negative temperature $-\theta$ and the time t, and its ultimate long-term value ϕ_{lt} must be used in ultimate-load calculations; p is the external pressure in kg/cm^2.

Thus, we have

$$\text{des } \tau_{ul} = c_{lt} + 1 \tan \phi_{lt} p \qquad \text{(IV.4')}$$

As we noted earlier, cohesion is the dominant factor in the total shear strength of frozen and permanently frozen ground, especially clayey types, for which it is tens of times larger than it is for unfrozen soils; on the other hand, the coefficient of internal friction $\tan \phi$ is of considerably lesser importance for high-temperature frozen ground, especially under long-term load application. For this reason, dispersed frozen soils with angles of internal friction ϕ smaller than about 20° (see below) can be regarded as ideally cohesive materials, and their friction resistance can be neglected in determining the load limits, which gives us a certain margin and simplifies all calculations substantially without introducing any inadmissible errors. This is all the more legitimate because the cohesive force determined by the ball-test method takes account, as we indicated above, not only of cohesion, but also indirectly of the friction in the soil.

Two criteria should be distinguished in determining the critical resistances of frozen soils: (1) the initial critical load init p_{cr} at which no dangerous plastic flows (zones of limiting equilibrium limit) have as yet appeared in the soil under the foundation and (2) the ultimate critical load on the soil ult p_{cr}, which exhausts the bearing capacity of the frozen soils and on attainment of which the ground goes into progressive flow, which leads to its failure or to complete loss of stability.

The initial critical load (critical pressure on the ground) is determined on the basis of the following considerations in the case of the two dimensional problem with a load uniformly distributed on a strip.

The limiting-equilibrium condition for an ideally cohesive soil ($c \neq 0$ and $\phi = 0$) at any point in that soil will be: that the maximum shear stress max τ, be equal to long-term cohesion c_{lt} of the frozen soil at this point, i.e.,

$$\max \phi \approx c_{lt} \qquad (c_1)$$

But if the shear stress comes to exceed the cohesion of the soil at a given point, an elementary shear phane will appear at that point, and a series of successive elementary planes will form a shear zone (a region in which the ultimate stressed state prevails).

For the case of a two-dimensional problem we have:

$$\max \tau = \frac{\sigma_1 - \sigma_2}{2} \qquad (c_2)$$

where σ_1 and σ_2 are the principal maximum and minimum stresses.

When a uniformly distributed strip load p acts on the ground, the principal stresses will be determined from the familiar expressions:

$$\sigma_1 = \frac{p}{\pi} (2\beta + \sin 2\beta)$$

$$\sigma_2 = \frac{p}{\pi} (2\beta - \sin 2\beta) \qquad (c_3)$$

where 2β is the so-called angle of visibility (Fig. 92).

Applying also (c_2) and denoting by p_{cr} the critical pressure that satisfies condition (c_1), we obtain

$$\frac{p_{cr} \sin 2\beta}{\pi} = c_{lt} \qquad (c_4)$$

from which

$$p_{cr} = \frac{\pi c_{lt}}{\sin 2\beta} \qquad (c_5)$$

Obviously, the critical pressure will have its minimum value for this case when sin $2\beta = 1$. This pressure corresponds to the initial appearance of slip

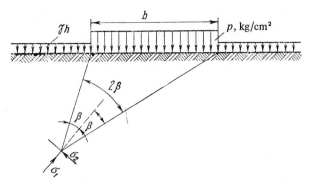

Fig. 92. Diagram showing action of strip load in the case of a two-dimensional problem.

zones in the soil under the margin of the loading area, which we denote by init p_{cr}.

We have then[28]

$$\text{init } p_{cr} = \pi c_{lt} \tag{IV.10}$$

Equation (IV.10) is the simplest and most convenient expression for determining the initial critical load on frozen ground and is widely used in practice, since it gives quite accurate results (for internal-friction angles $\phi \leqslant 20°C$).

Considering a side load increment $q = \gamma h$ (see Fig. 92), we obtain finally

$$\text{init } p_{cr} = \pi c_{lt} + \gamma h \tag{IV.10'}$$

where h is the depth of the foundation below grade.

The critical load init p_{cr} determined from Eq. (IV.10) should be regarded as a perfectly safe load on frozen soils when their negative temperatures are preserved; its value is near that of the limit of proportionality between load and settlement for these soils.

As this author reported at a conference of the Hungarian Academy of Sciences on 12 October 1955[29] and in a later paper written jointly with S. S. Vyalov (1956),[23] it is also permissible to regard frozen and permanently frozen ground as ideally cohesive bodies ($\phi_\infty = 0$) in determining the second critical loads on these soils—the limiting load lim p_{cr}.

Thus, Table 25* presents the results of some of our analytical calculations of the limiting critical load lim p_{cr} on soils, which were performed both using only the composite characteristic c_e (equivalent cohesion from ball-plunger test) with ϕ put equal to zero and using separate values for the cohesion (with the aforementioned friction corrections, i.e., the reduction coefficients M, applied to c_e) and the internal-friction coefficient ϕ found by an independent test.

In the former case (with consideration only of the cohesion c_e), the limiting load was calculated by the Prandtl equation[30] as for ideally cohesive

*Table 25 was taken from the article by the author that was published in the Transactions of the Hungarian Academy of Sciences, 1956.

Table 25.   Values of lim p_{cr} at various c_e and a side load
$q = 0.25$ kg/cm^2

ϕ, deg	Values of lim p_{cr}, kg/cm^2, as functions of c_e, kg/cm^2		
	0.75	2.3	4.0
0	4.1	12.1	20.8
10	4.4	12.3	21.1
20	4.8	11.3	18.5
30	–	13.0	19.3

bodies ($\phi \approx 0$), a formula first used to determine the limiting loads on cohesive frozen ground back in 1937.[15]

On the basis of the above, and using the ultimate long-term value of the equivalent cohesion c_{lt}, we have

$$\lim p_{cr} = (\pi + 2) c_{lt} + q \qquad (IV.11)$$

or

$$\lim p_{cr} = 5.14 c_{lt} + \gamma h \qquad (IV.11')$$

In the second case, i.e., with consideration of the friction ϕ and cohesion c, the limiting load lim p_{cr} was calculated from the well-known Novotortsev-Sokolovskiy equation for the two-dimensional problem[31]

$$\lim p = q \frac{1 + \sin \phi}{1 - \sin \phi} e^{\pi \tan \phi} + c \cot \phi \left(\frac{1 + \sin \phi}{1 - \sin \phi} e^{\pi \tan \phi} - 1 \right) \qquad (IV.12)$$

Comparison of the data in Table 25 for $\phi = 0$, $\phi = 10$, and $\phi = 20°$ shows that determination of the limiting loads on ground with consideration only of the cohesion c_e at internal-friction angles $\phi \leqslant 20°$ (and somewhat larger angles when c_e is large) yields values accurate enough for practical purposes; this greatly simplifies the calculations, the more so since it is necessary in complex cases (with consideration of both friction and cohesion) to use very cumbersome equations or to resort to numerical integration methods involving the use of computers.

Knowing the ultimate long-term equivalent cohesion c_{lt} determined by the ball-test method, we can also make sufficiently accurate determinations of the limiting load lim p_{cr} on frozen ground for the various other cases indicated in Table 26.

Extensive experiments to determine the limiting loads of permafrost using a flat circular plunger were performed at the Igarka Scientific Research Station of the USSR Academy of Sciences (by S. S. Vyalov and others), and the results were compared with analytical calculations based on equivalent cohesion for the same soils. The results of these comparisons appear in Table 27.

Table 26. Equations for determination of limiting loads on frozen ground as ideally cohesive bodies

No.	Shape of loaded area	Limiting load $\lim p_{cr}$	Author of solution
1	Strip	$\lim p_{cr} = (\pi + 2)\,c_{lt} + q$	Prandtl, 1920; N. A. Tsytovich and M. I. Sumgin, 1937
2	Sunken strip	$\lim p_{cr} = 8.3 c_{lt} + q$	Meyerhof, 1950
3	Circular, at the surface	$\lim p_{cr} = 5.68 c_{lt} + q$	A. Yu. Ishlinskiy, 1944 and V. G. Berezantsev, 1952
4	Square, at the surface	$\lim p_{cr} = 5.71 c_{lt} + q$	V. G. Berezantsev, 1952; Schild, 1958
5	Rectangular, at the surface	$\lim p_{cr} = \left[5.14 + 0.66\dfrac{b}{l}\right]c_{lt} + q$ (at $b/l \leqslant 0.53$)	Schild, 1958
6	Same	$\lim p_{cr} = \left[5.24 + 0.47\dfrac{b}{l}\right]c_{lt} + q$	Schild, 1958

Note: q is the side load, b is the width of the loaded area, and l is its length.

These data indicate good agreement between the test-load results and analytically calculated limiting loads for permafrost based on the cohesion c_{lt} determined by the ball-test method.

If both shear-strength parameters, i.e., the cohesion c_{lt} and the friction resistance of the frozen soils $p \tan \phi$, are taken into account separately in determining the critical loads on frozen ground and permafrost (init p_{cr} and lim p_{cr}), as is absolutely necessary when the internal-friction angle is large (on the

Table 27. Comparison of experimental data with analytical determinations of limiting loads on permafrost

Experiment no.	Designation of soils	Cohesion c_{lt}, kg/cm²	$\lim p_{cr}$, kg/cm²	
			According to test load	According to c_{lt}
19	Cover clay, dense	1.8	10.0	10.2
15	Same, mineral layers . .	1.8	9.0	10.2
17	Light silty sandy loam .	1.0	5.3	5.7
16	Same	1.2	6.0	6.8
14	Same	1.2	6.3	6.8
7	Heavy silty sandy loam, $i_{vol} > 50\%$	0.88	4.7	5.0

order of $20°$ and more; for example, for frozen sands or hard-frozen clayey soils), but the soils are at temperatures below the range of substantial water-to-ice phase transitions, it is not sufficient to use only the solutions of ideally cohesive body theory ($\phi \approx 0$). In this case it is necessary to resort to solutions that take account of both friction resistance and cohesion.

With consideration of friction and cohesion, the initial critical load init p_{cr} should be determined using the well-known equation of Prof. N. P. Puzyrevskiy[32] for the two dimensional problem, with consideration of the specifics of frozen ground:

$$\text{init } p_{cr} = \frac{\pi(\gamma h + c_{lt} \cot \phi_{lt}}{\cot \phi_{lt} + \phi_{lt} - \dfrac{\pi}{2}} + \gamma h \qquad (IV.13)$$

where h is the depth of the foundation below grade and c_{lt} and ϕ_{lt} are the long-term resistance parameters of the frozen ground that correspond to their ice contents and negative temperatures.

In determining init p_{cr} for frozen soils, we may make use of the interpretation of the Puzyrevskiy formula given in the SNiP II-B.1-62, Sec. 5.10, which was designed to permit the development of limiting-equilibrium zones under the edges of the strip load to a depth equal to $1/4$ of the width of the foundation; the calculations are made by the equation

$$\text{init } p_{cr} = R^s = (Ab + Bh)\gamma + Dc \qquad (IV.14)$$

where A, B, and D are the coefficients defined by Table 7 of SNiP II-B.1-62 as functions of the standardized angle of internal friction.

We note that ϕ_{lt} and c_{lt} must be introduced into Eq. (IV.14) in determining safe loads on frozen ground and permafrost.

The initial critical load determined from Eqs. (IV.10) and (IV.13) should be regarded as a perfectly safe load on permafrost if their negative temperature is preserved.

To determine the limiting critical load on frozen ground and permafrost, i.e., to evaluate their maximum bearing capacity, use should be made of the convenient relationships by V. G. Berezantsev[33] (which have also been tabulated) obtained under consideration of friction and cohesion in both the two-dimensional and three-dimensional problems of the limiting equilibrium theory of the soils.

The following equations were derived from the strip-line contours (obtained by exact solution of the differential equations of limiting equilibrium for soils), the approximation of these lines by simpler functions, and analysis of the equilibrium of the densified rigid core nucleus formed in the soil directly under the bases of loaded foundations on attainment of the limiting load, i.e., on exhaustion of the load-bearing capacity of the soils:

(1) for the case of continuous foundations (two dimensional) and foundations with rectangular base areas with ratios between the length l and the

width b; larger than three:

$$\lim p_{p1} = A_p \gamma b + B_p q + C_p c_{1t} \tag{IV.15}$$

(2) for the case of foundations with base areas circular or square but of equal area:

$$\lim p_c = A_c \gamma b_1 + B_c q + C_c c_{1t} \tag{IV.16}$$

where b is the width of the strip load (strip footings), b_1 is the half-width of a square or the radius of a circular area, q is the side load increment (usually $q = \gamma h$), and A, B, and C are the soil bearing-capacity coefficients calculated as functions of the theoretical internal-friction angles (for frozen soils, from the values of ϕ_{1t} corresponding to the ice contents and negative temperatures of the given soils) from the closed solutions derived by V. G. Berezantsev, which are, however, of very complex form.

To simplify calculations made with Eqs. (IV.15) and (IV.16), Berezantsev's tables of values of the coefficients A_p, B_p, C_p and A_c, B_c, C_c, which he compiled with the aid of a computer, can be used (Table 28).

The design resistances for frozen ground and permafrost are determined by the following relations:

(1) the design resistance des p, which corresponds, according to the GOST [All-Union State Standard], to the standardized resistance R^s, may not be larger than the initial critical pressure init p_{cr}, i.e.,

$$\text{des } p = R^s \leqslant \text{init } p_{cr} \tag{IV.17}$$

(2) further, the design resistance may not exceed a certain fraction of the limiting critical pressure, i.e.,

$$\text{des } p = R^s \leqslant km \, (\lim p_{cr}) \tag{IV.18}$$

where k and m are, respectively, the coefficient of homogeneity and the coefficient of working conditions for the frozen soil, and SNiP P-B.1-62, Sec.

Table 28. Values of bearing-strength coefficients

Coeffi-cient	Value of angle internal-friction for grozen ground, ϕ in degrees													
	16	18	20	22	24	26	28	30	32	34	36	38	40	42
A_p	1.7	2.3	3.0	3.8	4.9	6.8	8.0	10.8	14.3	19.8	26.2	37.4	50.1	77.3
B_p	4.4	5.3	6.5	8.0	9.8	12.3	15.0	19.3	24.7	32.6	41.5	54.8	72.0	98.7
C_p	11.7	13.2	15.1	17.2	19.8	23.2	25.8	31.5	38.0	47.0	55.7	70.0	84.7	108.8
A_c	4.1	5.7	7.3	9.9	14.0	18.9	25.3	34.6	48.8	69.2	97.2	142.5	216.0	317.0
B_c	4.5	6.5	8.5	10.8	14.1	18.6	24.8	32.8	45.5	64.0	87.6	127.0	185.0	270.0
C_c	12.8	16.8	20.9	24.6	29.9	36.4	45.0	55.4	71.5	93.6	120.0	161.0	219.0	300.0

5.30 permits the assumption (e.g., for rocky soils) that the product of the coefficients $km = 0.5$. For frozen ground, we can recommend the value $km \approx$ 0.6-0.7 for calculations based on the equivalent cohesion and smaller values (on the order of 1/3) for complete allowance for friction and cohesion and large ϕ.

7. Examples of Strength Calculations for Frozen Soils

Strength calculations for frozen and permanently frozen ground are of enormous practical importance, since they permit correct determination of the safe loads on such soils with allowance for their special properties and primarily for their composition (ice content), the magnitude of the negative temperature, and their rheological properties (stress relaxation and creep with time).

Strength calculations for frozen ground become particularly essential when the artificial ground-freezing technique is applied in construction and mining—a technique that is being used more and more frequently in excavation of deep unsupported construction trenches (e.g., for subway underground stations and inclined ramps), to protect hydraulic-engineering trenches from influx of ground water, and in sinking deep mine shafts under complex geological conditions through unstable ground.

During the past 30 years, according to Ya. A. Dorman, about 70 inclined tunnels for subway systems (10 of them of the underpass type), over 30 construction excavations, and about 100 deep mine shafts have been driven by artificial soil freezing.[34]

The successful large-scale use of artificial soil freezing, which became possible only on the basis of new methods developed for calculating the strength and stability of frozen soils with consideration of their rheological properties, has permitted substantial cost reductions while still guaranteeing safety and convenience in operations. Thus, the use of artificial ground freezing made it possible to save 700 tons of metal and 500 cubic meters of shoring timber in construction of a deep trench for the Moscow Subway (at the Red Gates), and to complete the project 11-12 months earlier than would otherwise have been the case.

This construction trench, which was built with the present author serving as consultant and on the basis of his strength calculations for the frozen soils (Engineer Dorman supervised the operations), was circular in section, having a diameter of 40 m and a depth of about 20 m (some of the freezing wells were as deep as 27 m). The thickness of the frozen retaining wall (composed of a quick sandy-loam ground), as obtained from a strength calculation under consideration of stress relaxation (with a 14 kg/cm² long-term strength of the frozen soil and an average temperature $\theta = -10°C$), was 5.6 meters, which was found to be quite acceptable and made it possible to dispense with all types of shoring inside the trench (Fig. 93).

Successful excavation of this enormous trench within walls of artificially frozen soils,[35] which was abutted (at a distance of about 1-2 meters) by a

Fig. 93. General view of frozen walls of the deep subway entrance trench (the walls were covered with straw mats in summer to protect them from direct exposure to the sun).

high-rise building (height 138.5 m, weight about 27,000 tons), confirmed the correctness of the strength calculations performed for the frozen ground with consideration of relaxation of their stresses, and served as an example for later use of the artificial ground freezing technique in driving the entrances and tunnels for the subways.

Artificial ground freezing has been used no less widely in mining to sink deep shafts through quick ground; allowance for the rheological properties of the frozen ground has made it possible to economize on shoring.

Thus, a calculation of the thickness required in frozen ground walls for the deep shafts (depths on the order of 500 m) in the Kursk Magnetic Anomaly required according to elasticity theory (using Lame's formula) 12-16 meters while calculations for the frozen soils with consideration of creep on the basis of the closed solutions obtained in the rheology theory[36] made it possible to limit the wall thickness to about 3.4 meters, a solution that was applied with success in practice.

We present a number of examples of strength calculations for frozen ground.

Example 1. Determine the safe load and limiting pressure on permanently frozen loam having the following mechanical properties: with insignificant friction resistance, the ultimate-long-term cohesion at $\theta = -0.4°C$ is $c_{lt} = 0.9$ kg/cm², but $c_{lt} = 2.0$ kg/cm² at $\theta = -4.0°C$; foundation depth below grade $h = 3$ m, bulk weight of soil above base of footing $\gamma = 1.8$ tons/m³, shape of footing base square.

The perfectly safe load on a frozen soil, assuming no change in its negative temperature, will correspond to init p_{cr} and will be determined from Eq. (IV.10′), i.e.,

$$\text{init } p_{cr} = \pi c_{lt} + \gamma h$$

at $\theta = -0.4°C$

$$\text{init } p_{cr} = 3.14 \times 0.9 + 0.0018 \times 300 = 3.4 \text{ kg/cm}^2$$

at $\theta = -4.0°C$

$$\text{init } p_{cr} = 3.14 \times 2 + 0.54 = 6.8 \text{ kg/cm}^2$$

The values obtained for the critical pressure should be regarded as the perfectly safe load for this particular permafrost.

The maximum pressure, corresponding to total utilization of the bearing ability of the permanently frozen loam as an ideally cohesive soil ($c_{lt} \neq 0$ and $\phi_{lt} \approx 0$) will be determined from the equation for a footing with a square loaded area (Table 26, line 4):

$$\lim p_{cr} = 5.71 c_{lt} + q$$

where q is the side load increment and is equal in this case to $q = \gamma h = 0.0018 \times 300 = 0.54 \text{ kg/cm}^2$.

Then for $\theta = -0.4°C$

$$\lim p_{cr} = 5.71 \times 0.9 + 0.54 = 5.7 \text{ kg/cm}^2$$

at $\theta = -4.0°C$

$$\lim p_{cr} = 5.71 \times 2 + 0.54 = 12.0 \text{ kg/cm}^2$$

If we put the product of the soil homogeneity coefficient by the coefficient of working conditions equal to $km = 0.6$, we have for the design load:

at $\theta = -0.4°C$

$$\text{des } p = km (\lim p_{cr}) = 5.7 \times 0.6 = 3.42 \text{ kg/cm}^2$$

at $\theta = -4.0°C$

$$\text{des } p = km (\lim p_{cr}) = 12 \times 0.6 = 7.2 \text{ kg/cm}^2$$

which is close to the values obtained previously for the perfectly safe pressure on the soil.

Example 2. Determine the initial (init p_{cr}) and limiting (lim p_{cr}) critical loads on a permanently frozen sandy loam for the following specifications (see, e.g., Table 20): $\theta = -0.4°C$, $\phi_{lt} = 26°$ (i.e., $\phi \neq 0$), $c_{lt} = 1.0 \text{ kg/cm}^2$, $q = \gamma h = 0.54$ kg/cm^2.

In this case, it is necessary to use solutions that allow both for the cohesion of the frozen soil and for its resistance to friction in shear.

To determine init p_{cr} according to Eq. (IV.13)

$$\text{init } p_{cr} = \frac{\pi(\gamma h + c_{1t} \cot \phi)}{\cot \phi_{1t} + \phi_{1t} - \dfrac{\pi}{2}} + \gamma h$$

For $\gamma h = 0.54$ kg/cm^2 and $\cot \phi = \cot 26° = 2.05$

$$\phi = \frac{26\pi}{180} = 0.453 \qquad c_{1t} = 1 \text{ kg/cm}^2$$

$$\text{init } p_{cr} = \frac{3.14\,(0.54 + 1 \times 2.05)}{2.05 + 0.453 - 1.57} + 0.54 \approx 9.3 \text{ kg/cm}^2$$

This pressure should be regarded as perfectly safe as long as the permafrost remains at its negative temperature.

Expression (IV.16) can be used (for a square base area) to determine $\lim p_{cr}$:

$$\lim p_{cr} = A_c \gamma b_1 + B_c q + C_c c_{1t}$$

For a footing width of 1 meter, the half-width $b_1 = 0.5$ m. Also given are $\gamma = 1.8$ tons/m^3, $\phi = 26°$, $c_{1t} = 1$ kg/cm$^2 = 10$ tons/m^2, and $q = \gamma h = 0.54$ kg/cm$^2 = 5.4$ tons/m^2. From Table 28 we find for $\phi = 26°$ that $A_c = 18.9$, $B_c = 18.6$, and $C_c = 36.4$. Then

$$\lim p_{cr} = 18.9 \times 1.8 \times 0.5 + 18.6 \times 5.4$$
$$+\, 36.4 \times 10 = 481 \ t/\text{m}^2 = 48.1 \text{ kg/cm}^2$$

To determine the design load, the resulting value of $\lim p_{cr}$ must be multiplied by the reduction factor km, which equals (as noted above) approximately 0.3 with both friction and cohesion considered.

Example 3. Use a strength calculation to determine the thickness required in frozen ground cylinder to secure a deep mine shaft by artificial soil freezing.

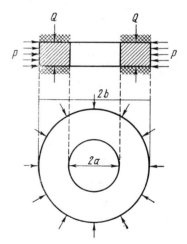

Fig. 94. Diagram of pressures on bilaterally restrained cylinder of frozen ground in shaft excavation by the method of artificial ground freezing.

We have the following equation[36] for a case with bilateral restraint of a frozen ground cylinder, which had a good performance record in application of artificial ground freezing in the Kursk Magnetic Anomaly (Fig. 94):

$$\phi = b - a = \frac{\sqrt{3}}{2} \times \frac{ph}{\sigma_t^{cm}} \tag{IV.19}$$

where δ is the thickness of the frozen ground cylinder, p is the external radial pressure, h is the height for a single cycle of artificial ground freezing, and σ_t^{cm} is the compressive strength for the time t of load application, i.e., considering relaxation during the time t.

For sandy loam, $\sigma_{t=1\ day}^{cm} = 30\ kg/cm^2$ at $\theta_{av} = -10°C, h = 2.4\ m, a = 4\ m$, and $p = 40\ kg/cm^2$.

Substituting the above into Eq. (IV.19) we obtain

$$\phi = b - a = \frac{\sqrt{3}}{2} \times \frac{40 \times 240}{30} = 277\ cm \approx 3\ m$$

Then $b = a + \delta = 4 + 3 = 7\ m$.

Example 4. Determine the limiting load (pressure) as related to a plastic deformation perm u_a of a frozen ground cylinder that is acceptable under operational conditions for artificial freezing of soils.

The initial equation for the state of stress and strain of the soil, the equation used to derive the equation given below, was Eq. (III.3), i.e., the shear stress intensity T is a nonlinear function of the shearing strain intensity Γ, namely

$$T = A(t)\Gamma^m$$

where $A(t)$ and m are the parameters of the frozen ground determined in the usual way. (See Chap. III, Sec. 5.)

By closed solution of the axisymmetric problem in limiting-equilibrium theory on the basis of the parametric creep equation (III.3), Yu. K. Zaretskiy obtained the following expression for the design load in terms of the specified acceptable plastic displacement u_a:

$$p = \frac{A(t)}{m}\left(\frac{2u_a}{a}\right)^m \left[1 - \left(\frac{a}{b}\right)^{2m}\right] \qquad (IV.20)$$

By specification, we have for the sandy loam at an average temperature $\theta = -10°C$: the parameter $A(t)$ at $t = 1$ day is $A(t) = 290\ kg/cm^2$; $m = 0.5, a = 4$ m, $b = 7$ m. The permissible deformation of the cylinder is $u_a = 5$ cm.

Substituting the numerical values of the quantities in Eq. (IV.20), we have

$$p = \frac{A(t)}{m}\left(\frac{2u_a}{a}\right)^m \left[1 - \left(\frac{a}{b}\right)^{2m}\right] = \frac{290}{0.5}\left(\frac{2 \times 5}{0.5}\right)^{1/2} \left[1 - \left(\frac{400}{700}\right)^1\right]$$

and, carrying out the calculations, we get $p = 39.2\ kg/cm^2$.

References

1. Tsytovich, N. A.: Mechanical Property Instability of Frozen and Thawing Soils, "Trudy I Mezhdunarodnoy konferentsii po merzlotovedeniyu" (Transactions of First International Converence on Geocryology USA), 1963.
2. See Ref. 1, this chapter and N. K. Pekarskaya: Prochnosty merzlykh gruntov pri sdvige i yeye zavisimost' ot tekstury (Shear Strength of Frozen Soils and its Dependence on Texture), *Izd-vo Akad. Nauk SSSR*, 1963.

3. Vyalov, S. S.: Zakonomernosti deformirovaniya l'da. Sb. II "Kontinental'naya ekspeditsiya 1956-1958 gg. Glyatsiologicheskiye issledovaniya" (Laws of Ice Deformation. Collection II. "Continental Expedition of 1956-1958. Glaciological Studies"), *Izd-vo Mortrans*, 1960.
4. Tsytovich, N. A.: Osnovaniya i fundamenty na merzlykh gruntakh (Bases and Foundations on Frozen Soils), *Izd-vo Akad. Nauk SSSR*, p. 117, 1958.
5. Orlov, V. O.: "Kriogennoye pucheniye tonkodispersnykh gruntov" (Cryogenic Heaving of Fine-Particle Soils), USSR Academy of Sciences Press, 1962.
6. Tsytovich, N. A.: On the Theory of the Equilibrium State of Water in Frozen Soils, *Izv. Akad. Nauk SSSR, Ser. Geog. Geof.*, no. 5–6, vol. 9, 1945.
7. Tsytovich, N. A.: Permafrost as a Base for Structures. "Materialy KEPS No. 80, Sb. Vechnaya merzlota" (Materials of KYePS. No. 80, Collection "Permafrost") *Izd-vo Akad. Nauk SSSR*, 1930.
8. Pekarskaya, N. K.: Ultimate Strength of Frozen Soils in Uniaxial Compression and Tension, "Materialy VIII Vsesoyuznogo soveshchaniya po geokriologii" (Materials of 8th All-Union Conference on Geocryology), No. 5, 1966.
9. Zaretskiy, Yu. K.: Rheological Properties of Plastic Frozen Ground, "Osnovaniya, fundamenty i mekhanika gruntov," 1971, No. 2.
10. Shusherina, Ye. P. and S. S. Vyalov: Long-Term Strength Study of Frozen Ground in Uniaxial Compression, Moscow State University, Collection "Merzlotnyye issledovaniya" (Frost Research), No. III, *Izd-vo MGU*, 1963.
11. Shusherina, Ye. P. and Yu. P. Bobkov: Dependence of Strength of Frozen Ground and Ice upon Temperature, "Trudy V Vsesoyuznogo soveshchaniya po stroitel'stvu" (Transactions of Fifth All-Union Conference on Construction), vol. VI, Krasnoyarsk, 1968.
12. Shusherina, Ye. P. and Yu. P. Bobkov: Influence of Moisture Content of Frozen Soils on Their Strength, *In* "Merzlotnyye Issledovaniya" (Frost Studies), no. IX, *Izd-vo MGU*, 1969.
13. Grechishchev, S. Ye.: Creep of Frozen Soils in a Complex State of Stress, *In* "Prochnost' i polzuchest' merzlykh gruntov" (Strength and Creep of Frozen Soils), Siberian Division, USSR Academy of Sciences, USSR Academy of Sciences Press, 1963.
14. Mel'nikov, P. I., S. Ye. Grechishchev, *et al.*: Fundamenty sooruzheniy na merzlykh gruntakh v Yakutii (Foundations of Structures on Frozen Soils in Yakutiya), *Sib. Otd. Akad. Nauk SSSR, Izd-vo "Nauka,"* 1968.
15. Tsytovich, N. A., and M. I. Sumgin: "Osnovaniya mekhaniki merzlykh gruntov" (Fundamentals of Frozen Ground Mechanics), p. 99, USSR Academy of Sciences Press, 1937.
16. Pekarskaya, N. K. and N. A. Tsytovich: The Role of Friction and Cohesion in the Total Shear Strength of Frozen Ground under Rapidly Increased Loads, *In* "Materialy po laboratornym issledovaniyam merzlykh gruntov" (Materials from Laboratory Studies of Frozen Soils), no. 3, *Izd-vo Akad. Nauk SSSR*, 1967.
17. Tsytovich, N. A.: Issledovaniye uprugikh i plasticheskikh deformatsiy merzlykh gruntov (Investigation of Brittle and Plastic Deformation in Frozen Soils), *Trudy KOVM Akad. Nauk SSSR*, vol. X, 1940.
18. Tsytovich, N. A.: Instruktsiya po opredeleniyu sil stsepleniya merzlykh gruntov (Instructions for Determination of Cohesive Forces in Frozen Soils), Institut merzlotovedeniya Akad. Nauk SSSR, 1947.
19. Tsytovich, N. A.: Instruktivnyye ukazaniya po opredeleniyu sil stsepleniya merzlykh gruntov, Sb. 2 "Materialy po laboratornym issledovaniyam merzlykh gruntov pod rukovodstvom N. A. Tsytovicha" (Instructive Notes on the Determination of Cohesive Forces in Frozen Soils, Collection 2 of "Materials from Laboratory Studies of Frozen Soils under the Direction of N. A. Tsytovich), *Izd-vo Akad. Nauk SSSR*, 1954.
20. Ishlinskiy, A. Ya.: The Axisymmetric Problem of Plasticity Theory and the Brinell Test,

In "Prikladnaya matematika i mekhanika" (Applied Mathematics and Mechanics), vol. III, no. 3, 1944.

21. Tsytovich, N. A.: A Study of the Coherent Forces of Cohesive Clayey Soils by the Ball-Test Method, *Trans. Czechoslovak Academy of Sciences*, vol. V, no. 3, Prague, 1956.

22. Berezantsev, V. G.: Limiting Equilibrium of a Cohesive Medium under Spherical and Conical Punches, *Izvestiya Akad. Nauk SSSR*, OTN, 1955, no. 7.

23. Vyalov, S. S. and N. A. Tsytovich: Evaluation of the Bearing Ability of Cohesive Soils from the Impression Depth of a Spherical Punch, *Doklady Akad. Nauk SSSR*, vol. III, no. 6, 1956.

24. Vyalov, S. S.: "Reologicheskiye svoystva i nesushchaya sposobnost' merzlykh gruntov" (Rheological Properties and Bearing Strength of Frozen Ground), USSR Academy of Sciences Press, 1959.

25. Tsytovich, N. A., I. S. Vologdina, M. L. Sheykov, *et al.*: "Laboratornyye issledovaniya mekhanicheskikh svoystv merzlykh gruntov" (Laboratory Investigations of the Mechanical Properties of Frozen Ground), Collections 1 and 2, *Izd-vo Akad. Nauk SSSR*, 1936.

26. Gol'dshteyn, M. N.: "Deformatsii zemlyanogo polotna v osnovanii sooruzheniy pri promerzanii i ottaivanii" (Earth-Bed Deformations in Base Fills in Earthwork Structures on Freezing and Thawing), Transzheldorizdat Press, 1948.

27. The present section is based on A. N. Zelenin's Osnovy razrusheniya gruntov mekhanicheskimi sposobami (Fundamentals of Ground Disintegration by Mechanical Methods), *Izd-vo "Mashinostroyeniye,"* 1968, and V. P. Bakakin and A. N. Zelenin's article, Development of Frozen Soils, Sec. 2, in "Doklady na Mezhdunarodnoy konferentsii po merzlotovedeniyu" (U.S.A., 1963) [*Papers at International Conference on Geocryology* (U.S.A., 1963)], *Izd-vo Akad. Nauk SSSR*, 1963.

28. Berezantsev, V. G.: Resistance of Soils to Local Loading at Constant Negative Temperature, Collection 1 of "Materialy po laboratornym issledovaniyam merzlykh gruntov pod rukovodstvom N. A. Tsytovicha" (*Materials from Laboratory Studies of Frozen Soils Under the Supervision of N. A. Tsytovich*), *Izd-vo Akad. Nauk SSSR*, 1953.

29. Tsytovich, N. A.: Theoretical Problems of Soil-Mechanics in Major Construction (Paper at the 12 October 1955 session of the Hungarian Academy of Sciences), *Trudy Akad. Nauk Vengrii*, vol. XIX, no. 1-3, 1956.

30. Prandtl, L.: The Hardness of Plastic Bodies, Gött. Nachrichten, 1920.

31. Sokolovskiy, V. V.: Statika Sypuchei sredy (Statics of Incoherent Media), Gostekhizdat, 1954.

32. Tsytovich, N. A., V. G. Berezantsev, *et al.*: Osnovaniya i fundamenty (Bases and Foundations), *Izd-vo* "Vysshaya skola," 1970.

33. Berezantsev, V. G.: Raschet osnovaniy sooruzheniy (Design of Bases for Structures), Stroyizdat, 1970.

34. Dorman, Ya. A.: Iskusstvennoye zamorazhivaniye gruntov pri stroitel'stve metropolitenov (Artificial Freezing of Soils in Subway Construction), *Izd-vo "Transport,"* 1971.

35. Tsytovich, N. A. and Kh. R. Khakomov: Use of Artificial Soil Freezing in Construction and Mining, *In* "Doklady k V Mezhdunarodnomu kongressu po mekhanike gruntov" (*Papers at Fifth International Congress on Soil Mechanics*), N. A. Tsytovich (ed.), Gostroyizdat, 1961.

36. Vyalov, S. S., S. E. Gorodetskiy, Yu. K. Zaretskiy, *et al.*: "Prochnost' i polzuchest' merzlykh gruntov i raschety l'dogruntovykh ograzhdeniy" (Strength and Creep of Frozen Soils and Calculations for Frozen Ground Barriers), USSR Academy of Sciences Press, 1962.

Chapter V

DEFORMATION OF FROZEN GROUND AT NEGATIVE TEMPERATURES

1. Types of Deformation of Frozen Ground without Temperature Change

When frozen soils are acted upon by external loads (in either a simple or complex state of stress), various types of deformation occur, depending on both the time of load application and the magnitude and nature of the applied load. The load may be applied once (instantaneously) and then removed; it may be cyclic, it may increase progressively with time in accordance with any of various laws and then, on reaching a certain value, become constant, or it may be constant to begin with—and the deformation of the frozen ground will take a wide variety of forms depending on all of the above circumstances.

For convenience, all types of deformation of frozen ground can be treated under the following three basic headings:

    I) Instantaneous deformations
   II) Long-term deformations
  III) Destructive deformations

The first class (I) of frozen-ground deformations includes (1) adiabatic and (2) elastic deformations.

The following should be distinguished in the second class (II): (1) compacting deformations (structural-migrational deformations, irreversible and partly reversible); (2) damping-creep (viscous) deformations; and (3) plastic flows (irreversible viscoplastic deformations), which evolve into progressive flows under certain conditions with the passage of time.

The third class (III) has two subclasses: (1) brittle destructive deformations (failure of continuity) and (2) excessive (inadmissible) plastic shape changes (loss of stability).

All of the above classes and types of deformations of frozen and permanently frozen ground must be treated as interrelated, but the various types will vary in importance in specific problems of frozen-ground mechanics, a matter which we shall discuss briefly.

Among the instantaneous (class I) deformations; elastic deformations are of fundamental practical importance; adiabatic deformations, on the other hand, arise at the first instant of loading without the development of dangerous shear stresses and are of virtually no practical importance. The magnitude of the elastic deformations strongly influences the work of frozen ground under dynamic load (impact, explosion, seismic oscillations, vibrations), and knowledge of the elastic-property characteristics of frozen ground and permafrost is absolutely necessary for prediction of the behavior of structures and their foundations under dynamic load and for seismological and ultrasonic engineering-geological surveys of the frozen strata.

The deformations of densification in frozen ground will be decisive in strain-limit calculations for foundations to be built on high-temperature frozen ground with maintenance of its negative temperature, in exactly the same way as damping-creep deformations. Foundations for structures to be built on plastically frozen soils with preservation of their negative temperature cannot be designed rationally without knowledge of the compacting deformations and damping creep, especially if the temperature of permafrost will be equal to or above the limit of substantial (rapid) water-to-ice phase transitions during the structures useful life.

Prediction of viscoplastic flow deformations of permafrost and ice will be necessary when structures are to be built with a specific time of existence in mind, e.g., for artificial freezing of soils, in driving horizontal or inclined excavations in viscoplastic clayey frozen soils or in ice for use for a specified time, and so forth.

But for structures that are to be used continuously for a considerable time (on the order of several tens or even hundreds of years), it is, of course, not admissible to neglect viscoplastic flows of permafrost base soils, especially if this flow is nonuniform (as a result of inhomogeneity of the soils or differences in their negative temperatures) or substantial in magnitude (over the limits permitted by the norms).

As for the third class of deformations, those which cause failure of frozen ground and permafrost (brittle failure or loss of stability), it is necessary, as we showed in the preceding chapter, to specify a load at the base of the structure that will be perfectly safe and constitute a certain fraction of the maximum load, which causes progressive flow (failure).

2. Elastic Deformations of Frozen Ground and Their Characteristics

Elastic deformations of frozen ground are governed by purely reversible changes in the crystal lattices of the mineral particles and the ice, by the elastic properties of the thin films of unfrozen water, and by the elastic properties of enclosed air bubbles that are present in some quantity in frozen and permanently frozen ground.

As the present author showed in 1940,[1] frozen soils retain their elastic properties throughout the entire plastic range of deformation, and these properties are especially strongly in evidence under cyclic loading and unloading.

The author and his coworkers have studied the elastic deformations of frozen ground and determined its characteristics (under laboratory and field conditions) in experiments on an enormous number of frozen and permanently frozen ground specimens, for the most part from 1935 to 1940 with minor supplements from later studies.*

As for any other material, the basic indicator to the elastic properties of homogeneous frozen soils is composed of the modulus of normal (longitudinal) elasticity (Young's modulus E, kg/cm^2) and the coefficient of transverse elasticity (the Poisson coefficient μ).

The normal elastic moduli of the frozen soils were determined under laboratory conditions (in the LISI Refrigerated Laboratory) in a special heat insulator (Fig. 95) on specimens of cubic shape (20 cm on a side) and in field excavations using a Tsytovich spring press (see Fig. 63), with which permafrost specimens can be compression-tested and their elastic properties determined, along with the compressibility characteristics of the soils in the frozen and thawed states. In addition, specimens of frozen ground (artificially frozen and undisturbed permafrost specimens) were torsion-tested for determination of the shear modulus and Poisson's coefficient, with the latter also calculated from results of direct and quite accurate measurements of the transverse and longitudinal relative elastic deformations of the frozen soils under longitudinal compressive forces.

The author performed or assisted in about 20,000 individual measurements to investigate the elastic and plastic properties of the frozen soils and established the value of their characteristic coefficients. The reduction results of these measurements and general conclusions drawn from them are set forth concisely below.

The modulus of normal elasticity E, kg/cm^2, was determined with the frozen ground specimens under a cyclic load that was repeated until the elastic deformations had become constant; the averages from no fewer than five individual determinations were used in the analysis.

The experiments showed that frozen soils have normal elastic moduli E [kg/cm^2] tens and hundreds of times larger than those of unfrozen soils ($E \approx$ 3000-300,000 kg/cm^2) and that the value of the modulus depends on a number of factors: the composition of the frozen soils, their ice contents, their negative temperatures, and the external pressure.

The basic experiments were performed with three types of icy soils: frozen sand (93.0% content of particle fraction > 0.25 mm, 5.6% from 0.25 to 0.05 mm; 1.4% 0.05 mm, and averaged total moisture content $W_d = $ 17-19%; a frozen silty soil (35.6% content of fraction > 0.05 mm, 9.2% < 0.005 mm and $W_d = $ 26-29%), and a frozen clay (more than 50% < 0.005 mm and $W_d = $ 46.56%), and also with structurally undisturbed specimens of disperse permafrost.

*Some of these studies were published in the collections of the SOPS AN SSR (see N. A. Tsytovich, I. S. Vologdina, M. L. Sheykov, et al., Laboratory Investigations of the Mechanical Properties of Frozen Ground, Collections I and II, SOPS AN SSSR, 1936), but most of the material remained unpublished due to the beginning of the 1941-1945 war.

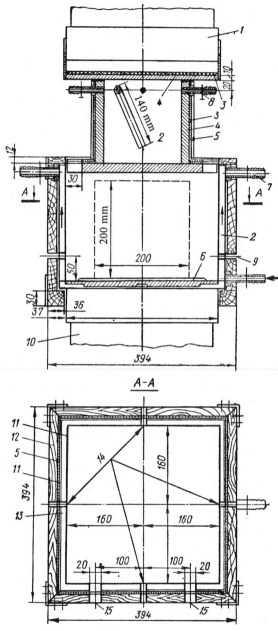

Fig. 95. Thermoinsulator for 150-ton press used in determining the normal elastic moduli of frozen ground. (1) Plunger of 150-ton press; (2) cooling alcohol; (3) ebonite; (4) steel; (5) asbestos; (6) shelf; (7) connection with valve for discharge of alcohol (on all four walls); (8) valve; (9) adjusting screw; (10) bedplate of press; (11) brass; (12) wood; (13) airspace; (14) 4 holes in Martens mirror for insertion of screws; (15) 2 holes for Martens mirror rods.

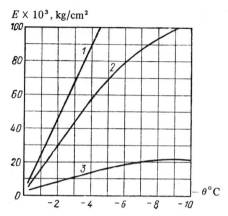

Fig. 96. Modulus of normal elasticity E, kg/cm², of frozen ground, dependent upon the negative temperature $-\theta°$, at constant pressure $\sigma = 2$ kg/cm². (1) Frozen sand; (2) frozen silty soil; (3) frozen clay.

Frozen sand has the largest modulus of normal elasticity (the experiments yielded values from 8200 kg/cm² at $\theta = -0.2°$C to 225,000 kg/cm² at $\theta = -10.2°$C), and the frozen clays had the smallest values (from 6800 kg/cm² at $\theta = -1.2°$C to 27,800 kg/cm² at $\theta = -8.4°$C); the values for the silty loams and sandy loams were intermediate.

The elastic moduli of frozen ground are most strongly influenced by their negative temperature $-\theta°$. The average results from a large number (more than 200) determinations of the normal elastic moduli for frozen sands, a silty soil, and clay are given in the form of averaged curves in Fig. 96. Generalizing on the results of these experiments and on other data set forth in detail in the aforementioned works of the author on the influence of the negative temperature $-\theta°$ on the moduli of normal elasticity E of frozen soils, we conclude that in the most general case the $E = f(-\theta)$ relation is curvilinear (see Fig. 96) and can be approximated, as suggested earlier,[1] by a complete polynomial of third degree at strongly negative temperatures (on the order of -10 and $-30°$C), i.e.,

$$E = \alpha + \beta\theta + \gamma\theta^2 + \delta\theta^3 \qquad \text{(V.1)}$$

where α, β, γ, and δ are experimentally determined parameters and θ is the absolute value of the negative temperature of the frozen soil in °C; or it must be approximated by a power-law function

$$E = \alpha + \beta\theta^n \qquad \text{(V.2)}$$

where n is the nonlinearity parameter.

At not too low temperature of frozen ground (above -5 to $-7°$C for clays and silty soils and, in the extreme case, above $-10°$C or slightly lower for sands), the exponent n can be put equal to unity ($n \approx 1$), i.e., it can be assumed within these limits that the elastic moduli of frozen soils depend linearly on the negative temperature; then

$$E = \alpha + \beta\theta \qquad (V.3)$$

We obtained for the above frozen soils at a compressive stress $\sigma = 2$ kg/cm^2 : frozen sand (temperatures above $\theta = -10°C$)

$$E = [0.5 + 2.1\theta] \times 10^4 \text{ kg/cm}^2 \qquad (V.3')$$

frozen silty soil (above $\theta \approx -5°C$)

$$E = [0.4 + 1.4\theta] \times 10^4 \text{ kg/cm}^2 \qquad (V.3'')$$

frozen clay (above $\theta \approx -5°C$)

$$E = [0.5 + 0.23\theta] \times 10^4 \text{ kg/cm}^2 \qquad (V.3''')$$

The values of the initial parameter α of Eq. (V.3) for the normal elastic modulus of frozen sand are practically constant as the external pressure varies through a factor of 5–10, while the parameter β, which evaluates the influence of negative temperature, depends on the applied compressive stress σ [kg/cm^2] at which the normal elastic modulus was determined.

Thus, we obtained for frozen sand (at $W_d = 16$–20%):

at $\sigma = 0.5$ kg/cm^2 .... $\beta = 3.3$
at $\sigma = 1.0$ kg/cm^2 .... $\beta = 2.3$
at $\sigma = 2.0$ kg/cm^2 .... $\beta = 2.1$
at $\sigma = 4.0$ kg/cm^2 .... $\beta = 2.0$

An obvious conclusion to be drawn from the above data is that the influence of negative temperature on the elastic properties of frozen sand is stronger the lower the external pressure, i.e., negative temperature and external pressure work in opposite directions.

To explain the established effect of external pressure on the elastic properties of frozen soils, we must assume that, even at soil temperatures below the limit of substantial water-to-ice phase transitions (for example, for frozen sands), as a result of the sharp angularity of the mineral particles and the transformation of pressures into enormous compressive stresses at contact points between particles, external pressure has a substantial influence on the elastic properties of frozen ground. As the temperature is lowered, of course, the pore ice becomes stronger, but this would hardly have such a strong effect on the elastic modulus at moderately low temperatures.

The results for frozen clayey soils were different: the coefficient β (the slope of the $E = f(\theta)$ curve with respect to the θ axis) was found to be practically constant as the external pressure was doubled from 1 to 2 kg/cm^2 (it was independent of external pressure within this range) at a value $\beta = 0.23$–0.24, while the initial coefficient α in expression (V.3''') decreased by a factor of 1.4

(from 0.7 at $\sigma = 1$ kg/cm^2 to 0.5 at $\sigma = 2$ kg/cm^2), indicating a different type of effect of external pressure on clayey soils, which contain much larger numbers of flat scaly mineral particles than sandy soils—particles that apparently account for the higher initial ice contents of these soils. We note also that longer freezing of the clay produced only an increase in α, while β remained unchanged.

The influence of external pressure on the normal elastic modulus E of frozen ground is particularly strongly felt at small negative temperatures (above the limit of the water-to-ice phase transitions) and preferentially in soils with a rigid mineral component. This is confirmed by the following data: for frozen sand ($W_d = 16$-20%), the normal elastic modulus decreased as follows as the external pressure (the compressive stress at which the elastic modulus was determined) varied from 1 to 3 kg/cm^2, i.e., by 2 kg/cm^2:

at $\theta = -\ 0.5°C$ ... from 13.2×10^3 to 8.0×10^3 kg/cm^2, or by 40%
at $\theta = -\ 2.1°C$ ... from 71.4×10^3 to 45.0×10^3 kg/cm^2, or by 37%
at $\theta = -10.0°C$ ... from 200×10^3 to 190×10^3 kg/cm^2, or by 5%

These data pertain even to a wider range of external pressure variation, from 2 to 10 kg/cm^2.

We note that field experiments with structurally undisturbed permafrost specimens (Yakutsk area) confirmed the relationships obtained in the laboratory for the variations of the normal elastic moduli of frozen ground and also the absolute values of the elastic modulus.

Thus, the elastic modulus of a specimen of silty permafrost with undisturbed structure varied from 2,780 to 5,100 kg/cm^2 at $\theta = -0.1°C$, and the elastic modulus of the same soil after artificial freezing at $\theta = -0.1°$ to $-1.4°C$ was found equal to 3,200-5,900 kg/cm^2.

In exactly the same way, the normal elastic modulus of a clean sand determined in the field was 29,500 kg/cm^2 at $\theta = -1°C$, while the value obtained in the laboratory for the artificially frozen specimen at the same temperature was about 30,000 kg/cm^2.

Thus, the normal elastic modulus was found to be of the same order for specimens of frozen soils with the natural undisturbed structure and in the artificially frozen state; the moduli for permafrost decreased with increasing external pressure and increased with decreasing temperature, especially rapid in the case of frozen sandy soils.

It is also interesting to compare the normal elastic moduli of frozen soils with the elastic modulus of ice. According to experiments conducted by the author at $\theta = -1.5°C$ and a compressive stress $\sigma = 2$ kg/cm^2, the value of the latter was $E_{ice} = 24,500$ kg/cm^2.

At the same temperature and stress, the elastic moduli of the frozen soils were:

for sand ($W_d = 18.9\%$) ... $E_{sd} = 36,500$ kg/cm^2
for silty soil ($W_d = 30.0\%$) ... $E_{si} = 28,600$ kg/cm^2
for frozen clay ($W_d = 55.1\%$) ... $E_{cl} = 73,000$ kg/cm^2

Table 29. Values of Poisson's coefficient for frozen soils

Designation of soil	W_d, %	$\theta°C$	σ, kg/cm²	Poisson's coefficient μ
Frozen sand	19.0	− 0.2	2	0.41
	19.0	− 0.8	6	0.13
Frozen silty loam	28.0	− 0.3	1.5	0.35
	28.0	− 0.8	2	0.18
	25.3	− 1.5	2	0.14
	28.7	− 4.0	6	0.13
Frozen clay	50.1	− 0.5	2	0.45
	53.4	− 1.7	4	0.35
	54.8	− 5.0	12	0.26

It follows from comparison of the above data that the elastic modulus of ice is smaller than those of soils with a rigid mineral skeleton and substantially greater than that of frozen clay; this can be explained only in terms of different contents of unfrozen water in the frozen soils, the frozen clays naturally containing the largest amount, so that they are substantially less elastic.

The coefficient of transverse elasticity (Poisson's coefficient) was determined for the same three basic types of frozen soils: sand, silty soil (loam), and clay. The value of this coefficient (which equals the ratio of the transverse relative elastic deformation to the longitudinal elastic deformation at the same axial stress) was calculated from direct measurements of the elastic relative deformations, longitudinal and transverse, and also from the equation known from elasticity theory:

$$\mu = \frac{E}{2G} - 1 \qquad (V.4)$$

where G is the shear modulus determined in torsion tests of cylindrical frozen ground specimens and E is the normal elastic modulus determined in a compression test.

The values obtained for the Poisson coefficient by the two methods were closely similar.

The results of direct determinations of the Poisson coefficient for frozen ground appear in Table 29.

These data indicate substantial influence of the temperature θ on the Poisson's coefficient μ for frozen soils, which approaches 0.5 as $\theta \to 0°$ (as for ideally plastic bodies), and the values for solids at lower temperatures.

3. Elastic Deformations in Layered Strata of Frozen Ground

When structures are to be built on permafrost under natural conditions, a stratified bed is almost always found below the base of the footings. The individual layers of frozen ground can be tested to obtain coefficients that characterize their deformation properties. However, the question arises as to how to convert to a determination of the total deformability of the frozen stratum from known characteristics of the individual layers. Special experimental studies were set up to solve this problem.

The purpose of the experimental deformability studies made on stratified frozen ground layers was to establish the basic premises for conversion from the deformability indicators of individual layers to an estimate of the deformability of the entire layered stratum under both continuous and localized loads.

We shall confine ourselves in the present section to considering the influence of viscoplastic clay and ice interlayers on the elastic deformations of hard-frozen soil beds.

Several series of experiments were set up to investigate the elastic properties of the layered frozen soils:

Series 1—determination of the elastic modulus in tests of prisms of frozen soil composed of individual layers with large differences in deformability;

Series 2—tests of prisms of frozen water-saturated sand with various thicknesses of pure ice in continuous interlayers;

Series 3—impression of plungers (application of local loads) into frozen soils.

1. The specimens (I, II, and III) of the series 1 experiments are represented schematically in Fig. 97.

In scheme I was studied the elastic compressive deformations of a two-layer prism of frozen ground consisting of frozen sand and frozen clay; in scheme II the prism of frozen sand had a central layer of frozen clay, and in scheme III the frozen-clay prism had a middle layer of frozen sand.

Figure 97 indicates the dimensions of the prisms and interlayers and the placement of the strain indicators used to measure the deformation (on a 10-cm base).

The total elastic deformation of the frozen ground prisms was determined under cyclic loading halfway up their height (10 cm), and the result was used to calculate the average normal elastic modulus E_m for the entire prism.

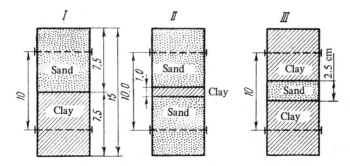

Fig. 97. Diagram showing arrangement of soil layers in a study of the elastic deformations of frozen-ground specimens (series 1).

Table 30. Values of normal elastic modulus E_m from experiments with stratified frozen ground

Scheme	Physical model of soil bed	$\sigma = 1$ kg/cm²		$\sigma = 2$ kg/cm²		$\sigma = 3$ kg/cm²	
		$\theta °C$	E_m, kg/cm²	$\theta °C$	E_m, kg/cm²	$\theta °C$	E_m, kg/cm²
I	Stratified	−1.4	14,300	−1.4	12,700	−	−
II	Same	−1.7	33,500	−1.7	30,100	−1.7	29,300
III	Same	−1.6	8,500	−1.6	8,200	−	−
−	Sand (uniform)	−	−	−1.7	40,700	−	−
−	Clay (uniform)	−	−	−1.7	8,900	−	−

The results of the experiments are given in Table 30.

If the elastic moduli of the individual frozen ground layers are known, the total elastic modulus of the stratified bed can easily be determined by equating the total elastic deformation of the stratified soil to the sum of the elastic deformations of its individual layers. Then the average normal elastic modulus of the stratified frozen-soil bed, E_m, can (in the case of an one-dimensional problem) be determined from the equation

$$E_m = \frac{\sum_1^n h_i}{\sum_1^n \dfrac{h_i}{E_i}} \tag{V.5}$$

where h_i are the thicknesses of the individual frozen ground layers and the E_i are the normal elastic moduli for the individual layers of the frozen ground stratum.

If we calculate E_m, for example for scheme II (for a prism of frozen sand with a 1-cm interlayer of frozen clay at $\sigma = 2$ kg/cm² and $\theta = -1.7°C$), from Eq. (V.5):

$$E_m = \frac{10}{\dfrac{9}{40,700} + \dfrac{1}{8,900}} \approx 30,300 \text{ kg/cm}^2$$

and compare it with the experimentally determined $E_{II} = 30,100$ kg/cm² (see Table 30), we see that they have very close values.

In exactly the same fashion, calculation of the average elastic modulus E_m for the other frozen ground stratification schemes (in the case of an one-dimensional problem) using Eq. (V.5) yields results sufficiently close to those obtained experimentally.

We note that the use of other relationships to calculate the average normal elastic moduli of stratified ground beds under a continuous, uniformly distributed load, e.g., the formulas for reducing the individual layers to the thickness of an "effective layer," such as are used in roadbuilding, yield much poorer (incompatible with experiment) results.

2. The schemes (I, II, and III) of the 2nd series of experimental specimens to determine the normal elastic modulus of frozen sand with continuous ice interlayers are shown in Fig. 98.

Also in the same series of experiments, we made numerous determinations of the normal elastic modulus of pure ice in compression perpendicular to the freezing surface at ice-specimen temperatures $\theta = -1.5$ and $\theta = -5.0°C$. In this series of experiments, we made about 300 determinations of the normal elastic modulus of frozen sand with ice interlayers and of pure ice alone; we also determined the steady-state strain rate as the average of a large number (about 30) of individual measurements.

It was found, for example, that the average steady-state deformation rate $\dot{e} = 0.0000054$ min^{-1} for specimens for frozen sand with a 10-mm ice interlayer (scheme I, Fig. 98) at $\theta = -1.5°C$ and $\sigma = 3$ kg/cm^2, but that $\dot{e} = 0.0000144$ min^{-1} at the same temperature with $\sigma = 4$ kg/cm^2, while $\dot{e} = 0.0000065$ min^{-1} for scheme III specimens (with two ice interlayers of 5 and 25 mm) at $\theta = -1.5°C$ and $\sigma = 2$ kg/cm^2, and $\dot{e} = 0.0000155$ min^{-1} at $\sigma = 3$ kg/cm^2 and the same temperature.

The given data indicate that the development of plastic deformations in stratified frozen soils depends almost entirely on the thickness of the ice inclusions.

Generalized (averaged) results from determination of the normal elastic moduli for frozen sand, ice, and frozen sand with ice interlayers appear in Table 31.

We note that if the average values found experimentally for the normal elastic moduli of the stratified frozen soils or are compared with the results calculated from the normal elastic moduli of the individual layers of frozen sand and ice according to Eq. (V.5), the disagreement ranges from 1 to 2.5 percent, indicating that Eq. (V.5) can be used in practice.

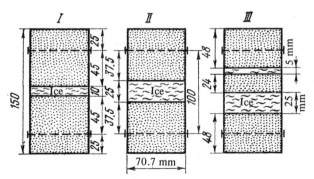

Fig. 98. Schemes (I, II, and III) of position of continuous ice
interlayers in frozen-sand prisms (series 2).

Table 31.   Results of experiments to determine the normal elastic modulus E (kg/cm^2)
of stratified frozen soils (series 2)

Temperature θ, °C	Compressive stress σ, kg/cm^2	Elastic modulus E, kg/cm^2				
		Homogeneous frozen sand	Sand with ice interlayers			Ice
			Scheme I	Scheme II	Scheme III	
− 1.5	1	39,500	34,500	32,600	31,300	37,000
− 1.5	2	32,000	–	29,000	28,600	24,500
− 5.0	1	120,000	111,000	92,800	74,200	45,500
− 5.0	2	110,000	89,000	70,000	65,000	34,000

3. The schemes of the series 3 experiments to investigate the elastic deformations of frozen soils (homogeneous and stratified) under indentation (localized load action) appear in Fig. 99; here experiments 1–3 were performed with homogeneous frozen soils (sand, clay, and ice), while Nos. 4 and 5 were performed on stratified frozen soil beds: experiment 4 (frozen sand with W_d = 23.5% with a layer of solid ice 10 mm thick at a depth equal to half the width of the rigid square-section plunger, i.e., at a depth of 25 mm, at $\theta = -1.8$ and $\theta = -1.6°$C) and experiment 5 with a two-layer frozen soil (upper layer of frozen sand with W_d = 25.1% with a thickness equal to half the width of the plunger and a lower layer of frozen clay with moisture content W_d = 53.8% at $\theta = -1.2°$C).

The elastic deformations of the frozen soil beds under application of a local load were determined at stresses from 4 to 50 kg/cm^2 on the soil.

Analysis of the experimental data showed that for the case of localized load action the relation between pressure and elastic deformation can be assumed linear with full confidence for frozen ground at pressures up to at least 10 kg/cm^2 and a temperature on the order of $-1°$C.

The experiments of this series also permitted experimental verification of the applicability of Tsytovich's equivalent-layer method (which was first proposed by the author in 1934 and underwent substantial development in the years that

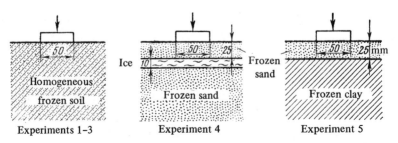

Fig. 99. Diagrams of experiments to determine elastic properties of frozen soils in indentation test (series 3).

followed)[2] to frozen and permanently frozen ground for calculation of the average modulus of a layered bed of frozen ground and the elastic deformations of such ground under the action of a local load.

Equating the elastic deformation of frozen ground to the average normal elastic modulus E_m for elastic deformation of a half-space under a local load (using the Boussinesq-Schleicher equation), we obtain an expression for the elastic equivalent layer $(h_e)_{el}$:

$$(h_e)_{el} = (1 - \mu^2)\omega \sqrt{F} \qquad (V.6)$$

where μ is Poisson's coefficient, ω is the shape and rigidity factor for the plunger (foundation),[2] and F is the basal area of the plunger.

If, for example, we assume $\mu = 0.3$ for the Poisson's coefficient of a stratum of frozen ground and take as ω the average depth of penetration of the rigid plunger, we have

$$(h_e)_{el} = 0.8645b \qquad (V.7)$$

where b is the width of the plunger.

To determine the normal elastic modulus E_m of a stratified bed of frozen ground for the entire depth of the active compression zone, whose maximum value, as the author showed earlier,* is twice the thickness of the equivalent layer in the case of local load application, i.e., $(2h_e)_{el}$, we equate the elastic deformation of the entire massif (which equals the elastic deformation of the equivalent layer) to the sum of the elastic deformations of the individual layers down to the depth $(2h_e)_{el}$, taking the decrease in pressure with depth from the equivalent curve* for calculation of E_m, i.e., approximately proportional to $z_i/(2h_e)_{el}$ (where z_i is the distance from the center of the layer under consideration to the depth $(2h_e)_{el}$).

As a result we obtain the following expression for the average elastic modulus of the stratified bed of frozen ground:

$$E_m = \frac{2(h_e)_{el}^2}{\sum\limits_{1}^{n} \frac{z_i h_i}{E_i}} \qquad (V.8)$$

Let us check the applicability of this relation for determining the elastic deformations of a stratified bed of frozen ground under application of a local load, e.g., for the conditions of experiments 4 and 5 (Fig. 99).

It was found as a result of direct measurements that the elastic modulus of frozen sand at $\theta = -1.7°C$ and a stress $\sigma = 6$ kg/cm^2 is $E_{sd} = 21,600$ kg/cm^2, while the elastic modulus of ice $E_{ice} = 12,500$ kg/cm^2.

---

*See, e.g., Fig. 32 in N. A. Tsytovich's book "Mekhanika gruntov" (Kratkiy kurs) [Soil Mechanics (A Short Course)], Izd-vo "Vysshaya Shkola," 1968.

For the conditions of experiment 4 (Fig. 100) with application of load $p = 6$ kg/cm² with a 5 × 5 cm plunger (which gives fully satisfactory results for highly viscous and tough solids such as frozen ground), it was found that the deformation of the layered frozen ground stratum was*

$$\exp s_{el} = 0.013 \text{ mm} = 0.00130 \text{ cm}$$

Let us calculate the amount of elastic deformation of the frozen stratum for the condition of experiment 4.

We have: $(h_e)_{el} = 0.8645 \times 5 = 4.32$ cm and $2(h_e)_{el} = 8.64$ cm, and we find from the stratification diagram (experiment 4) $h_1 = 2.5$ cm and $z_1 = 7.5$ cm (frozen sand); $h_2 = 1.0$ cm and $z_2 = 5.64$ cm (ice) and $h_3 = 5.14$ cm and $z_3 = 2.57$ cm (underlying bed of frozen sand). Then

$$E_m = \frac{2(h_e)_{el}^2}{\sum_1^n \dfrac{z_i h_i}{E_i}} = \frac{2(4.32)^2}{\dfrac{7.39 \times 2.5}{21,600} + \dfrac{5.64 \times 1}{12,500} + \dfrac{2.57 \times 5.14}{21,600}} \approx 19,600 \text{ kg/cm}^2$$

We find the magnitude of elastic settling from the well-known Boussinesq-Schleicher equation:

$$\text{calc } s_{el} = \frac{(1 - \mu^2)\omega p \sqrt{F}}{E_m} = \frac{0.9 \times 0.95 \times 6 \times 5}{19,600} = 0.00132 \text{ cm}$$

The calculated value found for the elastic deformation (0.00132 cm) is very close to that obtained from direct experiments (0.00130 cm).

For the conditions of experiment 5 (see Fig. 100), which was performed at a temperature $\theta = -2.1°C$ (we obtained $E_{sd} = 22,600$ kg/cm² for the frozen sand and $E_{cl} = 8,800$ kg/cm² for the frozen clay at this temperature), the average elastic modulus of the stratified frozen ground bed down to a depth $2 \times (h_e)_{el}$

---

*Detailed tabulated results of experiments on the elastic deformations of frozen ground strata under the action of a localized load are given in the papers of N. A. Tsytovich (see Chap. I, Refs. 1a, 1b, and 13).

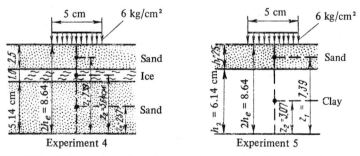

Fig. 100. Scheme of local load application for the calculation of average elastic modulus of stratified frozen-ground layer.

with $\mu = 0.3$ and $b = 5$ cm will be

$$E_m = \frac{2 \times (4.32)^2}{\dfrac{7.34 \times 2.5}{22,600} + \dfrac{3.07 \times 6.14}{8,800}} = 12,300 \text{ kg/cm}^2$$

Then the elastic deformation of the frozen ground under the plunger will be

$$\text{calc } s_e = \frac{0.9 \times 0.95 \times 6 \times 5}{12,300} \approx 0.0021 \text{ cm} = 0.021 \text{ mm}$$

which is very close to the value found experimentally: exp. $s_e = 0.022$ mm.

Thus, Eq. (V.8) gives average values of the elastic modulus for a stratified frozen ground bed *with sufficient accuracy for practical calculations.*

The data obtained on the normal elastic moduli and Poisson's coefficients of the frozen soils can be used successfully in seismic and ultrasonic engineering-geological surveys to determine the stratification conditions of permafrost beds and in dynamic-load calculations for structures and their foundations (heavy impact loads, construction blasting and other explosions, seismic shock, vibration of unbalanced machines, etc.). In addition, knowing the elastic constants (E and μ) of the frozen and permanently frozen soils, we also have a sound basis for calculation (design) of resiliant foundations as beams and slabs on an elastic half-space based on the "local elastic base" theory.

In the latter case, the coefficient of local (uniform) elastic compression of the frozen ground can be determined from the following relation, which arrives from the Boussinesq-Schleicher settlement equation:[3]

$$c_z = \frac{E_m}{(1 - \mu^2)\omega \sqrt{F}} \tag{V.9}$$

Recent studies of the author indicate that Eq. (V.9) is valid only if the footing area is not too large (approximately up to 50 m^2). At larger areas F, it is necessary to consider the decrease in soil compressibility with increasing depth.

4. Compacting Deformations of Frozen Ground

Among engineers, the notion is still prevalent that the frozen soils should be regarded as practically incompressible bodies at negative temperatures and at the pressures that prevail in the bases of structures. However, this position is approximately valid only for low-temperature frozen ground. But it is not valid for high-temperature ground (temperatures near $0°$ and not below the limit of substantial water-to-ice phase transitions).

It was shown as early as 1953 by the experiments of S. S. Vyalov[4] and independently by those of N. A. Tsytovich[5] that high-temperature frozen ground exhibits considerable compressibility (compactibility) under load, a fact that must be reckoned with in the erection of structures on high-temperature permafrost with maintenance of negative ground temperatures. Major studies of the compressibility of frozen and permanently frozen ground were carried out under the supervision of the author by A. G. Brodskaya.[6]

According to the most recent data, compacting of high-temperature frozen soils (reduction of their porosity) is a highly complex physicomechanical process governed by the deformability and displacements of all components—gaseous, liquid (unfrozen water), viscoplastic (ice), and solid (mineral particles).

The gaseous component (closed air and vapor bubbles) is governed to a substantial degree in the absence of structural-bond breakage by the elasticity of the frozen soils, and, after the mineral-particle aggregates that trap the gases have been broken down by the load, by rapid inelastic volume changes in the soil with extrusion of the gaseous component.

Unfrozen water—present in the frozen ground before loading and water formed at the points of contact between mineral particles—is pressed slowly out of the pores in the frozen soils, rendering them compressible; this is also clearly evident from Brodskaya's experiments on determination of changes in total moisture content (its reduction) in specimens of frozen ground after compacting under pressure.

Under the influence of pressures arising at the points of contact between mineral particles, the *ice* is partially melted, and, migrating toward less heavily stressed microscopic and macroscopic zones, it freezes again, which is a factor in compacting of frozen ground. In addition, ice is restructured under compression (into finer-grained ice) with surface melting of the sharp faces of the ice crystals (as has been established by direct crystal-optical studies), with substantial viscoplastic flows of the ice crystals and their aggregates.

When the shear stresses in the aggregates exceed a certain critical value, solid mineral particles undergo shear displacements that change the structure (texture) of frozen soils, giving rise to more compact packing of the individual particles and particle aggregates, another effect causing general compacting of frozen soils under load at constant negative temperatures.

The principal causes of compacting of frozen soils under load should be sought in displacements of their structural elements and disturbance of the equilibrium state between unfrozen water and pore ice.

The deformability of closed gaseous inclusions in frozen ground and the elasticity of all of its other components govern the magnitude of the instantaneous deformations of the soil.

Deformations caused by migration of unfrozen water (whether present in the frozen ground before loading or formed on melting of ice at the points of particle contact under load) are responsible for the filtration-migrational part of the compactional deformation of frozen ground; here the corresponding calculations indicate that the change in the porosity of the frozen ground as a result of melting of ice at contact points constitutes no more than 1/3 of the total compacting deformation, while the rest is due to attenuating-creep deformation (or secondary consolidation) as controlled by irreversible shears of particles and particle aggregates. If, for example, we assume in accordance with Hoekstra's experiments* that the increase in the unfrozen-water content on a 100-kg/cm^2 increase in the external pressure is $\Delta W_u = 0.07$ or, for 1 kg/cm^2, $\Delta W_u = 0.0007$, we obtain:

at $p = 1$ kg/cm^2, the porosity change in the frozen ground due to extrusion of water formed by the melting of ice at contact points will be $\Delta\epsilon = \gamma_{sp}\Delta W_u = 2.78 \times 0.0007 \approx 0.00195$;

at $p = 2$ kg/cm^2, $\Delta\epsilon = 0.0039$;

at $p = 8$ kg/cm^2, $\Delta\epsilon = 0.0156$, or approximately one-third of the change in the porosity coefficient obtained experimentally for loam with massive texture.** For example, with $W_d = 31.6\%$, $\theta = -0.4°C$, and $p = 1$ kg/cm^2, the result was $\Delta\epsilon_1 = 0.007$; with $p = 2$ kg/cm^2 on the same frozen-soil specimen, $\Delta\epsilon_2 = 0.0150$, and at $p = 8$ kg/cm^2 $\Delta\epsilon_8 = 0.0420$, in confirmation of the above statement.

We note that filtration-migrational deformation occurs primarily during the initial period after loading of the soil and, as time passes, composes a steadily smaller fraction of the total compacting deformation.

No method of predicting the changes in time (no analytical calculation) of the filtration-migrational consolidation of frozen ground has as yet been developed, and in forecasting the settling of structures on high-temperature frozen ground we must confine ourselves to determination of the total compacting (consolidation) settling without separating it into filtration-migrational (primary) and attenuating-creep (secondary consolidation) components and treat the settling of frozen ground versus time, beginning only after a certain time interval from the start of loading, as a damping-creep process, as will be set forth in the next section.

In the general case, the compression (compaction) curve of a frozen soil at constant negative temperature takes the form shown in Fig. 101.

Inspecting the compression curve of the frozen ground, we can distinguish three basic segments: aa_1, a_1a_2, and a_2a_3 (Fig. 101). Section aa_1 (as far as point a_1), which corresponds to the first maximum on the compression curve, characterizes the elastic and structurally reversible deformations of frozen ground under compression (without disturbance to structural bonds). In this pressure range the rate of deformation is very large and the deformation can be assumed practically instantaneous. The pressure corresponding to point a_1 is close to the structural strength of the frozen soil, which must be exceeded to begin the compaction (an irreversible porosity decrease) of the soil. At small stresses (about 1.5–1 kg/cm^2, the structurally reversible deformations (according to A. G. Brodskaya) may compose 100 percent of the total deformation, but at medium pressures (on the order of 4–10 kg/cm^2) and at not very low temperatures (say down to $-4°C$), the elastic and structurally reversible

*See Chap. I, Sec. 6.
**See Ref. 2, Table 8, Experiment 4.

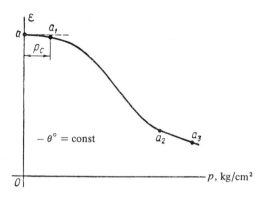

Fig. 101. Compression graph for higher content
of frozen ground.

deformations are indicated by the corresponding experiments to represent from
10 to 30 percent of the total deformation.

Section $a_1 a_2$ of the compression curve characterizes the structurally irreversible deformations of the frozen soil under compression, which amount to 70 to
90 percent of the total deformation. They are due principally to irreversible
shears of particles and particle aggregates (by creep of the skeleton and ice of the
frozen soil) and reach their largest values at pressure corresponding to the
inflection point on curve $a_1 a_2$; some frozen soils still exhibit substantial
structurally irreversible deformations (according to I. V. Boyko) at pressures in
the 200-kg/cm² range.

Finally the portion $a_2 a_3$ on the graph, which not always can be reached at
medium external-pressure magnitudes, characterizes the work hardening of
frozen ground, mainly owing to increased electro-molecular bonding between
the particles as the distances between them become shorter.

As we noted above, it is now necessary, in forecasting the settling of
foundations built on permafrost with preservation of negative temperatures, to
confine ourselves to determination of the resultant (stabilized) compaction
settling from the resultant relative compressibility coefficient of the frozen
ground, a specified external load, and the thickness of the active compression
zone, without separating the compacting deformation of the frozen soils into its
filtration-migration and creep-deformation components.

The magnitude of summons relative compressibility coefficient of frozen
ground (which we shall denote by a_r^Σ) is determined in compression experiments
performed on the soils (in thermally nonconductive odometers) or is calculated
from the results of field test-load measurements made with cold plungers.*

In the former case, the resultant relative compressibility coefficient is
determined by

*See Ref. 42, Chap. IV.

$$a_r^\Sigma = \frac{s_\infty}{hp} \qquad (V.10)$$

where s_∞ is the stabilized settling of the soil layer with preservation of its negative temperature, h is the thickness of the soil layer in the odometer, and p is the effective pressure.

In the latter case (for test loading with a plunger), we have in accordance with the formula of linearly-deformable-body theory

$$a_r^\Sigma = \frac{\beta}{E_t} \qquad (V.11)$$

where β is a coefficient characterizing the lateral expansion of the soil, which equals, as we know from general soil mechanics, $\beta = [1 - 2\mu_r^2/1 - \mu_r]$; μ_r is the coefficient of relative transverse deformation, analogous to the Poisson coefficient for the elastic state of solids; we may put $\mu_r \approx 0.1$-0.2 for hard-frozen soils and $\mu_r \approx 0.3$-0.4 for plastically frozen soils; E_t is the total strain modulus of the frozen soil, determined from test-load results using the equation

$$E_t = \omega(1 - \mu_r^2)\frac{pb}{s_\infty} \qquad (V.12)$$

where ω is a shape factor equal to 0.88 for a rigid plunger with a square base area, b is the width of the square plunger, and s_∞ is the stabilized settling of the plunger under the particular load p at constant negative temperature of the frozen ground.

Some values of summary relative compressibility coefficients of frozen ground according to original experiments are presented in Table 32,* and the results of special experiments to investigate the compressibility of frozen and permanently frozen ground in Table 33.**

We see on inspection of the above data on the summary relative compressibility coefficients of frozen soils (a_r^Σ) that this coefficient is variable and depends on the negative temperatures of the frozen ground, its composition, and the range of variation of the external pressure.

The data also indicate how large the compressibility coefficients of frozen soils can be; thus, in designing structures to be built on frozen ground it is necessary to consider the compressibility of plastically frozen soils, which governs their settling under load, even when the negative ground temperatures are preserved.

If the summary relative compressibility coefficients of the frozen soils, the dimensions of the footings (based on prior selection of the base area for safe pressure on the ground), and the magnitude of external load (the pressures from the structure to be built on the frozen soils in the base) are all known, the total stabilized settling of the foundations is determined from the known equations of

*See Ref. 1, Chap. IV.
**See Ref. 2, this chapter.

Table 32. Values of summary relative compressibility coefficient a_r^{Σ} for frozen ground

| Soil type | Physical properties | | | Summary relative compressibility coefficient a_r^{Σ}, cm^2/kg | Test conditions $(F, m^2; p, kg/cm^2)$ |
	Bulk density γ, g/cm^3	Moisture content W_d, %	Temperature θ, °C		
Sand*	1.99	13.20	−0.5	0.0010	Compression experiments
Fine-grained sand** ...	2.10	31.9	−2.0	0.0002	$F^{****} = 0.49; p = 8$
Medium sandy loam** .	2.11	43	−2.2	0.0014	$F = 0.49; p = 8$
Same**	–	24.30	−1.0	0.0011	$F = 1; p = 8$
Same**	–	24.30	−1.0	0.0017	$F = 0.5; p = 8$
Loam**	1.34–2.28	46	−2.0	0.0020	$F = 0.98; p = 8$
Same*	1.88	39.8	−1.0	0.0032	Compression experiments
Silty sandy loam*** ...	2.00	28.35	−3.0	0.0139	$F = 0.5; p = 3.75$
Same***	2.00	28.35	−1.0	0.0231	$F = 0.5; p = 2.5$

*N. A. Tsytovich, On the Compressibility (Compression) of Frozen Ground.

**Calculated from test-load results and field tests performed by G. N. Maksimov and L. P. Gavelis after a suggestion by G. Ya. Shamshur, 1952–1954.

***S. S. Vyalov, Long-Term Strength of Frozen Soils and Permissible Pressures on Them, Collection No. 1, TsNIMS, *Izd. Akad. Nauk SSSR*, 1954.

****Base area of test plunger.

Table 33. Summary relative compressibility coefficients for various frozen soils

Soil type	Physical indicators				Summary relative compressibility coefficient ($a_r^\Sigma \times 10^4$, cm²/kg) in the following pressure intervals, kg/cm²				
	Total moisture content W_d, %	Unfrozen water W_u, %	Bulk density γ, g/cm³	Soil temperature θ, °C	0-1	1-2	2-4	4-6	6-8
Medium-grained sand .	21	0.2	1.99	−0.6	12	9	6	4	3
Same	27	0.0	1.87	−4.2	17	13	10	7	5
Same	27	0.2	1.86	−0.4	32	26	14	8	5
Heavy sandy loam, silty, massive texture	25	5.2	1.90	−3.5	6	14	18	22	23
Same	27	8.0	1.88	−0.4	24	29	26	18	14
Medium loam, silty, massive texture ...	35	12.3	1.83	−4.0	8	15	26	28	24
Same	32	17.7	1.84	−0.4	36	42	37	21	14
Medium loam, silty, veined texture	42	11.6	1.71	−3.8	5	10	18	42	32
Same	38	16.1	–	−0.4	56	59	39	24	16
Medium loam, silty, stratified texture ..	104	11.6	1.36	−3.6	54	54	59	44	34
Same	92	16.1	1.43	−0.4	191	137	74	36	18
Varved clay	36	12.9	1.84	−3.6	15	22	26	23	19
Same	34	27.0	1.87	−0.4	32	30	25	20	16

Notes: 1. Values of the summary relative compressibility coefficients are given for frozen soils in their natural state (except for sand); 2. The accuracy of the deformation measurements was apparently inadequate for hard-frozen soils at $p \leqslant 2$ kg/cm².

general soil mechanics for the compaction settling of soils, but under consideration of the variability of a_r^Σ, as will be discussed in Sec. 7 of the present chapter.

5. Attenuating Creep Deformations in Frozen Ground

We noted in the preceding section that, according to experimental data, the structurally irreversible creep deformation of plastically frozen soils under compression constitutes from 70 to 90 percent of the total compaction deformation.

In consolidation by compression (with lateral spreading of the soil prevented) or in the case of operation of a continuous uniformly distributed load (one-dimensional problem), the creep deformation will always be attenuating.

As we showed in Chap. III, the later is accurately described by the equation of hereditary creep theory, and both the exponential (III.6) and hyperbolic (III.6″) creep factors are used in calculations for frozen ground and permafrost.

If we disregard the migrational filtration part of the total structurally irreversible compacting deformation of frozen soils, plastically frozen ground can be regarded as a composite quasihomogeneous body that is subject to hereditary creep theory.

Let us consider how the deformations are determined (in regard to their magnitude and variation in time) in the damped creep of frozen soils.

Since the creep factor in the theory of hereditary creep $[K(t - t_0)]$ is the creep rate of the soil under a constant unit stress, creep affects only the progress of settling in time, and the total stabilized settling of frozen ground in the case of the one-dimensional problem and a time-constant load (p = const) will be expressed in the same way as the stabilized settling of unfrozen ground:

$$s_t = ha_{r.c}p \tag{V.13}$$

where h is the thickness of the deformed soil, $a_{r.c}$ is the relative compressibility coefficient of the frozen soil in creep, the analytical expression for which is a function of time and depends on the form of the creep factor $K(t - t_0)$, and p is the external pressure.

As was proposed earlier,[8] the relative compressibility coefficient in creep, $a_{r.c}$, can be expressed with an exponential creep factor by the equation

$$a_{r.c} = a'_r + a''_r (1 - e^{-\delta't}) \tag{V.14}$$

where a'_r and a''_r are the primary and secondary relative compressibility coefficients of the soil.

Disregarding the primary compressibility (which represents approximately 10–20 percent of the total deformation), we have

$$a_{r.c} \approx a''_r (1 - e^{-\delta't}) \tag{V.14'}$$

where δ' is the creep parameter (the attenuation coefficient of the creep alone), which is determined from a curve constructed from experimental data in the coordinates $\ln (1 - s_t/s_\infty)$ and t (where s_t is the settling during time t and s_∞ is the stabilized settling) as the tangent of the slope angle of the rectified curve and the t axis.

Then the fading creep deformations (the settling due to creep) of a layer h of frozen ground under a continuous load p [kg/cm^2] will be determined by expression (V.13), into which the value of $a_{r.c}$ according to Eq. (V.14') is to be substituted. We then have

$$s_{t(t)} = ha''_r p (1 - e^{-\delta't}) \tag{V.15}$$

For sufficiently large times t_f, as indicated by Eq. (III.18), the magnitude of the secondary relative compressibility coefficient a''_r can be assumed to be equal to

$$a''_r \approx a^f_r - a^{in}_r \tag{V.16}$$

where a^f_r is the final relative compressibility coefficient and a^{in}_r is the initial relative compressibility coefficient (which corresponds to the beginning of the dominance of creep in the total deformation of the frozen ground, not including the instantaneous deformation).

We note that comparison of calculated results with observations indicates that the exponential relation for the fading creep (V.15) corresponds to the type of creep encountered in high ice content hard-frozen soils.

Example 5. Plot the creep-deformation curve for a layer of frozen soil with a high ice content and a thickness $h = 5$ m if $a''_{r.c} = 0.01$ cm^2/kg, $\delta' = 0.001$ day^{-1}, and $p = 4$ kg/cm^2.

We determine the creep deformation for various time intervals: $t_1 = 30$ days; $t_2 = 100$ days, $t_3 = 300$ days, $t_4 = 1,000$ days, and $t_5 = 3,000$ days.

At $t_1 = 30$ days, Eq. (V.15) yields $s_{t(30)} = ha''_{r.c}p\,(1 - e^{-\delta't}) = 500 \times 0.01 \times 4\,(1 - e^{0.001 \times 30}) = 0.6$ cm. In exactly the same way for $t_2 = 100$ days

$$s_{t(100)} = 20\,(1 - e^{-0.001 \times 100}) = 1.8 \text{ cm}$$

$$\text{at } t_3 = 300 \text{ days, } s_t = 5.2 \text{ cm}$$
$$\text{at } t_4 = 1,000 \text{ days, } s_t = 12.4 \text{ cm}$$
$$\text{at } t_5 = 3,000 \text{ days, } s_t = 19.0 \text{ cm}$$
$$\text{at } t = \infty, s_{t\infty} = 20 \text{ cm}$$

A theoretical attenuating creep curve for the layer of frozen ground is plotted from these data (Fig. 102).

For plastic frozen soils, better agreement with experimental data is obtained if the hyperbolic creep factor is used in analytical calculations of the attenuating creep deformation [Eq. (III.6")].

According to Eq. (III.5) of the hereditary-creep theory, we have the following expression for the relative fading creep deformation:

$$e_c = \frac{\sigma(t_0)}{E_{\text{inst}}} + \int_0^t K(t - t_0)\,f\,[\sigma(t_0)]\,dt_0$$

If we disregard the instantaneous deformation, i.e., if we put

$$\frac{\sigma(t_0)}{E_{\text{inst}}} = 0 \qquad \text{and} \qquad f[\sigma(t_0)] = a_{r\infty}\sigma$$

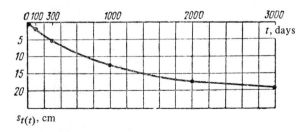

Fig. 102. Theoretical attenuating creep curve for a layer of frozen ground.

(where $a_{r\infty}$ is the stabilized secondary relative compressibility coefficient), substitution of the value of the hyperbolic creep factor from Eq. (III.6″), i.e.,

$$K_3(t - t_0) = \frac{T}{(T + t)^2}$$

into the expression for e_t yields

$$e_{t(t)} = a_{r\infty} T \int_0^t \frac{\sigma(t)}{[T + (t - t_0)]^2} \, dt_0$$

Yu. K. Zaretskiy[8] obtained the following very simple expression for the relative attenuating creep deformation of frozen ground in the case of a constant load $[\sigma(t) = p = \mathrm{const}]$:

$$e_{t(t)} = a_{r\infty} p \left(\frac{t}{T + t} \right) \tag{V.17}$$

where settling due to creep of the frozen ground $(s_t = e_{c(t)} h)$ will be

$$s_t = a_{r\infty} p h \left[\frac{t}{T + t} \right] \tag{V.18}$$

The value of the stabilized secondary relative compressibility coefficient $a_{r\infty}$ of a frozen soil and the value of the hyperbolic creep factor T are determined, as was shown in Sec. 5 of Chap. III, from a diagram construed on experimental data in the coordinates $t/[e(t)]$ vs. t.

However, as Zaretskiy showed in his paper,[8] the parameter T depends on the load and a more exact expression is

$$T = T_0 \frac{p_\infty}{p_\infty - p} \tag{V.19}$$

where T_0 is the value of the parameter, which does not depend on the external load at $p \ll p_\infty$; p_∞ is the ultimate long-term strength $(p_\infty = \sigma_{lt})$.

Like Eq. (IV.1), expression (V.19) enables us to determine the ultimate long-term value $p_\infty \approx \sigma_{lt}$ that causes creep of a frozen base from a few values of T for various load levels p.

We note that, according to calculations, very good agreement is observed between the theoretical magnitudes (based on the above theory of attenuating creep application of a hyperbolic factor) and the values measured on frozen ground in load tests at a constant negative temperature.

Example 6. Plot an attenuating creep deformation graph for a frozen ground layer of thickness $h = 4$ m if the steady-state (secondary) relative compressibility coefficient of the frozen soil $a_{r\infty} = 0.02$ cm^2/kg, the external uniformly distributed load on the layer is $p = 5$ kg/cm^2, and the parameter of the hyperbolic creep factor for the frozen ground (found from experimental results) is $T = 360$ hr $= 15$ days.

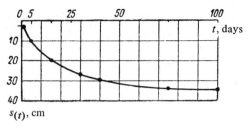

Fig. 103. Theoretical attenuating creep deformation curve of frozen soil layer.

Specifying various time intervals ($t_1 = 5$ days, $t_2 = 15$ days, $t_3 = 30$ days, and so forth), we find the corresponding creep-settling values for the soil from Eq. (V.18):

at $t_1 = 5$ days

$$s_{t_3} = a_{r\infty}ph\left(\frac{t}{T + t}\right) = 0.02 \times 5 \times 400\left(\frac{5}{15 + 5}\right) = 10 \text{ cm}$$

In exactly the same way, we find:

$$\begin{aligned}
\text{at } t_2 &= 15 \text{ days, } s_{t_{15}} = 20 \text{ cm} \\
\text{at } t_3 &= 30 \text{ days, } s_{t_{30}} = 26.7 \text{ cm} \\
\text{at } t_4 &= 40 \text{ days, } s_{t_{40}} = 39.1 \text{ cm} \\
\text{at } t_5 &= 75 \text{ days, } s_{t_{75}} = 33.3 \text{ cm} \\
\text{at } t_6 &= 100 \text{ days, } s_{t_{100}} = 34.7 \text{ cm}
\end{aligned}$$

The data obtained are used to plot the attenuating creep graph for the layer of frozen ground (Fig. 103).

It should be noted that use of the hyperbolic creep parameter gives a steeper descent of the creep curve (a higher creep rate at the earlier time intervals as compared with the curve plotted with the exponential creep factor), in closer conformity to the process of attenuating creep in plastically frozen soils.

6. Viscoplastic Flow Deformations of Frozen Ground

The development of rheological processes in frozen ground under the action of external loads was examined in detail in Chap. III, where it was established that steady state creep arises only when the frozen soil is subjected to stresses exceeding its long-term strength and has three stages: (1) nonsteady (with a gradually decreasing rate of deformation); (2) steady (or viscoplastic flow at a constant deformation rate), and (3) progressive (with a steadily increasing deformation rate); here the second stage, that of viscoplastic flow, occupies the longest time and is of the greatest interest for practice.

In analytical prediction of the viscoplastic flow deformations, the equation usually used is (III.7) of Chap. III:

$$\dot{e}_t = \frac{1}{\eta_{t,\theta}}(\sigma - \sigma_0)^n$$

The viscoplastic flow of ice is accurately described in more general form by Eq. (III.7'''):

$$\dot{e}_i = \frac{1}{\eta_{(t,\theta)}}\, \sigma_i^n$$

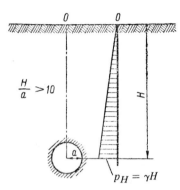

Fig. 104. Diagram of deep gallery (tunnel) in ice.

i.e., the deformation-rate intensity $\dot{e}_i$ in viscoplastic flow is proportional to the shear stress intensity σ_i to the nth power, with the proportionality coefficient (the viscosity η) dependent on both the negative temperature and the time of load action.

Example 7. Determine the viscoplastic flow deformations that cause closure of a deep circular-section gallery in a bed of ice under natural pressure (Fig. 104).

For a cylinder subjected to natural pressure $(p_H = \gamma H)$ by the weight of the ice bed, we have the following expressions for the stresses and displacements at the inner contour of the gallery.*

Stresses:

$$\left. \begin{array}{l} \sigma_z = p_H \left[1 - \left(\dfrac{a}{r} \right)^{2/n} \right] \\[3mm] \sigma_\theta = p_H \left[1 + \dfrac{2-n}{n} \left(\dfrac{a}{r} \right)^{2/n} \right] \end{array} \right\} \qquad (d_1)$$

where a is the radius of the gallery, r is the present radius, and n is the parameter of Eq. (d$_1$).

Based on the general equation of viscoplastic flow of ice (III.7''), the displacements in the direction of the gallery radius are determined by

$$\frac{u_a}{a} = \frac{3^{\frac{n+1}{2}}}{2n^n}\, p_H^n \int_0^t \frac{d\tau}{\eta(\tau,\theta)} \qquad (d_2)$$

The time dependence of the viscosity coefficient may be assumed to be in a form (a) from Eq. (III.21):

$$\eta_t = \eta_0 t^q \qquad (d_3)$$

where η_0 and q are parameters determined from the results of experiments:

(b) from N. N. Maslov's equation (III.22):

$$\eta_t = \eta_f - (\eta_f - \eta_0)e^{-\zeta t} \qquad (d_4)$$

*See Ref. 9, Chap. VII.

where η_f and η_0 are the final and initial viscosity coefficients and ζ is the nonlinearity parameter.

Substituting the expressions for the viscosity coefficients (d_3) or (d_4) into the equation of the viscoplastic displacement (d_2) after integration, we obtain: in the first case the viscoplastic flow (closure) deformation will be

$$\frac{u_a}{a} = \frac{3^{\frac{n+1}{2}}}{2n^n} p_H^n \left[\frac{1}{\eta_0} \times \frac{t^{1-q}}{1-q} \right] \qquad (V.20)$$

and in the second case

$$\frac{u_a'}{a} = \frac{3^{\frac{n+1}{2}}}{2n'^n} p_H^n \left\{ \frac{t}{\eta_f} + \frac{1}{\zeta\eta_f} \ln \left[\frac{\eta_f - (\eta_f - \eta_0) e^{-\zeta t}}{\eta_0} \right] \right\} \qquad (V.20')$$

Thus, suppose that a circular-section gallery (tunnel) is being driven through underground ice. Then, using Eq. (V.20) and assuming $m \approx 2$ and $q = 0.5$ for the ice, the pressure due to the weight of the ice at a temperature $\theta = -10°C$, $\eta_0 = 4 \times 10^{13} P \times 2.83 \times 10^{-10} = 11.32 \times 10^3$ kg hr/cm^2, that the depth of the center of the gallery $H = 30$ m, that the inside radius of the gallery $a = 2$ m, and that the ice has a bulk density $\gamma_{ice} = 0.9$ g/cm^3 is found to be

$$p_H = 0.0009 \times 3,000 = 2.7 \text{ kg/cm}^2$$

Specify time $t = 100$ days and determine the viscoplastic displacement as a factor of the radius a.

According to Eq. (IV.20)

$$\frac{u_a}{a} = \frac{3^{3/2}}{2 \times 2^2} 2.7^2 \left[\frac{2}{11,320} \times \frac{2,400^{1-0.5}}{1-0.5} \right] \approx 0.041$$

Thus, with $a = 2$ m $= 200$ cm, the viscoplastic displacement (closure) of the circular ice gallery after 100 days will be

$$u_a = 0.041a = 0.041 \times 200 = 8.2 \text{ cm}$$

It is seen from the above example that, by determining experimentally the parameters of the equation of the stressed and strained state in the phase of established creep (viscoplastic flow) and the viscosity as a function of negative temperature and time of load action, we can also easily determine, from existing solutions, the displacements of the ice or frozen ground in viscoplastic flow for any time interval following the beginning of loading.

7. Prediction of Foundation Settling on Plastic Frozen Ground at Constant Temperature

As we pointed out in Sec. 4 of the present chapter, it is sufficient to know only one deformability characteristic of frozen soil to calculate the total stabilized settling of a given layer (disregarding the increase in the settling with time): the summary relative compressibility coefficient a_r^Σ, besides the dimensions of the frozen ground layer as well as the pressure on it.

Thus, in the case of the one-dimensional problem (compression of a layer of soil under a continuous load or with lateral expansion prevented), the total stabilized settling of the frozen ground is determined by the equation known from general soil mechanics,* in which it is only necessary to replace the relative compressibility a_r by the summary relative compressibility a_r^Σ of the frozen ground:

$$s_\infty = h a_r^\Sigma p \qquad (V.21)$$

In Sec. 5, we also considered the settling increase with time due only to creep of the frozen ground, treating it as a quasihomogeneous single-component body under the conditions of the one-dimensional problem [relations (V.15) and (V.18), which can be used directly only if the working conditions of the particular foundation correspond practically adequately to those of a one-dimensional problem]. This will be the case only for very wide foundations (when the width is at least twice the thickness of the frozen ground layer), something that occurs very seldom. In other cases, the problem of determining the settling of foundations placed on permafrost must be treated as a three-dimensional problem.

In the case of a three-dimensional problem (localized load imposed by the foundations of structures upon frozen ground), it is also possible to use the known general methods of soil mechanics to determine the total stabilized settling of foundations on plastically frozen (high-temperature) soils from the summary relative compressibility coefficient a_r^Σ and the thicknesses of the individual layers: either the approximate method of layer-by-layer summation (according to SNip II-B.1-62) or the engineering method of the equivalent layer, which was developed by the author.

If the frozen ground is homogeneous to a sufficient depth (several times the width of the footing base), the thickness of the soil is broken up (in accordance with the SNiP) into layers of thickness $0.2b$, where b is the width of the footing base (or in accordance with the boundaries between the different soil strata, provided that the thicknesses do not exceed $0.2b$), and the total stabilized settling of the foundations on the frozen bed is determined from the layer-by-layer summation formula

$$s_\infty = \sum_i^n h_i a_{ri}^\Sigma \sigma_{zi} \qquad (V.22)$$

where σ_{zi} is the vertical compressive stress determined from Table 8 of SNiP II-B.1-62; a_{ri}^Σ is the resultant relative compressibility coefficient for the

*See footnote on p. 193.

particular frozen ground sublayer (it is determined by experiment or taken from Tables 32 and 33 for preliminary calculations), and h_i is the thickness of the sublayer.

In the equivalent-layer method, which takes account of all normal-stress components (σ_z, σ_y, and σ_x), the lateral expansion of the ground, and the dimensions and rigidity of the foundation, the thickness of an equivalent ground layer whose settling is equal to that of a foundation of the specified dimensions and rigidity is first determined from the author's formula

$$h_e = A\omega b \qquad (V.23)$$

where $A\omega$, the coefficient of the equivalent layer, is taken from the tables* as a function of the ratio of the foundation's length l to its width b, i.e., as a function of $\alpha = l/b$ (the rigidity of the foundation) and the coefficient of lateral expansion of the soil, μ_r, which is analogous to Poisson's coefficient.

Then, without carrying out the summation according to (V.22) or calculating the compressive stresses σ_z, we have a simpler and more accurate expression for the total stabilized settling of foundations on a bed of homogeneous frozen soils:

$$S_\infty = h_e a_r^\Sigma p \qquad (V.24)$$

where p is the pressure exerted by the structure at the level of the footing base.

If the individual layers of frozen ground have different compressibilities through the entire active compression zone, whose maximum thickness (according to the solution that we obtained earlier) $H = 2h_e$, it is first necessary to calculate the average summary relative compressibility coefficient of the frozen soils from the equation

$$a_{rm}^\Sigma = \frac{\sum h_i a_{ri}^\Sigma z_i}{2h_e^2} \qquad (V.25)$$

where z_i is the distance from the level corresponding to the depth $2h_e$ to the midpoint of each frozen ground layer considered.

A solution was obtained in 1972 by Yu. K. Zaretskiy** on the basis of hereditary-creep theory with a hyperbolic coefficient of the form of (III.6″) for computation of the increase in settling solely due to creep of frozen soils homogeneous to a sufficient depth under foundations with rectangular base areas. In the case of a load increase from the start of loading the foundation to t_1 according to a linear law $p = kt$, followed during $t > t_1$ by a constant load ($p = $ const), this solution takes the following form:

(1) Using the method of layer-by-layer summation to determine the stabilized settling due to attenuating creep:

*See footnote, p. 193.
**See Ref. 9, this Chap.

$$s_{tt} \approx \left(\sum_1^n a_{r\infty} \alpha_z h_i \right) kt \left[1 - \frac{T}{t} \ln \left(1 + \frac{t}{T} \right) \right]$$

at $t < t_1$

(V.26)

$$s'_{tt} \approx \left(\sum_1^n a_{r\infty} \alpha_z h_i \right) t_1 \left[1 - \frac{T}{t} \ln \left(1 + \frac{t_1}{T - t_1 + t} \right) \right]$$

at $t > t_1$

where $a_{r\infty}$ is the steady-state secondary compressibility coefficient (for creep); this parameter is determined from the curve of Fig. 68 and equation (III.19'); T is the parameter of the hyperbolic creep coefficient (see Sec. 5 of Chap. III), and α_z is the compressive-stress dissipation coefficient (Table 8, SNiP II-B.1-62);

(2) Using the equivalent-layer method, the solution is simpler:

$$s_{tt} = h_e a_{r\infty m} kt \left[1 - \frac{T}{t} \ln \left(1 + \frac{t}{T} \right) \right]$$

at $t < t_1$ and

(V.27)

$$s'_{tt} = h_e a_{r\infty m} kt_1 \left[1 - \frac{T}{t_1} \ln \left(1 + \frac{t_1}{T - t_1 + t} \right) \right]$$

at $t > t_1$

where $a_{r\infty m}$ is the average stabilized secondary compressibility coefficient for attenuating creep for the frozen ground stratum as determined from Eq. (V.25), in which the resultant relative compressibility coefficient a_{ri}^{Σ} is replaced by the secondary compressibility coefficient $a_{(r\infty)i}$.

In the case of constant load ($p = $ const), Eq. (V.27) become much simpler, and the expression for the attenuating creep deformation assumes the following form with a hyperbolic creep coefficient:

$$s_{tt} = h_e a_{r\infty m} p \left(\frac{t}{T + t} \right)$$

(V.28)

This equation can be recommended for practical use.

With the exponential creep coefficient, we have

$$s_{tt} = h_e a_{r\infty m} p (1 - e^{-\delta' t})$$

(V.29)

Equations (V.22) to (V.29) determine the major part of the settling of frozen ground under load, due to its creep; however, the total settling will be somewhat larger, since the initial settling s_0 due to filtrational-migration deformation must be added to the results, which is as we noted earlier, about 10 percent but no more than 30 percent of the total deformation according to experimental data.

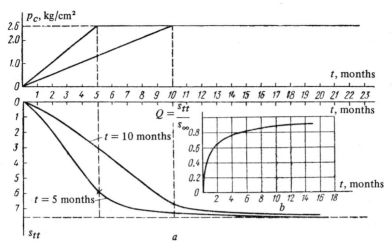

Fig. 105. Illustrating increase of attenuating creep deformation of a plastic frozen ground under variable load with the passage of time.

Example 8.* Find the time variation of creep deformation of a foundation base with a footing area 3×4.8 m^2, to be built on plastic frozen soils ($\theta = -0.5°C$).

From the surface to a depth of 5.6 m, the base soil is a silty sandy loam (bulk density $\gamma = 1.85$ ton/m^3, characterized by a hyperbolic creep coefficient parameter $T' = 0.4$ month and $a'_r = 0.01$ cm^2/kg); from 5.6 to 11 meters it is a clay of massive texture ($\gamma = 1.7$ tons/m^3, $T'' = 0.2$ month, $a''_{r\infty} = 0.006$ cm^2/kg); the pressure on the soil under footing level is $p = 2.6$ kg/cm^2 and the rate of load increase $k = 0.26$ kg/cm^2 month over 10 months.

The calculation was made by Eq. (V.26), with the depth of the active compression zone under a footing of the specified dimensions is taken to be equal to 660 cm according to SNiP.

The results were used to plot a creep attenuation curve of the plastic frozen base of this foundation (Fig. 105a). The calculation showed that the fully stabilized attenuated creep deformation will be 7.6 cm and that about 88 percent of it will take place by the time the construction is completed (within 10 months of its onset).

But if construction proceeds at a more rapid pace, e.g., is complete within five months, the development of the settling in time will be somewhat different in nature, and only 78 percent of the total stabilized settling will have occurred by completion.

Figure 105b shows a graph of the degree of settling stabilization ($a = s_{tt}/s_\infty$) plotted against time during construction.

---

*Example taken from the paper by Yu. K. Zaretskiy, Rheological Properties of Plastically Frozen Ground, 1972.

References

1. Tsytovich, N. A.: "Issledovaniye deformatsiy merzlykh gruntov (A Study of the Deformations of Frozen Soils)," vol. 2 of Doctoral thesis, pp. 45, etc., Leningrad, 1940; N. A. Tsytovich: "Construction Under Permafrost Conditions," abstracts of papers at Conference of USSR Academy of Sciences Council for the Study of Productive Resources (SOPS), USSR Academy of Sciences, 1941; N. A. Tsytovich: Issledovaniye uprugikh i plasticheskikh deformatsiy merzlykh gruntov (Investigation of Brittle and Plastic Deformations in Frozen Soils), *Trudy KOVM Akad. Nauk SSSR*, vol. X, 1940.

2. Tsytovich, N. A.: Calculation of the Settling of Foundations as Function of Time, Soil Properties, and Foundation Dimensions, *Izd. LISI*, 1934; see also N. A. Tsytovich: "Principles of the Mechanics of Frozen Soils," USSR Academy of Sciences Press, 1952; N. A. Tsytovich and I. I. Cherkasov: Determination of the Coefficients of Compressibility of Soils from Plunger-Indentation Tests, "Osnovaniya fundamenty i mekhanika gruntov," no. 6, 1970; and Ref. 1.

3. Tsytovich, N. A.: On the Coefficient of Elastic Compression of Soils, *Byull, LIS*, no. 46, 1932.

4. Vyalov, S. S.: Report of the Igarka Scientific Research Station for 1953, Institute of Permafrostology, USSR Academy of Sciences.

5. Tsytovich, N. A.: Compressibility (Compression) of Frozen Ground, Report of the Institute of Permafrostology, USSR Academy of Sciences, 1953; see also N. A. Tsytovich: Osnovaniya i fundamenty na merzlykh gruntakh (Bases and Foundations on Frozen Soils), *Izd-vo Akad. Nauk SSSR*, 1958, p. 117.

6. Brodskaya, A. G.: Compressibility of Frozen Ground, *Izd-vo Akad. Nauk SSSR*, 1962.

7. Ter-Martirosyan, Z. G., and N. A. Tsytovich: On Secondary Consolidation, "Osnovaniya, fundamenty i mekhanika gruntov," 1965, no. 5.

8. Zaretskiy, Yu. K.: Rheological Properties of Plastic Frozen Ground, "Osnovaniya, fundamenty i mekhanika gruntov," 1971, no. 2.

Chapter VI

THAW SETTLING OF FROZEN GROUND

1. Importance of Predicting the Settling of Frozen Ground on Thawing

If the settling that occurs on thawing of permafrost at the bases of structures is not provided for by the design, and the amount of settling exceeds the permissible limit for the particular structure, unacceptable deformations and failure of foundations and superstructures are inevitable.

Most of the failures that were encountered in construction on permafrost were due precisely to the fact that no account was taken for settling of the frozen bases when thawing was permitted, or of the drastic changes that take place in the strength properties of the base soils: frozen soils, whose design strength is of the order of 5-20 kg/cm^2 when their negative temperatures are maintained, undergo partial conversion to liquefied masses that are incapable of bearing the load imposed by the structure when thawing occurs (this applies in particular to soils with high ice content).

If the deformations of frozen ground on thawing are due to abrupt (avalanche-like) changes in ice-cement structure, with bond loss of local collapse nature (for example, under the influence of local heat sources—hot furnaces, steam boilers, etc.) and take place rapidly, usually with extrusion of the thawed soils; they are referred to as subsiding.

But if general compacting deformation (which may vary under different foundations) occurs on thawing of the permafrost in the bases of the structures, such deformation is called settlement.

The extent of settling of frozen ground and the course that it takes in time depend not only on the properties of the frozen soil (its structure, the presence of ice inclusions, etc.) and the acting load, but also on the temperature regime of the soils in the process of thawing. In addition, frozen ground will settle on thawing under its own weight, since it is undercompacted under natural conditions owing to the loosening of the ground during freezing and the presence of ice crystallization bonds (ice-cement bonds).

Below we shall examine the settling of permafrost on thawing with simultaneous compacting by the load imposed by the foundations of structures.

The Construction Norms and Rules of the USSR State Committee on Construction (SNiP II-B.6-66, Sec. 3.25) recommend the use of soil bases in the thawing and thawed states "in the presence (to the calculated thawing depth) of permafrost whose settling on thawing *does not exceed the limiting values* (underscoring ours–N.Ts.) if technological or design features of the buildings and structures render it economically inexpedient to preserve the frozen state of the base soils," or if there is bedrock not far below the surface. Further, ". . . the possibility of allowing gradual thawing of the base soils during use of the building or structure, or the need for preconstruction thawing of the ground, is established on the basis of a preliminary evaluation of the *possible settling of the permafrost base soils* (underscoring ours-N.Ts.) when they thaw."

The tremendous importance of correct prediction of settling of permafrost during thaw under structures is clear from the above quotations.

Study of the settling of frozen soils during thawing was begun back in 1933, when a method was proposed for testing frozen ground in a special device to determine its settling on thawing under load.[1]

This method made it possible to determine only the total compacting settling of the thawing soils, without resolving it into its components.

Beginning in 1937, and especially during 1939-1940, a great deal of attention was given to the settling of frozen ground on thawing in view of the special actuality of the problem and practical requirements. Here we should note the work of G. I. Lapkin (1939), who suggested on the basis of an experiment at Noril'sk that the settling of frozen ground on thawing be divided into two components: the "conditional thaw settling" (which includes not only the settling due to thawing, but also the part of the pressure settling that is constant at the given pressure) and the variable pressure settling, which is assumed proportional to the pressure increase above that at which the frozen ground was tested; the work of A. Ye. Fedosov (1942 and 1944), who proposed a prediction method for water-saturated clay soils based on the known pressure dependence of the moisture content in clayey soils (a method that did not, however, come into practical use); the work of M. N. Gol'dshteyn (1942), who investigated the compression of frozen ground under load and upon unloading, and our own papers (1937-1939) on the compression of frozen ground in the thawing process with simultaneous compaction of thawing soils under load.*

We note that research on compression of thawing ground has resulted in the detailed development not only of a method of computing the final stabilized settling of foundations on thawing of soils (with rigorous separation into a component that does not depend on the external pressure—the so-called thaw settling—and a second component, the compacting settling, which is a direct function of the normal pressure), but also a method for predicting the course taken by settling of the thawing ground in time.

The cited studies on settling of frozen ground during thaw (dating from 1937-1942) have been developed further both in the work of the present author and in that of various other investigators (V. P. Ushkalov, M. F. Kiselev, and others). The work done during the last decade by a group of scientific workers at NIIosnovaniy working under the author's supervision (Yu. K. Zaretskiy, M. V. Malyshev, and, in the experimental aspect, V. G. Grigor'yeva, V. D. Ponomarev, and others) has made a valuable contribution to the theory and practice of

---

*See Chap. I, Refs. 1a, 1b, 13, and 14.

prediction of foundation settling on thawing and thawed soils; on the basis of more than five years of detailed experimental testing (in the laboratory and in the field) and analytical calculations, the "Instructions for the Calculation of Settling of Thawing and Thawed Soils in Time" were compiled (1967–1970), and found application in practice.

2. Changes in Soil Compressibility on Freezing and Subsequent Thawing

A number of papers by M. N. Gol'dshteyn, H. F. Winterkorn, A. M. Pchelintsev, N. K. Zakharov, Ye. P. Shusherina, and the author in collaboration with V. G. Grigor'yeva, V. D. Ponomarev, and others have been devoted to study of the effects of the freeze-thaw cycle on the properties of soils.

Here we shall discuss only the results of research on the deformation properties, and chiefly the compressibility, of soils subject to freezing and subsequent thawing, as being of greatest practical significance for the prediction of the thaw settling of frozen soils.

As was shown earlier (Chaps. I and II), freezing of moist soils is accompanied by very significant changes in their texture and structure and by the formation of a new and extremely complex cryogenic texture, associated with the migration of water and disperse mineral particles in the process of ground freezing and subsequent freezing of water with an increase in its volume, with differentiation of the ground by formation of ice inclusions, with compression of mineral aggregates and individual layers by ice crystals, and so on and so forth. Frozen ground exhibits particularly complex structure when there is water influx from the outside with excessive ice segregation in the form of ice lenses, interlayers, etc., and a stratified texture forms in the frozen ground, also at nonlinear freezing of water-saturated ground, when a reticular (cellular) texture forms, with numerous thin interlayers and cells of ice oriented in various directions. At rapid freezing, moist soils acquire a massive texture in the frozen state with more or less uniform distribution of the pore ice throughout its volume of the ground. It should also be noted that the freezing of liquefied clayey soils is accompanied by a series of physicochemical processes, such as: coagulation of soil colloids, aggregation of clay particles, etc., and that the density of the soil aggregates may become quite high, equaling the density (consistency) of clays at the plastic limit W_p.

The constitution of frozen soils (their texture and structure) has a substantial influence on the properties of the thawing soils (their shear resistance, as will be discussed in Part Two of the present book when we evaluate the bearing capacities of thawing soils, and especially on the values of their deformability indicators).

As we noted earlier (Chap. I), ice in the ground pores begins to melt at any increase of temperature, with the cohesion between mineral particles, which is governed by ice-cement bonds, decreasing as the ground temperature approaches zero Centigrade and finally dropping suddenly virtually to zero when the ground is completely thawed.

When frozen soils thaw, two opposite processes occur in them:[2] compacting, i.e., a decrease in porosity due to extrusion of melt water, and swelling of aggregates, which may, at light loads on the soil, cause an increase in its porosity after thawing. The resultant effect may be either an increase in ground porosity after freezing and thawing (something that is observed specially frequently in disperse soils of stratified structure) or (occasionally) a decrease.

If, after freezing and subsequent thawing, the ground porosity increases over the original (before freezing), the compressibility coefficient of the thawed soil (which equals the ratio of change in the soil's void ratio to the pressure exerted) increases substantially over the value for the same soil which has not been subjected to freezing; with a decrease in the porosity of the soil (when water is extruded from excessively moist soil under special conditions during freezing), we observe a decrease in compressibility (Table 34). But the most characteristic phenomenon, the one nearly always observed in nature, will be a substantial increase in the compressibility coefficient of the thawed soil (up to several times its original value), especially at the first steps of the load application on the ground; but after the thawed ground is compacted by a pressure of approximately 2 to 7 kg/cm^2, the changes in compressibility become insignificant.

According to experiments of the author* designed specifically to study compressibility variations of soils on freezing and subsequent thawing, an increase in the void ratios was invariably observed on freezing and a substantial

*See Chap. I, Ref. 13.

Table 34. Effects of freeze-thaw cycle on compressibility of cover loam (after Ye. P. Shusherina's experiments)

Load increment, kg/cm^2	Compressibility coefficient, cm^2/kg		Remarks
	Before freezing	After the freeze-thaw cycle	
0–0.5	0.040	0.120	Void ratio before freezing 0.71; after
0.5–1	0.02	0.050	freezing and thawing 0.75
1–2	0.025	0.040	
2–4	0.018	0.022	
4–8	0.021	0.014	
0–0.5	0.426	0.233	Void ratio before freezing 0.99; after
0.5–1	0.120	0.087	freezing and thawing 0.83
1–2	0.067	0.040	
2–4	0.030	0.037	

Table 35.   Changes in compressibility of soils on freezing and thawing

Designation of soil	On thawing at a pressure $p = 1$ kg/cm^2			At positive temperature and pressure $1 = k$ kg/cm^2		
	ϵ_0	ϵ_p	$a = \dfrac{\Delta\epsilon}{p}$	ϵ_0	ϵ_p	$a = \dfrac{\Delta\epsilon}{p}$
Sand (93% content of 1-0.25-mm fraction)	0.649	0.606	0.043	0.628	0.619	0.009
Silty loam (72% content of fraction from 0.05 to 0.005 mm)	0.710	0.619	0.091	0.703	0.688	0.030
Clay (88% content of 0.005 mm fraction)	1.160	0.901	0.259	1.128	1.047	0.081

Note: ϵ_0 is the initial void ratio, ϵ_p is the void ratio under pressure, and a is the compressibility coefficient.

increase in compressibility on thawing, as compared to the compressibilities of the same soils under the same compacting load when they had not been frozen (Table 35), which is explained by the drastic change in the constitution (texture) of the frozen soils on thawing.

The largest increase in porosity of a frozen ground and, consequently, its highest compressibility upon thawing are observed for soils frozen under open-system conditions with influx of water into the soils and formation of a stratified and reticular texture with substantial amounts of ice inclusions. Frozen soils of these types always settle substantially on thawing, and slump at correspondingly high moisture contents and pressures, while overcompacted soils with massive texture or individual mineral interlayers between compressed ice interlayers (for example, such as have been under glacier pressure in the past and froze in the compacted state under pressure) may swell, i.e., increase in volume, instead of being compacted, especially during the first stages of loading. But at a certain pressure (larger than the "swelling pressure") they will also be compacted and settle upon thawing.

Thawing is accompanied by a drastic change in the texture of the soils, and this affects not only their compressibility, but also, and primarily, their water permeability. Thus, according to Shusherina's experiments, the rate of compaction (which depends on the filtration capacity of the soils and their compressibility) increases by a factor of 7 to 10 for topsoil loam in the first loading stages after freezing and thawing. According to field observations at the Vorkuta Scientific Research Station of the NIIOSP, the filtration coefficient of thawed ground is many times larger than that of unfrozen ground of the same composition.

The results of laboratory and field studies of consolidation of thawing ground made over the past decade (1960-1970) by the author and his coworkers[3] have shown that the ice saturation of thawing ground is a very important indicator for evaluation of its process of settling as a function of time: the consolidation process will take totally different directions depending on whether the ground has a high ice content ($i_{vol} \geqslant 0.5$) and fluid or fluid-plastic during thawing or slightly icy ($i_{vol} \leqslant 0.25$) and semisolid or solid on thawing. While the theory of filtrational consolidation is applicable to ground with high ice content (in a special elaboration that takes account of the distinctive properties of the thawing soils), the creep theory will be preferentially applicable for ground with low ice content.

Beside this, a reticular and stratified texture forms in frozen ground when clayey soils of a fluid and fluid-plastic consistency are frozen; this endows them with special properties on thawing, among which the following may be regarded as most important (based on tests performed):

1) the change in the void ratio on compacting under load is much larger for frozen ground of this type than for unfrozen soils;

2) the water permeability of thawing ground with a high ice content is tens and hundreds of times greater than that of the same soils after thawing, but decreases with time, dependent upon the change in its porosity coefficient due to compression;

3) the pore pressure in thawing high ice content ground remains constant throughout the entire thawing process;

4) dense soils (of semisolid and solid consistency) have a definite structural strength on thawing, and the pore-pressure coefficient at the thaw boundary is always much smaller than unity.

Below we shall consider in detail the settling of frozen soils on thawing, its extent and the course that it takes in time based on results of direct tests performed to investigate the compression of frozen soils on thawing and their settling under conditions of localized load application when lateral expansion is impossible. In analyzing the results of direct studies of thaw settling of frozen ground, it is necessary to bear in mind the qualitative changes that take place in the soils on freezing and thawing, as described in the present section, and the fundamental physical premises that proceed both from the material set forth above and from special experiments to be described below.

3. Compression of Ground During Thawing

A compression-testing method for thawing ground (compaction under uniform load with lateral expansion of the ground prevented) was developed back in the 1930's. The author proposed a special device—a thermally nonconductive odometer (Fig. 106) that would ensure plane-parallel thawing of frozen ground specimens by heat from an element with a filtering floor both without a load on the soil and under uniform compacting loads up to 7-8 kg/cm^2. This device is now widely used in practice.

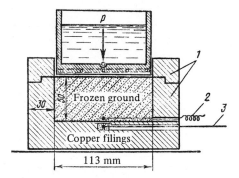

Fig. 106. Tsytovich odometer for testing of frozen ground for settling during thawing. (1) Thermally nonconductive plastic; (2) thermocouple; or (3) thermometer.

We should note that observance of plane-parallel thawing is mandatory, since otherwise the conditions of a one-dimensional problem will be violated (uniform compression with lateral expansion of the ground prevented). Unfortunately, this is overlooked by certain specialists who test frozen ground for thaw settling in metal rings, which, of course, creates conditions that violate the one-dimensional problem posed and leads to results that depend on the irregularity of thawing from the sides and top of the specimen, i.e., to a three-dimensional thaw case very difficult to analyze.

As we noted earlier, the object of the author's first experiments (1933) was to determine only the total stabilized settling of thawing soils that were not allowed to expand laterally. Naturally, in the case of compression of frozen ground during thawing with simultaneous compaction and observance of the boundary conditions described above, this settling followed exactly the relation known from general soil mechanics for the one-dimensional problem:

$$s = \frac{h}{1 + \epsilon_0} \ (\Delta\epsilon)_p \qquad\qquad (e_1)$$

where h is the total thaw depth (of the entire soil layer tested) in cm, ϵ_0 is the initial void ratio of the frozen soil, and $(\Delta\epsilon)_p$ is the compression change in the void ratio at a uniform pressure p [kg/cm^2] on the ground.

After complete thaw of the soil layer tested, h and ϵ_0 will be constant; the quantity $(\Delta\epsilon)_p$, which is a function not only of the properties of the soil, but also of the external compacting pressure p, remains as a variable on which the total stabilized settling of the thawing soil depends.

In further experimental studies of changes in void ratio $(\Delta\epsilon)_p$, special attention was also given to its values during thawing of frozen ground without load and under various degrees of loading.

Detailed compression experiments on thawing ground were carried out in the second stage of the studies (1937-1940) and included determination of the compaction deformations (and, from them, of the variation of the soil's void ratio), not only in the frozen state, but chiefly during the thawing process and thereafter upon subsequent loading of the thawed soil.

In addition, we made the first specifically designed studies of the increase in the settling of frozen ground (sandy and clayey) during the time in which the soils reached the stabilized state; we shall return to this somewhat farther on.

In confirmation of what was said concerning the importance to study the changes in the void ratio $\Delta\epsilon$ of ground under compression, Fig. 107 shows compression curves obtained in our experiments of positive ground temperatures (unfrozen soils) and on thawing of identical frozen soils under the same compacting load (on the thawing and unfrozen soils) of 1 kg/cm^2.

It is obvious from comparison of the compression curves for the unfrozen and frozen soils during thawing that the largest changes in the porosity coefficient occur during the process of thawing and that the quantity determining the settling of the thawed soils is the change $\Delta\epsilon$ in the void ratio during the thawing process.

Complete compression studies of thawing soils were carried out both by the author and his staff on a whole series of characteristic frozen and permanently frozen (structurally undisturbed) soils and by other investigators (M. N. Gol'dshteyn, V. F. Bakulin, V. F. Zhukov, M. F. Kiselev, V. P. Ushkalov, and others). Here we shall confine our statement to a description of the results of our most typical experiments with clean frozen sand (93 percent content of fraction from 1 to 0.25 mm) and disperse clay (50 percent content of

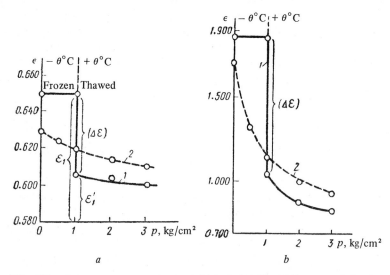

Fig. 107. Compression curves for sand (*a*) and clay (*b*). (1) Frozen and thawing; (2) unfrozen (at positive temperature).

Table 36. Compression changes of void ratios of thawing frozen sand and clay

Test conditions	Initial void ratio ϵ_0	Void ratio during thawing with simultaneous compaction ϵ_p	Change in void ratio $\Delta\epsilon = \epsilon_0 - \epsilon_p$
Sand			
Without load (at constant heater temperature $\theta = +15°C$)	0.6600	0.6325	0.0275
Under load $p = 1$ kg/cm^2 (heater temperature $\theta = +15°C$)	0.6490	0.6055	0.0435
Under load $p = 3$ kg/cm^2 (heater temperature $\theta = +15°C$)	0.6496	0.5757	0.0746
Clay			
Without load (at constant heater temperature $\theta = +15°C$)	1.166	0.937	0.229
Under load $p = 1$ kg/cm^2 (heater temperature $\theta = +15°C$)	1.160	0.901	0.259
Under load $p = 3$ kg/cm^2 (heater temperature $\theta = +15°C$)	1.161	0.840	0.321

fraction < 0.005 mm), conducted sufficiently rigorously, with observance of the boundary conditions of thawing and compaction.

The thaw compression of the frozen sand and clay was investigated at various external pressures: without load (under only the weight of the heating plunger, $p < 0.002$ kg/cm^2) and at two different compacting pressures ($p = 1$ kg/cm^2 and $p = 3$ kg/cm^2).

After thaw and stabilization of settling, the thawed soil was placed under several additional load steps, in which the compaction settling was also observed until full stabilization. For each experiment, we determined the characteristics of the initial and final physical states of the specimen (bulk density, moisture content, initial and final specimen heights), from which we calculated the corresponding soil void ratios. We obtained data for plotting the complete compression curves of the thawing soils, which were similar to those shown in Fig. 107, but for different values of the compacting load during thawing, for which we also determined the variations of the void ratio during thawing with simultaneous compacting under load, i.e., $(\Delta\epsilon)_p$. Table 36 gives averaged results from three determinations of the void ratio changes in frozen sand and clay thawed under compression and at various pressures.

Figure 108 shows graphs relating compressional load to changes of the void ratio with simultaneous densification during thaw of frozen ground.

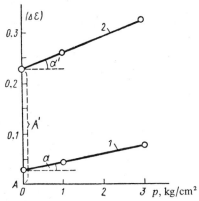

Fig. 108. Variation $\Delta\epsilon$ of void ratios of frozen soils on thawing as functions of external pressure p. (1) Sand; (2) clay.

According to Fig. 108, the change $\Delta\epsilon$ in the void ratios of frozen soils are strictly linear functions of the pressure p within the range studied (at least up to 3 kg/cm²). This extremely important relationship is also observed on direct comparison of the numerical data in Table 36.

Field studies (for example, those of V. P. Ushakov at Petrovsk-Zabaykalsk in 1944) have shown that a linear relation between the change in the void ratios of frozen ground on thawing and the external pressure will also hold for permafrost with natural undisturbed structure.

Using the terminology of Fig. 108, the equation of the straight line $\Delta\epsilon = f(p)$ is written in the form:

$$\Delta\epsilon = A + \tan\alpha p \qquad (e_2)$$

We denote

$$\tan\alpha = a \qquad (e_3)$$

The quantity $\tan\alpha = a$ might be called the coefficient compaction of thawing soils, since the compactness of the soils increases with increasing angle α. Then

$$\Delta\epsilon = A + ap \qquad (VI.1)$$

Relation (VI.1) indicates that the change in the void ratios of thawing frozen soils with simultaneous compaction consists of two parts: A, which does not depend on the external pressure, and ap, which is directly proportional to the external pressure within the range studied.

Equation (e_1) for the stabilized settling of the soil layer with lateral expansion prevented can be rewritten

$$\frac{\Delta\epsilon}{1 + \epsilon_0} = \frac{s}{h} = e_{th} \qquad (e_1')$$

The quantity $e_{th} = \Delta\epsilon/(1 + \epsilon_0)$ is the relative settling of the soil layer under continuous load (under the conditions of the one-dimensional problem). Substituting the $\Delta\epsilon$ relation from Eq. (VI.1) into Eq. (e_1'), we obtain

$$e_{th} = \frac{A}{1 + \epsilon_0} + \frac{a}{1 + \epsilon_0} p \qquad (e_4)$$

or (Fig. 109)

$$e_{th} = \overline{A} + \overline{a}p \qquad (VI.2)$$

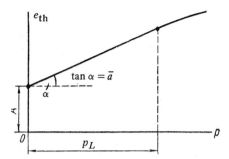

Fig. 109. Relative settling e_{th} of thawing frozen ground plotted against external pressure p.

where $A/(1 + \epsilon_0) = \bar{A}$ is the thawing coefficient (an abstract number, since e_{th} is an abstract number and ϵ_0, the initial void ratio of the frozen soil, is also an abstract number); $a/(1 + \epsilon_0) = \bar{a}$ is the coefficient of relative (referred only to the volume of the soil mineral particles) compaction of the frozen ground during thawing, which has the dimensions cm^2/kg.

Then the equation for the stabilized settling of the layer of thawed ground ($s = eh$) assumes the following form in the one-dimensional problem:

$$s_{th} = \bar{A}h + \bar{a}hp \tag{VI.3}$$

In this equation, which is the most important relationship for calculation of settling of thawing ground and was the basis for development of the methods now in use for prediction of the settling of foundations on thawing soils, the first term ($\bar{A}h$) is the so-called thaw settling, which does not depend on the external pressure, and the second term ($\bar{a}hp$) is the expression known from general soil mechanics for the stabilized settling of a soil layer under a continuous load and depends directly on the value of the compacting load and is known as the compaction settling during thawing.

In later experiments, the external pressures on thawing soils were higher than those considered above; it was found that the $e_{th} = f(p)$ relation is curvilinear only at pressures above 3 kg/cm^2 for clayey soils and 5 kg/cm^2 for coarse-skeleton and sandy soils.

It should be noted that the high pressures indicated here are very seldom permitted in practice for thawing soils; smaller pressures are encountered much more frequently, so that it is possible to assume a linear $e_{th} = f(p)$ relationship.

However, for cases in which high pressures are exerted on thawing ground, we proposed the following relation for the relative settling e_{th} of thawing frozen soils:

$$e_{th} = \bar{A} + \bar{a}p_i - \bar{b}\,(p_i - p_L)^m \tag{VI.4}$$

where $\bar{b}$ is a parameter that we called the work-hardening coefficient of the thawing soil, p_L is the pressure corresponding to the linear part of the equation $e_{th} = f(p)$, and m is an experimentally determined nonlinearity parameter.

To establish the most workable procedure for determining the design coefficients of the basic equation (VI.2) or (VI.3), namely the thawing coefficient $\bar{A}$ and the relative compaction coefficient on thawing $\bar{a}$, and to improve the general relation $e_{th} = f(p)$, Ye. P. Shusherina[4] designed special experiments whose results were as follows.

1. On the basis of rigorous statistical reduction of a significantly large number of experiments performed to investigate the settling of frozen soils during thawing, it was found that the linear relation $e_{th} = f(p)$ will very probably (with a statistical correlation coefficient $R = 1.00$ to 0.96) be fully valid for thawing soils up to pressures of 3–4 kg/cm².

Thus, the correlation coefficient $R = 1.000$ at pressures up to 4.0 kg/cm² for a frozen varved clay and $R = 0.960$ for a cover loam at the same pressure, but $R = 1.000$ also at lower pressures (up to about 2.5 kg/cm²); $R = 1.000$ for frozen sand at pressures up to 4–5 kg/cm². On the other hand, determination of the statistical "reality" (from the equation $R\sqrt{m-1}$ where m is the number of observations) of the established linear relationship invariably yielded values larger than three, indicating it to be fully legitimate.

For practical purposes, therefore, we may confine ourselves to the first two terms of Eq. (VI.4) in computing the relative settling of thawing ground and still obtain satisfactory accuracy.

2. In experimental determination of the thaw coefficient $\bar{A}$ and the coefficient $\bar{a}$ of compaction during thawing, it is necessary to remember that both of these quantities appear as unknowns in the same equation (VI.3), and that their rigorous determination requires settlement tests of two identical blocks of permafrost during thawing under different external pressures.[5] We then have

$$\left. \begin{array}{l} s_1 = \bar{A}h + \bar{a}hp_1 \\ s_2 = \bar{A}h + \bar{a}hp_2 \end{array} \right\}$$

from which we obtain at $p_2 > p_1$

$$\left. \begin{array}{l} a = \dfrac{s_2 - s_1}{h(p_2 - p_1)} \\ \bar{A} = e_1 - \bar{a}p_1 \end{array} \right\} \tag{VI.5}$$

But in practice it is sometimes very difficult to find even two identical specimens of frozen soil with natural structure. The author therefore proposed the following procedure, verified by Shusherina's experiments, for approximate determination of the coefficients $\bar{A}$ and $\bar{a}$ in this case. The frozen soil is first tested for settling while thawing under a very light pressure (for example, $p \leqslant 0.1$ kg/cm²). Then the second term in the right-hand side of Eq. (VI.3) can be neglected, i.e.,

$$s'_{0.1} \approx \bar{A} h$$

from which

$$\bar{A} \approx \frac{s'_{0.1}}{h} \qquad (VI.6)$$

where $s'_{0.1}$ is the settling of the thawing soil at insignificant external pressure.

On the other hand, the thawing coefficient $\bar{a}$ can be determined with accuracy sufficient for practical purposes from the results of a settling determination under the first load after thawing, when the soil is still uncompacted.

We then have the following expression for the relative thawing coefficient:

$$\bar{a} \approx \frac{\Delta s_1}{h p_1} \qquad (VI.7)$$

where Δs_1 is settling increase under the action of the first load step p_1 after thawing.

4. Determination of Total Stabilized Settling of Foundations on Thawing Soils

In this section, we shall discuss only the settling of frozen ground on thawing, i.e., the compacting deformation, and not plastic extrusion, such as may occur when high ice content ground sags with thawing under the weight of structures built upon them.

To prevent sag of thawing soils in the bases of structures, the pressure from the foundations of the structure should not exceed init p_{cr} [Eq. (IV.11), (IV.13), or (IV.14)] and must be substantially smaller than lim p_{cr} [Eqs. (IV.15) to (IV.18)] at the shear strength indicator values determined experimentally for thawing soils (from cohesion of thawed soil c_{th} and the angle of internal friction of the thawed soil ϕ_{th}).

It has been shown by a number of studies (those of Zakharov, Shusherina, the author, and others) that the shear strength (that basic strength indicator of soils) may decrease by as much as five times of that for soils after freezing and subsequent thawing, especially if they acquired a stratified and network texture on freezing, but there may be no decrease at all in the shear strength of frozen soils with massive texture.

In any event, the cohesion of the soils decreases sharply (by many times) on thawing, and $c_{th} \ll c_{fr}$ always, while the angle of internal friction may change only slightly upon thawing, especially for coarse-skeleton and sandy soils, with $\phi_{th} \leqslant \phi$.

In predicting foundations settling on thawing soils, therefore, it is first necessary to determine the safe pressure (from the values of c_{th} and ϕ_{th} found by direct experiments and from the equations for init p_{cr} and lim p_{cr}), at which the thawing soils in the bases of the foundations will be loaded by the foundations only in the compacting phase, i.e., they will be compacted but not squeezed out to the sides.

In the statement to follow, we shall also predict the settling of foundations on thawing soils based on the fundamental relation (VI.3) for the stabilized compaction settling of thawing frozen soils with simultaneous loading.

In addition to our procedure for computing the settling of thawing frozen soils, which we shall set forth below (and which has been improved by V. P. Ushkalov, Yu. K. Zaretskiy, and others), a contact-pressure method (G. I. Lapkin, 1939-1947)[6] and a method for determining the settling from the elementary physical characteristics of the thawing soils (M. F. Kiselev, 1952-1957)[7] have also been proposed. Without dwelling in detail on these studies, since a detailed contemporary analysis was published, we note only the following.

In support of the method of settling calculation that he later (1943-1947) referred to as the "contact pressure method," Lapkin proceeds from the experimental deformation curve of a permafrost specimen as recorded during thawing in an experiment at Noril'sk (Fig. 110); here the settling is determined from the expression

$$s = s_0 + s'_{zh} + s''_{zh} \qquad (f_1)$$

where s_0 is the "thaw settling" found by extrapolation (Fig. 110), s'_{zh} is the "pressure settling constant," which is determined graphically, and s''_{zh} are the "variable pressure settling."

Then, replacing the sum of the first two terms in the righthand side by the "conditional thaw settling" s_{01}, assuming in approximation that the curve of soil settling at pressures greater than the first load step (p_1 in Fig. 110) was a straight line and thenceforth referring to tan $\alpha = a$ as the "compression coefficient," Lapkin obtained the equation

$$s = s_{01} + ap_i \qquad (f_2)$$

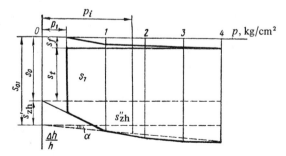

Fig. 110. Deformation graph of frozen ground specimen during thawing (according to G. I. Lapkin). s_f is the residual settling in the frozen state.

This expression is quite similar in form to our initial equation for the settling of a thawing soil layer under continuous load [Eq. (VI.3)], but the quantities that appear in it have a totally different meaning. Thus, the first term s_{01} in the right-hand side includes both the thaw settling and the constant part (for a given pressure) of the compression settling, which depends on the pressure; it will therefore be effective only at a pressure equal to or higher than the pressure at which the frozen ground was tested. In exactly the same way, $a = \tan \alpha$ is a characteristic of the straightened segment of the compression curve of the already thawed soil, and has a different meaning and a totally different value from the compacting coefficient $\bar{a}$ in Eq. (VI.3), which characterizes the rate of change of the porosity of the ground during thaw and not after thawing of the soil is complete. It is clear from the above that Lapkin's initial equation pertains only to the particular case of ground thawing at a certain definite pressure and subsequent load application only after it has been completely thawed. In addition, Lapkin recommends in a later paper (1947) that the design pressure p_i be determined on the basis of the so-called contact-pressure method, i.e., by regarding the underlying frozen ground as totally incompressible, a procedure that gives excessively large values for the compacting pressures (since it is known, for example, that plastically frozen soils are quite compressible at temperatures near $0°$), and assuming that the total compacting of the thawed ground layer occurs instantaneously.

However, field experiments and observations by V. P. Ushkalov[8] showed that even frozen sands continue to settle after thawing, and that the actual value of the compacting pressure is much smaller (sometimes by 40–60 percent) than for a layer of soil on an incompressible base and, in some cases, even smaller than for the homogeneous half-space. The latter was conclusively proven by experiments performed by the author with V. D. Ponomarev in a study of the pressure distribution in a soil layer under compaction by a local load (when its porosity increases with depth) by the electrohydrodynamic analogy (EGDA) method, the results of which appear in Fig. 111.

Thus, there is no longer any adequate basis for recommending the "contact-pressure method" for prediction of the settling of foundations on thawing soils.

Concerning the settling determination method of thawing ground from elementary physical characteristics, which is recommended by SNiP II-B.6-66 for preliminary calculations, we consider its use to be possible for these calculations only for thawing sandy soils; for thawing clays, however, determination of the settling requires knowledge of four characteristic moisture contents (W_i, W_d, W_p, and W_r) and the empirical coefficient K_d, the use of which does not always yield reliable results (especially at low pressures), making it difficult to use this method in practice.

Based on the above, we consider it possible to recommend Eq. (VI.3) as the basic starting relation in predicting the total stabilized compaction settling of thawing soils:

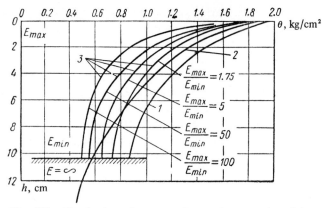

Fig. 111. Distribution of sum Θ of normal stresses in soil layer under local load. (1) On incompressible base; (2) in homogeneous half-space; (3) for variable strain modulus.

$$s_{th} = \bar{A}h + \bar{a}hp_e$$

where $\bar{A}h$ is the thaw settling, which is independent of the external pressure (and is composed of the volume change of ice on its conversion to water and the volume changes due to a certain closure of macrofissures in the ground during thawing), $\bar{a}hp_e$ is the compaction settling for not very large pressure variations (up to 3–4 kg/cm^2) and is directly proportional to the effective pressure p_e, and h is the thawing depth.

The effective pressure p_e can be assumed to be equal to:

a) in the case of one-dimensional problem (with a continuous load p [kg/cm^2] on the soil)

$$p_e = p + \gamma_{th}\,(H + z_i) \tag{VI.8}$$

where H is the depth of the foundation below grade, γ_{th} is the specific gravity of the thawed soil, and z_i is the distance from the base of the footing to the middle of the soil layer under consideration;

b) in the case of a local load without consideration of lateral expansion of the ground (for example, when s_{th} is calculated by layerwise summation)

$$p_e = \sigma_z + \gamma_{th}\,(H + z_i) \tag{VI.8'}$$

where σ_z is the vertical compressive stress at the middle of the soil layer considered.

In the case of stratified ground beds, relation (VI.3) assumes the form

$$s_{th} = \sum_1^n \bar{A}_i h_i + \sum_1^n \bar{a}_i h_i p_{ei} \tag{VI.9}$$

where the sign $\sum\limits_{1}^{n}$ is to be extended over all ground layers from the loaded surface to the total depth of thawing.

Expression (VI.9) for the thaw settling s_{th} will be fully applicable to the one-dimensional problem, for example, to compression of a ground layer under a continuous load (when the width of the load is several times the thickness of the layer) or compression of ground when lateral expansion is prevented.

Keeping in mind that the product of the thickness h_i of a single soil layer by the mean value of the compacting pressure p_i acting on this layer is the area under the compacting-pressure curve, we obtain

$$s_{th} = \sum_{1}^{n} \bar{A}_i h_i + \sum_{1}^{n} \bar{a}_i (F_{\gamma i} + F_{p i}) \qquad (VI.10)$$

where $F_{\gamma i}$ is the area under the compacting-pressure curve due to the weight of the soil (Fig. 112) and $F_{p i}$ is the corresponding area due to the external load.

We note that in approximate determinations of thaw settling of ground under the foundations of structures by layer-by-layer summation, which is permitted by the SNiP, only the maximum compressive stress $\sigma_{z i}$—the average for each layer of thawing soil—is taken into account in calculating F_p (see Fig. 112).

Relations (VI.9) and VI.10) will be valid for frozen soils without significant ice interlayers. In the presence of ice inclusions (several mm thick), the SNiP II-B.6-66 recommend that their settling be added separately, but with a reduction factor K_m that accounts for incomplete closure of cavities in the ground after the ice has melted out.

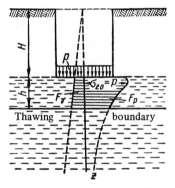

Fig. 112. Diagram showing distribution densifying pressures in a layer of thawing ground under the action of a local load.

The settling of individual foundations built on permafrost that thaw during operation of the buildings and structures is determined from the formula (SNiP II-B.6-66, Sec. 5.35):

$$s_{th} = \sum_{1}^{n} (1 - I_{ii}) e_{th} h_i + \sum_{1}^{n} K_m I_{ii} h_i \qquad (VI.11)$$

where I_{ii} is the ice content due to ice inclusions and is determined from the results of special experiments or calculated from relation (I.6) and e_{th} is the relative compression of the ground between the ice interlayers (notation ours.—N. Ts.); according to the SNiP, $e_{th} = \delta_i$.

The value of the coefficient K_m is chosen in accordance with the thickness Δ_{ice} of the ice inclusions: when $\Delta_{\text{ice}} = 1$ cm, $K_m = 0.4$; for $\Delta_{\text{ice}} > 1 < 3$ cm, $K_m = 0.6$, and for $\Delta_{\text{ice}} \geqslant 3$ cm, $K_m = 0.8$. These values of K_m are, of course, quite tentative, requiring further improvement.

However, it should be noted that because of the great practical complexity and possible significant errors associated both with the determination of the ice inclusion content I_i and with direct measurement of ice inclusions during study of the permafrost (as permitted by the SNiP), considerable difficulty may be encountered in predicting the settling of frozen ground when it thaws under the foundations of structures based on the relation (VI.11), especially for permafrost with reticulate texture and ice lens inclusions.

When only continuous interlayers of ice more than several mm in thickness extending over the area that influences settling of the foundations are present, and their dimensions can be gauged in surveys, it can be recommended that the settling determined from Eq. (VI.10) be increased by the settling due to complete thaw of the ice interlayers (of thickness Δ_{ice}), since experiments have shown that cavities in the ground formed by melting-out of ice interlayers are gradually completely closed, while small lenses and films of ice and the like are taken into account together in experimental determination of the thaw coefficient $\bar{A}_i$.

Then relation (VI.10) for the settling of the frozen soils on thawing under structures becomes

$$s'_{\text{th}} \approx \sum_1^n \bar{A}_i(h_i - \Delta_{\text{ice-}i}) + \sum_1^n \bar{a}_i(F_{\gamma i} + F_{p i}) + \sum_1^n \Delta_{\text{ice-}i} \qquad \text{(VI.12)}$$

It is, of course, possible to introduce a coefficient of incomplete closure of the soil cavities resulting from melting out of the ice, K_m, into the last term of this expression, but the author considers this unnecessary.

A method of the author, the equivalent-layer method, which is known from general soil mechanics, gives more accurate results in determination of the compaction settling of thawing soils in the presence of local loads (imposed by structural foundations). For foundations with specified base-area dimensions (determined from a preliminary calculation) and a known relative lateral deformation coefficient μ_0, the equivalent-layer coefficient $A\omega_m$ is determined (as for the average settling of rigid foundations) and used with the known foundation base width b to determine the depth h_e of the equivalent soil layer, whose settling is exactly equal to the compaction settling of the soils under the foundation. According to (V.23)

$$h_e = A\omega_m b$$

Certain values of the equivalent-layer coefficient $A\omega_m$ for $\mu_r = 0.3$ are given in Table 37.

Table 37. Values of the equivalent-layer coefficient $A\omega_m$

Value of μ_r	Ratio of length of footing area to its width, $\alpha = l/b$				
	1	2	3	5	10 and more
0.3	1.17	1.60	1.89	2.25	2.77

Note: At other values of μ_r, the value of $A\omega_m$ is multiplied by a coefficient M with the following values: $M = 0.83$ for $\mu_r = 0.1$, $M = 0.87$ for $\mu_r = 2$, and $M = 1.47$ for $\mu_r = 0.4$.

The average coefficient of compaction of a stratified bed of thawing ground (down to the maximum depth of the active compacting zone, which equals $H_a = 2h_e$) is determined from the equation given earlier (V.25), in which a_{ri} is replaced by $\bar{a}_i$:

$$\bar{a}_m = \frac{\sum h_i \bar{a}_i z_i}{2h_e^2} \tag{V.25'}$$

where z_i is the distance from the center of each layer to the depth $2h_e$.

Then the compaction settling of the thawed ground due solely to the external load p will be determined by Eq. (V.24):

$$s_p = h_e \bar{a}_m p_e \tag{V.24'}$$

Or, allowing for the thaw settling and the compaction settling of the thawing ground under its own weight γ_i, and applying the equivalent-layer method, we finally obtain

$$s_{th} = \sum_1^n \bar{A}_i(h_i - \Delta_{ice-i}) + \sum_1^n \bar{a}_i F_{\gamma i} + \frac{p}{2h_e}\sum_1^n \bar{a}_i z_i h_i + \sum_1^n \Delta_{ice-i} \tag{VI.13}$$

For vertically homogeneous frozen ground (without continuous ice layer), Eq. (V.13) assumes the simpler form

$$s_{th} \approx \bar{A}h + \frac{\bar{a}\gamma}{2}(h^2 + 2Hh) + \bar{a}hp\left(1 - \frac{h}{4h_e}\right) \tag{VI.14}$$

where H is the depth of the foundation placement.

We note that for thaw depths $h \geqslant 2h_e$, the last term in Eq. (VI.14) assumes the constant value $\bar{a}h_e p$.

Equations (VI.13) and (VI.14) take account of the largest number of factors influencing the total stabilized compaction settling of thawed soils under the foundations of structures.

We made laboratory control determinations of the settling of frozen ground on thawing for a homogeneous frozen sand (experiment 1) and sand in the presence of a thick (25-mm) interlayer of pure ice (experiments 2, 3, and 4) at a

Table 38. Values of calculated and actual settling of frozen soils on thawing
in experiments 2, 3, and 4

Experi- ment no.	Conditions of experiment	Thaw depth, mm	Settling, mm	
			Calculated	Actual (measured)
1	Homogeneous frozen sand	14	–	3.48
2	Frozen sand with 25-mm ice interlayer at depth $b/2$	26	7.2	7.3
3	Same	34	15.2	14.7
4	Same	39	20.2	21.9

depth of 25 mm under a 50 × 50-mm plunger under a pressure of 2 kg/cm² at temperatures from + 10 to + 16°C. The coefficient of compaction upon thawing was determined in a compression test and found equal to $\bar{a} = 0.01$ cm² /kg; the thawing coefficient $\bar{A}$ was calculated from the results of experiment 1 and had the value $\bar{A} = 0.23$ for the sand and was assumed equal to $\bar{A} = 1$ for the pure ice.

The results of comparison of the calculated [from Eq. (VI.13) with consideration of the ice interlayer] and actually measured settling are given in Table 38.*

The resulting data indicate quite good agreement between the calculated and experimental data even in the case of conspicuous stratification in frozen ground (when they contain a thick ice interlayer).

We note in conclusion that the determination of foundation settling on frozen ground upon thawing set forth in the present section pertains, as we noted earlier, only to the total stabilized compaction settling of the ground under the foundations at the full thaw depth; determination of the time variation in settling, which is also important, requires a separate analysis.

5. Prediction of the Settling Course in Time of Thawing
of Coarse Skeleton Ground

The settling of thawing ground takes a different course in time as compared with the settling of unfrozen soils; firstly in that the settling of thawing ground continues after thawing is complete (especially in the case of frozen clayey soils) and, secondly, in that the nature of the time variations of settling depends not

---

*A detailed table of the settling depths of the plunger on the thawing stratified frozen soil in the described experiments is given in the book by N. A. Tsytovich, "Printsipy me- khaniki merzlykh gruntov" (Principles of Frozen Ground Mechanics), pp. 145–147, *Izd. Akad. Nauk SSSR*, 1952.

only on the properties of the thawing soils material, but also on the rate and conditions of thawing.

Beginning in 1938-1940, special experiments were performed to study the settling of thawing ground as a function of time under various thawing conditions. Without describing these experiments in detail,* we present only some of the results in the form of time graphs of settling, plotted as averages of several settling measurements made on ground during thawing with strict observance of parallelity of the thawing process (in odometers, see Fig. 106) of the ground and the boundary conditions of the one-dimensional problem.

Figure 113 shows a settling graph of a sand thawed at a constant odometer heater temperature ($\theta = $ const), as an average of five experiments (the standard deviation ranged up to 7.8 percent, indicating the difficulty of preparing identical specimens), and Fig. 114 shows a similar curve for a thawing clay as an average of three experiments (standard deviations up to 3.7 percent). All of the frozen ground specimens were originally $\sim$ 5.17 cm high, the external load was $p = 3$ kg/cm^2, and the thawing time was $t_{av} = 52$ min for the sand specimens and $t_{av} = 100$ min for the clay specimens.

The unbroken curves in Figs. 113 and 114 are the s_t settling curves, whose ordinates are proportional to $\sqrt{t}$, i.e., $s_t = \alpha\sqrt{t}$, which agree well with the average experimental data. This indicates that under the conditions considered, the settling is proportional to the thaw depth h both for frozen sand and for frozen

*See Chap. I, Refs. 1a, 1b, 13 and 14.

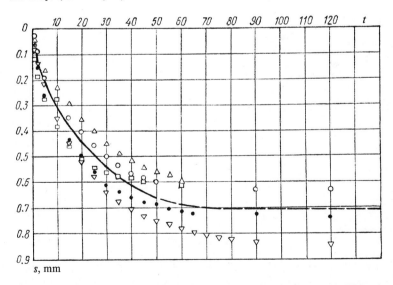

Fig. 113. Time variation of settling of a frozen sand during thaw with simultaneous compacting by a pressure $p = 3$ kg/cm^2. o—Experiment 1; ∇—experiment 2; △—experiment 3; □—experiment 4; ●—experiment 5; ——analytical curve.

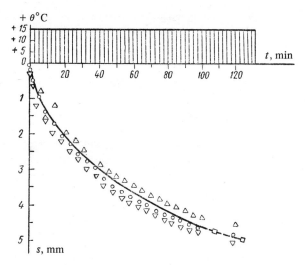

Fig. 114. Time variation of settling of frozen clay during thaw.
o–Experiment 1; ∇–experiment 2; ∆–experiment 3; □–points
of the analytical curve $s = \alpha\sqrt{t}$.

clay. As we know, this depth is determined from the well-known thermophysical solution for the one-dimensional problem by the expression

$$h = \beta_t \sqrt{t} \qquad\qquad (VI.15)$$

where β_t is a thermal coefficient.

In this expression, $\beta_t = $ const at $\theta = $ const; when the ground surface heater temperature varies differently, say, linearly ($\theta = Bt$), $\beta_t' = \alpha_1 \sqrt{t}$, and for parabolic variation ($\theta = Ct^2$), which is very closely similar to the temperature rise at the base of a foundation as it heats frozen ground, $\beta_t'' = \alpha_2 t$. Then the settling of the frozen ground on thawing will also depend linearly on time in the case of linear heating-temperature variation, since $h = \beta_t \sqrt{t}$, or, substituting $\beta_t' = \alpha_1 \sqrt{t}$, we obtain

$$h' = \alpha_1 t \qquad\qquad (VI.15')$$

On the other hand, if the heating temperature of the ground varies parabolically, the settling curve (whose ordinates are proportional to h) will be curvilinear, since with $\beta_t'' = \alpha_2 t$ we obtain

$$h'' = \alpha_2 t^{3/2} \qquad\qquad (VI.15'')$$

These last relationships (VI.15′) and (VI.15″) are fully confirmed by direct experiments, the results of which appear in Figs. 115 and 116 for a thawing sand with the soil-surface heater temperature varied during a certain time t (increasing linearly in Fig. 115 to $t' = 30$ min and curvilinearly in Fig. 116 up to $t' = 50$ min).

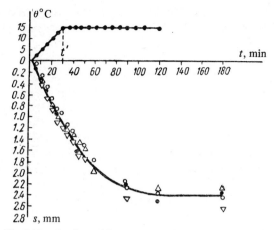

Fig. 115. Settling of frozen sand in time during linear increase of heater temperature to $t = 30$ min. ○—Experiment 1; ▽—experiment 2; △—experiment 3; ●—experiment 4; ——averaged curve.

It follows from the above theoretical arguments and results of direct experiments and from observations in nature that the contours of the time curves of the settling of frozen ground during thawing and compacting depend not only on the physical properties of the thawing ground and the magnitude of the external load, but also on the temperature regime in the heated soil surface.

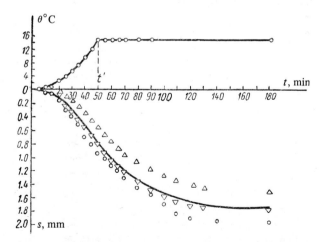

Fig. 116. Settling of frozen sand (1–0.25 mm) on thawing with simultaneous compaction as the heater temperature θ varies curvilinear up to $t = 50$ min. ○—Experiment 1; ▽—experiment 2; △—experiment 3; ——averaged curve.

Turning to calculation of the settling of thawing coarse-skeleton and sandy (water-permeable) soils as a function of time, we may assume that the settling during a time t is proportional to the depth of thawing h of frozen coarse-skeleton ground and will stabilize during the thawing process, although (strictly speaking) it will increase to some degree (sometimes very slightly) even after thawing, an effect that can be neglected in practical calculations.

Thus, the settling curve of the thawing sand (see Fig. 113) indicates that at $t = 50$ min, i.e., after complete thawing of the sand, the settling increases (though only slightly, by no more than about 10 percent), presumably as a result of creep of the mineral skeleton of the frozen sand. In that case, the equations derived earlier for the stabilized settling of thawing soils will also be valid for prediction of the settling of coarse-skeleton frozen soils as a function of time on thawing with simultaneous compacting by a constant load, requiring only substitution of the time-variable (determined by a thermal-engineering calculation) thawing depth h_t for the constant total-thawing depth h. For example, relation (VI.14) then assumes the following form for the settling of foundations on thawing homogeneous coarse-skeleton soils:

$$s_t = \bar{A}h_t + \frac{\bar{a}\gamma}{2}\,(h_t^2 + 2Hh_t) + \bar{a}h_t p\left(1 - \frac{h_t}{4h_e}\right) \qquad (\text{VI.14}')$$

In the other settlement formulas for foundations on thawing water-permeable soils (sandy, coarse-skeleton, pebbly, etc.) e.g., in Eq. (VI.9) or (VI.13), it is necessary to extend the summation sign from zero to the thawing depth h_t.

We note that according to SNiP II-B.6-66, it is permissible in preliminary calculations to determine the settling of thawing coarse-skeleton soils from the total relative strain* e_i:

$$s_t \approx \sum_0^{h_t} e_i h_i \qquad (\text{VI.16})$$

Here e_i is determined from test-load results or is calculated in approximation from N. F. Kiselev's equation

$$e_i = \frac{\gamma_{(sk)t} - \gamma_{(sk)f}}{\gamma_{(sk)t}} \qquad (\text{VI.17})$$

where $\gamma_{(sk)t}$ is the bulk weight of the skeleton of the thawed soil (which may be assumed equal to the bulk weight of the skeleton of the dry soil at maximum density), $\gamma_{(sk)f}$ is the bulk weight of the skeleton of the frozen soil $[\gamma_{(sk)f} = \gamma_f/(1 + W_d)]$, where γ_f is the bulk weight of the frozen soil).

Example 9. Determine the settling of the foundation of a steam boiler on permanently frozen sand inside a heated building.

*The relative deformation is denoted by δ_i in the SNiP.

Given: the base dimensions of the foundation area $l = 8$ m, $b = 4$ m; the pressure on the soil $p = 1.5$ kg/cm^2 $= 15$ tons/m^2; depth of foundation placement $H = 2$ m. Properties of the thawing sand: $\gamma = 1.85$ tons/m^3, thaw coefficient $A = 0.015$, coefficient of compacting during thawing $a = 0.008$ m^2/ton; thermal coefficient (found from solution of the thermal-engineering thawing problem) $\beta_t = 5$ m $\times$ yr$^{1/2}$.

Since the foundation is placed in the middle of the heated building, it can be assumed within certain limits that the thawing depth will be

$$h_t = \beta_t \sqrt{t}$$

We determined the settling of the foundation for a series of values of t, e.g., for $t = 0.3$ year, $t = 1$ year, $t = 2$ years, and $t = 5$ years.

We use Eq. (V.23) to determine the depth h_e of the equivalent layer for the sides ratio of the footing-base area, $\alpha = l/b = 8/4 = 2$, and the relative lateral expansion coefficient for sand, $\mu_r = 0.2$.

From Table 37 for $\mu_r = 0.2$ with the coefficient $M = 0.87$ and $\alpha = 2$ for the average settling of the foundation

$$A\omega_m = 1.60 \times 0.87 = 1.39$$

Then the thickness of the equivalent ground layer is

$$h_e = A\omega_m B = 1.39 \times 4 = 5.56 \text{ m} \qquad \text{and} \qquad 2h_e = 11.12 \text{ m}$$

We determine, for example, the settling after 0.3 year:

$$\text{at } t = 0.3 \text{ year, } h_t = 5\sqrt{3} = 2.74 \text{ m}$$

Then the settling of the foundation [according to Eq. (VI.14′)] is

$$s_{0.3} = 0.015 \times 2.74 + \frac{0.0008 \times 1.85}{2} (2.74^2 + 2 \times 2.74 \times 2) + 0.0008$$

$$\times 15 \times 2.74 \left(1 - \frac{2.74}{4 \times 5.56} \right) = 0.041$$

$$+ 0.0137 + 0.0288 = 0.0836 \text{ m} \approx 8.4 \text{ cm}$$

At $t = 1$ year, we obtain $h_t = 5\sqrt{1} = 5$ meters and

$$s_1 = 0.015 \times 5 + \frac{0.0008 \times 1.85}{1}$$

$$(5^2 + 5 \times 2 \times 2) + 0.0008 \times 15 \times 5$$

$$\left(1 - \frac{5}{22.24} \right) = 0.1548 \text{ m} \approx 15.5 \text{ cm}$$

In exactly the same way, at $t = 2$ years, $s_2 = 22.2$ cm and at $t = 5$ years, $s_5 = 35.9$ cm.

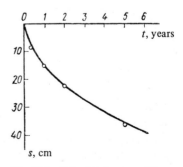

Fig. 117. Calculated curve of foundation settling on thawing sand.

The data obtained in the above example are used to plot a settling graph for the foundation of the steam boiler on thawing sandy permafrost over 5 years (Fig. 117).

We note that when foundations are built on frozen clay, calculation of their settling in time as a result of thawing will be much more complicated, since it is necessary to consider the consolidation (compacting in time) of clay in the thawing process and an additional compaction after complete thawing.

6. Predicting the Course of Settling in Time of Thawing Disperse (Clayey) Soils

In computing the settling of thawing clayey soils as a function of time, it is necessary to consider the possibility of incomplete consolidation during the process of thawing with simultaneous loading, since the rate of compaction under an external load and under the ground's own weight may be lower than the thaw rate for soils of this type. Depending on the water permeability of the thawing ground, the time to its complete compaction may exceed the thaw time, and the shear strength of the ground in the lower part of the thawed zone will be lower, since cohesion drops to the insignificant value c_{th} on thawing of the soils, while the friction resistance drops to values equal $\tan \phi_{th}(p - u)$, where u is the pressure developed in the pore water (and not in the mineral skeleton of the soil), the so-called pore pressure.

All of these factors create a quick-thaw hazard, and may cause instability of the thawing and thawed ground strata.

Since permafrost of natural bedding usually has pores filled with ice and, on thawing, saturated with water, we shall discuss below only the thaw consolidation of disperse soils that have low water permeability but are saturated with water.

The first studies of the consolidation of thawing disperse (clayey) soils with simultaneous compaction by a load were conducted back in 1938-1940,[9] and an engineering method was developed in the same period for prediction of settling of frozen ground as a function of time; it was based on the assumption of a linear vertical distribution of the compacting pressures through the thawing ground layer during the entire thaw process, for both the one-dimensional and three-dimensional problems (by reducing the solution of the spatial problem to an equivalent one-dimensional solution), with adoption of the postulate that the theory of filtrational consolidation could be applied to thawing water-saturated soils.

Later (in 1955-1965), extensive experimental studies of the consolidation of high ice content ground[10] during thawing with simultaneous compaction were organized under the author's supervision, with strict observance of boundary conditions, with measurement of pore pressures, relative settling, etc.

Certain results of these experimental studies are represented graphically in Fig. 118.

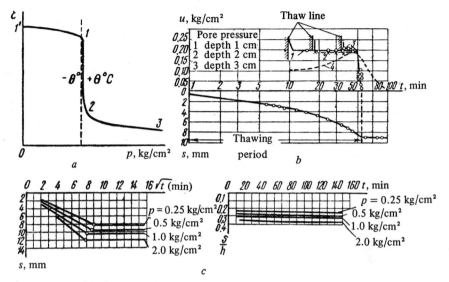

Fig. 118. Experimental data on investigation of the consolidation of high ice content frozen soils during thaw. (*a*) Compression curve; (*b*) pore-pressure curves (1, 2, and 3 for thawing; 4 for unfrozen soil); (*c*) settling vs. time (○—complete thawing).

These detailed experiments made it possible to establish a number of premises that were basic to further development of the prediction theory of the settling of thawing disperse soils in time.

1) At least three characteristic segments should be distinguished on the compression graph of thawing ground (Fig. 118*a*): segment 1′–1 compacting of the soil in the frozen state due to migrational-viscous deformation; segment 1–2 with a sharp change in the structure of the ground during the process of thawing, compaction, and filtration (for high ice content ground) extrusion of water and, finally, segment 2–3 with additional post-thaw compaction due to residual-filtrational consolidation and creep of the mineral skeleton of the thawed soil.

2) The assumed validity of the filtrational-consolidation theory for description of settling of high ice content ground during thawing was confirmed.

3) Also confirmed was the proportionality of the depth of thawing of high ice content ground and its settling up to the time of complete thaw (in the one-dimensional problem) to the square root of time.

4) It was established that the consolidation of thawing high ice content ground proceeds, practically, at a constant pore pressure u (Fig. 118*b*), although its initial value represents a certain fraction of the external compacting pressure.

5) Throughout the entire time of thawing, the relative settling of the thawing ground layer remains constant (Fig. 118*b*, inset at right); after thaw, however, the settling rate decreases and the thawed soil acquires a post-thaw compaction increment.

Then, on the basis of experimental results, a number of analytical solutions were proposed for the problem of consolidation of thawing soils.

On the basis of the assumed linear vertical distribution of the pressure in the thawing soil, an attempt was made to solve the complete differential equation of the one-dimensional problem of filtration consolidation by introducing a computing scheme with two moving boundaries, which made it possible to determine analytically the fraction of the total pressure to which the pore water in the thawing zone of the ground was subject.* However, the resulting solution was found to be highly complex and to require experimental determination of many characteristics of the thawing ground.

A solution to the same problem, but with consideration of the variability of the filtration and compressibility coefficients, was formulated by G. M. Fel'd-man,[11] but also in a highly complex form that requires further development.

An interesting approximate solution of the three-dimensional consolidation problem for thawing ground with consideration of only normal compressive stresses in the soil was developed on the basis of the theory of linearly deformable bodies by M. V. Malyshev,[12] who prepared special graphs of the rather complex functions appearing in the solution in order to facilitate calculations.

Based on a more natural formulation of the one-dimensional problem of the filtration theory of consolidation of thawing soils, i.e., with the variation of soil moisture content on displacement of the thaw front by an amount Δh equal to the amount of water filtered out across the thaw boundary during the time Δt, Yu. K. Zaretskiy[13] obtained a rigorous though complex solution of the consolidation differential equation, but with tabulation of the functions composing the solution, making it possible to use very simple equations to compute the settling of frozen ground during thaw with simultaneous compacting by a load and during post-thaw compaction of the soils.

We shall now dwell on the engineering solution of the one-dimensional and three-dimensional problems and more rigorous solutions of the same problems, considering also their use in practice.

The engineering solution of problems of time variation of foundation settling on icy thawing soils with low water permeability is based, as we noted above, on the approximate assumption that the compacting pressures in the skeleton of the soil are linear over the thaw depth for any time interval during the thawing process.

Under the action of a continuous uniformly distributed load, p_1 (from the weight of the soil layers above the bottom of the foundation and the weight γz of the thawing soil) will be represented in the general case of the vertical distribution graph of the compacting pressures by a trapezoid *abcd* (Fig. 119). The case in which a local uniformly distributed load also acts on the foundation of the structure also produces the same kind of trapezoidal compacting-pressure

*See Ref. 3, this chapter.

diagram, since, according to the known principles of general soil mechanics, this diagram can be assumed triangular in outline in the prediction of foundation settling with time (with accuracy sufficient for practical purposes), with the base of the triangle equal to the specific pressure p_2 exerted on the soil by the structure, and its vertex at the depth $2h_e$. Then the overall compacting-pressure curve (*aekl* in Fig. 119) will have one slope down to $2h_e$ and a different slope at greater depths. The total compacting pressure at any depth $z < 2h_e$ (where h_e is the thickness of the equivalent soil layer according to Tsytovich) can be expressed by the equation

$$p_{z_i} = p + \gamma' z_i$$

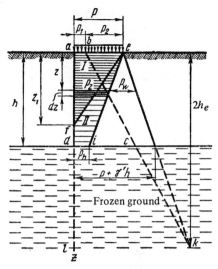

Fig. 119. Calculated graphs of compacting pressure in thawing soil beds under foundations.

where $p = p_1 + p_2$, $\gamma' = \gamma - (p_2/2h_e)$ is the angular coefficient of line *ek* (Fig. 119).

The volume of the water removed from the thawed soil layer (disregarding the small decrease in the height of the ice interlayers on transformation to water), referred to the unit area, will be equal to the settling of the soil, i.e.,

$$Q = \bar{A}h + \bar{a}hp \tag{g_1}$$

In general form, the volume of water per unit time per unit area will be

$$Q_z = \int_z^{z_1} \bar{A}\,dz + \bar{a}\int_z^{z_1} \frac{\partial p_z}{\partial t}\,dz \tag{g_2}$$

Assuming further a linear vertical distribution of the compacting pressures p_z (which, compared to the exact solution of K. Terzaghi for unfrozen soils, gives acceptable results for up to 75 percent of compacting), we have two basic cases:

I—rapid thawing of clayey ground, when compacting does not reach the thaw depth limit (see Fig. 119, graph I);

II—Sufficiently slow thawing, when compacting reaches the thaw depth but is incomplete (Fig. 119, graph II).

Case I. The settling will be proportional at any moment of time to the area *aef* of the compacting-pressure graph, which can be found if we know the time-varying quantity z_1:

$$p_z = p - p\frac{z}{z_1} \tag{g_3}$$

Differentiating this expression with respect to z_1, we obtain

$$\frac{dp_z}{dz_1} = p\,\frac{z}{z_1^2} \tag{g_4}$$

According to Darcy's water-permeability law

$$\frac{k\,(p + \gamma z_1)}{\gamma_w}\,t = \int_0^{z_1} Q_z dz \tag{g_5}$$

Using expressions (g_2) and (g_5), we obtain

$$\frac{k\,(p + \gamma z_1)}{\gamma_w}\,t = \int_0^{z_1} dz \int_z^{z_1} \overline{A} dz + \overline{a} p \int_0^{z_1} dz \int_z^{z_1} z dz \int_z^{z_1} \frac{dz_1}{z_1^2} \tag{g_6}$$

where k is the filtration coefficient of the thawed soil and γ_w is the specific gravity of water.

Integrating,

$$\frac{kt\,(p + \gamma z_1)}{\gamma_w} = \frac{\overline{A} z_1^2}{2} + \frac{\overline{a} p z_1^2}{6} \tag{g_7}$$

Solving this equation for the compacting depth z_1, we obtain

$$z_1 = \frac{B\gamma}{2} + \sqrt{\frac{B^2 \gamma^2}{4} + Bp} \tag{VI.18}$$

where

$$B = \frac{6kt}{\gamma_w (3\overline{A} + \overline{a} p)} \tag{g_8}$$

Equation (VI.18) determines the height of the compacting-pressure triangle for the pressures corresponding to time t; this height cannot exceed the thaw depth, i.e.,

$$\max z_1 = h \tag{g_9}$$

It is Eq. (g_9) that will be the condition for validity of Eq. (VI.18) for determination of the compacting depth of the soil during thawing.

Substituting the value of $\max z_1$ into Eq. (g_7) and solving it for the filtration coefficient of the thawed ground, which is denoted by k_∇ for the extreme case under consideration here. we obtain

$$k_\nabla = \frac{(3\overline{A} + \overline{a} p)\,\gamma_w h^2}{6t\,(p + \gamma h)} \tag{VI.19}$$

When

$$k \leqslant k_\nabla \tag{VI.20}$$

we have compacting case I; case II obtains when $k > k_\nabla$.

The settling s_t for the given time t in the present case will be determined on the assumption that the compacting settling of the ground equals the area of the graph of the compacting pressures transmitted to the soil skeleton (p_z) multiplied by the relative compressibility (densification) coefficient of the soil $(\bar{a})$:

$$s_{tI} = \bar{A}h + \frac{\bar{a}pz_1}{2} \qquad (VI.21)$$

Together with Eq. (VI.18) for z_1, relation (VI.21) determines the settling during time t for our case I.

Case II. Ground compaction during thaw time t may reach the base of the thawed layer during slow thawing, but will be incomplete. Then the diagram of the compacting pressures p_z will be the trapezoid *aeid* (Fig. 119).

The pressure in the ground skeleton at any depth $z < h$ will be

$$p_z = p - \frac{z}{h}(p - p_h) \qquad (h_1)$$

where p_h is the pressure in the soil skeleton at the thawing limit. Differentiating expression (h_1) and assuming p_h and h to be variable, we obtain

$$dp_z = \frac{z}{h^2}(p - p_h)\,dh + \frac{z}{h}\,dp_h \qquad (h_2)$$

The discharge of melt water pressed out of the pores of the soil is

$$Q_z = \int_z^h \bar{A}\,dz + \bar{a}\int_z^h \frac{z}{h^2}(p - p_h)\,dh + \bar{a}\int_z^h \frac{z}{h}\,dp_h \qquad (h_3)$$

and since, according to Darcy,

$$\frac{kt}{\gamma_w}(p - p_h + \gamma h) = \int_0^h Q_z\,dz \qquad (h_4)$$

substitution of the value of Q_z from (h_3) yields

$$\frac{kt}{\gamma_w}(p - p_h + \gamma h) = \bar{A}\int_0^h dz \int_z^h dz + \bar{a}(p - p_h)\int_0^h dz \int_z^h z\,dz \int_z^h \frac{dh}{h^2}$$
$$+ \frac{\bar{a}}{h}\int_0^h dz \int_z^h z\,dz \int_0^{p_h} dp_h \qquad (h_5)$$

Integrating the right-hand side of this equation and substituting the limits, we obtain after some manipulation

$$-\frac{3kt}{\gamma_w \bar{a}h^2} + \frac{1}{2} = \frac{3kt\gamma h}{\gamma_w \bar{a}h^2(p - p_h)} - \frac{3\bar{A}}{2\bar{a}(p - p_h)} + \ln\left(\frac{p - p_h}{p}\right) \qquad (VI.22)$$

Denoting for brevity

$$\frac{p - p_h}{p} = X \qquad (VI.23)$$

we finally obtain an expression for X in closed form:

$$\frac{\gamma_w \bar{a} h^2 - 6kt}{2\gamma_w \bar{a} h^2} = \frac{6kt\gamma h - 3\bar{A}\gamma_w h^2}{2\gamma_w \bar{a} h^2 p X} + \ln X \qquad (VI.24)$$

This equation is solved quite easily by successive approximations.

Knowing X from expression (VI.23), we find the pressure p_h in the soil skeleton at the thawing depth:

$$p_h = p(1 - X) \qquad (VI.23')$$

Then, as above, we obtain the compaction settling s_{tII} corresponding to given time t as the product of the area under the compacting-pressure graph by the compaction coefficient for thawing:

$$s_{tII} = \bar{A}h + \frac{p + p_h}{2} \bar{a}h \qquad (VI.25)$$

At $t = \infty$, we have $p_h = p$ for ground with very high water permeability (coarse-skeleton, sandy, and similar soils); as would be expected, Eq. (VI.24) reverts to Eq. (VI.3):

$$s_t = \bar{A}h + \bar{a}hp$$

A particular case. If we put $\bar{A} = 0$ and $\gamma h = 0$ in Eq. (VI.22), we have the case of compacting of ordinary unfrozen soil under the action of a local uniformly distributed load:

$$- \frac{3kt}{\gamma_w \bar{a} h^2} + \frac{1}{2} = \ln \frac{p - p_h}{p} \qquad (h_6)$$

from which, remembering that according to Eq. (h_1) $p_z = p - z/h\,(p - p_h)$, we obtain

$$p_z = p \left(1 - \frac{z}{h} e^{-\frac{3kt}{\gamma_w \bar{a} h^2} + \frac{1}{2}} \right) \qquad (h_7)$$

Expression (h_7) is exactly the same as Terzaghi's equation[14] (he omits $\gamma_w = 1$ ton/m^3), which was derived only for soils at positive temperature, while our equation (VI.22) gives the general solution for both frozen and unfrozen ground.

Example 10. Determine the settling of a layer of thawing clayey soil one year after the thawing begins, given: the thermal coefficient $\beta_t = 4.0$ m $\times$ yr$^{-1/2}$, filtration coefficient $k = 0.02$ m/yr, specific load on the soil $p = 2$ kg/cm$^2 = 20$ tons/m^2, bulk density of thawed soil with consideration of buoyant effect of water $\gamma = 1$ ton/m^3, coefficient of relative compacting of soil during thawing $a = 0.001$ m^2/ton, thawing coefficient of soil $A = 0.02$.

We determine the depth of thawing from the equation

$$h = \beta_t \sqrt{t} = 4 \text{ m}$$

To establish which equations should then be used in calculating the settling of the thawing clayey soil, we find the criterion k_∇ of applicability of cases I and II and compare the resulting value with the filtration coefficient of the soil. According to Eq. (VI.19)

$$k_\nabla = \frac{(3\bar{A} + \bar{a}p)\gamma_w h^2}{6(p + \gamma h)t} \approx 0.01 \text{ m/yr}$$

Since the filtration coefficient $k = 0.02 > k_\nabla = 0.01$ m/yr, we have case II. Substituting the numerical values of the quantities in Eq. (VI.24), we obtain

$$- 3.27 + \frac{0.48}{X} - \ln X = 0$$

Solving the resulting equation by successive approximations, we find $X = 0.26$.

According to Eq. (VI.23′)

$$p_h = p(1 - X) = 0.74p = 0.74 \times 20 = 14.8 \text{ tons/m}^2$$

Substituting the values thus found into Eq. (VI.25), we finally obtain

$$s_{t=1} = \bar{A}h + \frac{(p + p_h)}{2}\bar{a}h = 0.02 \times 4 + \frac{20 + 14.8}{2} \, 0.001 \times 4 = 0.15 \text{ m}$$

Thus, the settling of the layer of frozen clayey soil one year after freezing will be 15 cm for the conditions considered.

Zaretskiy[15] obtained *a rigorous solution* of the differential equation of the one-dimensional consolidation theory problem for thawing ground with high ice content and low water permeability, assuming on the basis of experimental data that the compressional changes in porosity during thawing of clayey ground are determined by the expression $\epsilon_0 - \epsilon_i - f_1(\sigma_z - \gamma_w H)$, and that the filtration coefficient $k = f_2(\sigma_z - \gamma_w H)$ (where σ_z is the normal compressive stress and H is the pressure head); assuming further that the change in moisture content (porosity) at the thawing front is equal to the amount of water filtered off, which is mathematically expressed in the form:

$$f_1(\sigma_z - \gamma_w H)_{z = h} = \frac{\gamma_{sp}}{\gamma_{sk}} \left| \frac{f_2(\sigma_z - \gamma_w H)}{\left(\dfrac{dh}{dt}\right)} \times \frac{\partial H}{\partial z} \right|_{z = h} \quad (i_1)$$

obtained the solution of the differential equation of consolidation theory

$$\frac{\partial H}{\partial t} = c_v \frac{\partial^2 H}{\partial z^2} \quad (i_2)$$

where c_v is the coefficient of consolidation.

Having determined the pressure head function H from the resulting solution and putting the effective stress equal to $\sigma_{ef} = \sigma_z - \gamma_w H$, we have the following expression for the settling during thawing:

$$s_t = \int_0^h \sigma_{ef}(z,\ t)\,dz \tag{i_3}$$

Then the total compaction settling of the ground during thawing and the additional compaction after thaw will be determined by

$$s_t = s_{1t} + s_{2t} + s_{3t} \tag{VI.26}$$

where

$$s_{1t} = \bar{A} h_t \tag{VI.27}$$

is the thaw settling, h_t is the depth of thaw responding to time t, s_{2t} is the compaction settling of the ground during thawing due to the action of the external load p and the weight $\gamma' h_t$ of the soil (where γ' is the specific gravity of the soil with consideration of the buoyant action of the water), and s_{3t} is the subsequent compaction settling of the thawed soil.

The compaction settling s_{2t} during the thaw process will be determined by

$$s_{2t} = \bar{a}\left(\chi_1 h_t p + \chi_2 \frac{\gamma' h_t^2}{2}\right) \tag{VI.28}$$

where χ_1 and χ_2 are complex functions depending on the parameter

$$r_{th} = \frac{\beta_t}{2\sqrt{c_m}} \tag{VI.29}$$

where β_t is the thermal coefficient calculated for the design time t (by the thermal-engineering calculation), e.g., for the one-dimensional problem

$$\beta_t = \frac{h_t}{\sqrt{t}} \tag{VI.30}$$

c_m is the consolidation coefficient averaged over the depth of the thawed layer and is determined by the expression

$$c_m = \frac{k_m}{\gamma_w \bar{a}_m} \tag{VI.31}$$

k_m is the vertically averaged filtration coefficient of the soil and equals

$$k_m = \frac{\sum h_i}{\sum \dfrac{h_i}{k_i}} \tag{VI.32}$$

h_i are the thicknesses of the individual soil layers; $\bar{a}_m$ is the average value of the relative coefficient of compaction during thawing, which equals, for the

Table 39. Table of values of χ_1 and χ_2

$r_{th} = \dfrac{\beta_t}{2\sqrt{c_m}}$	χ_1	χ_2	$r_{th} = \dfrac{\beta_t}{2\sqrt{c_m}}$	χ_1	χ_2
0	1.0	1.0	1.5	0.38	0.22
0.1	0.99	0.98	2.0	0.28	0.14
0.3	0.93	0.90	2.5	0.22	0.10
0.5	0.84	0.72	3.0	0.19	0.07
0.7	0.73	0.57	5.0	0.11	0.03
0.9	0.61	0.45	10.0	0.06	0.007
1.0	0.56	0.40	∞	0	0

one-dimensional problem,

$$\bar{a}_m = \frac{\sum \bar{a}_i h_i}{\sum h_i} \tag{VI.33}$$

and can be determined from Eq. (V.25) in the case of the three-dimensional problem:

$$\bar{a}_m = \frac{\sum h_i \bar{a}_i z_i}{2h_e^2}$$

Values of the dimensionless coefficients χ_1 and χ_2 are given in Table 39.

The post compaction s_{3t} of the thawed soils is determined by

$$s_{3t} = s_{3\infty}^{(\gamma h)} U_3^{(\gamma h)} + s_{3\infty}^{(p)} U_3^{(p)} \tag{VI.34}$$

where the stabilized compaction settling of the thawed ground under its own weight is

$$s_{3\infty}^{(\gamma h)} = \frac{1}{2}\bar{a}(1 - \chi_2)\gamma' h_\infty^2 \tag{VI.35}$$

where h_∞ is the depth of the stabilized (maximum) thaw bowl of the ground, determined by the appropriate thermal-engineering calculation or from the diagram of SNiP II-B.6-66, and the stabilized settling due to continuous load p in the one-dimensional problem is given by

$$s_{3\infty}^{(p)} = \bar{a}_m(1 - \chi_1)h_\infty p \tag{VI.36}$$

The values of the degree of soil consolidation (after completion of thawing) under the weight of the soil $U_3^{\gamma h}$ and the action of a continuous load U_3^p are determined from Table 40, which was compiled for the parameters

$$r_{th} = \frac{\beta_t}{2\sqrt{c_m}} \tag{i_4}$$

and

$$N = \frac{\pi^2 c_m}{4h_\infty^2}(t - t_0)$$

Table 40. Degree of consolidation (residual compaction) of ground after thawing

N	Values of $U_3^{(\gamma h)}$ and $U_3^{(p)}$ at	Values of $U_3^{(p)}$ at		
	$0 \leqslant r_{th} \leqslant 0.3$	$r_{th} = 0.5$	$r_{th} = 1$	$r_{th} = 3$
0.1	0.082	0.084	0.090	0.230
0.2	0.161	0.165	0.174	0.321
0.3	0.238	0.242	0.252	0.393
0.4	0.309	0.313	0.323	0.454
0.5	0.374	0.379	0.387	0.507
0.6	0.434	0.437	0.445	0.555
0.7	0.488	0.490	0.498	0.597
0.8	0.536	0.539	0.546	0.636
1.0	0.620	0.622	0.628	0.702
1.5	0.770	0.771	0.774	0.819
2.0	0.860	0.861	0.863	0.890
2.5	0.915	0.916	0.917	0.933
3.0	0.949	0.949	0.950	0.960

where

$$\beta_t = \frac{h_\omega}{\sqrt{t_0}}$$

t_0 is the time to complete formation of the thaw bowl.

We have extended expression (VI.28) for the compaction settling of the soil in the thawing process under a continuous load (following the equivalent-layer method described previously) to the action of a local load imposed by structure foundations (three-dimensional problem) by reducing the compaction-pressure distribution diagram to an equivalent diagram with a base equal to the external load p from the foundations and with a height equal to twice the equivalent soil layer (i.e., $2h_e$).

Then the compacting pressure at any depth z_i will be determined by the expression

$$p_{z_i} = p + \gamma'' z \qquad (k_1)$$

where γ'' is the slope of the equivalent compacting-pressure diagram where

$$\gamma'' = \gamma' - \frac{p}{2h_e} \qquad (k_2)$$

γ' is the bulk density of the ground lightened by the weight of the displaced water, and p is the external load imposed by the structure foundations. The compaction settling of the frozen soil in the process of thawing with simultaneous action of the constant local load, i.e., with

$$p_{z_i} = p + \gamma' - \frac{p}{2h_e} \qquad (k_3)$$

will be equal to

$$s'_{2t} = \bar{a}\left(\chi_1 ph_t + \frac{1}{2}\chi_2\gamma'h_t^2 - \frac{1}{2}\chi_1\frac{ph_t^2}{2h_e}\right) \qquad (VI.37)$$

When the thawing depth of the frozen soil under the foundations $\geqslant 2h_e$, the compaction settling under the constant load from the foundations assumes a constant value $\bar{a}\chi_1 ph_e$. Then

$$s'_{2t} = \bar{a}\left(\frac{1}{2}\chi_2\gamma'h_t^2 + \chi_1 ph_e\right) \qquad (VI.37')$$

We note that for the middle of a building, it is possible to assume the thawing depth of the soils to be equal to

$$h_t = \beta_t \sqrt{t}$$

where β_t is the thermal coefficient determined by a thermal-engineering calculation.

Relations (VI.37) and (VI.37') can also be used in calculating the compaction settling of frozen ground under the foundations of structures in the thaw process.

Example 11. Determine the settling of the thawing ground layer (which was calculated earlier in Example 10 by the author's method) for the same data as before.

We calculate the settling during the time t from (VI.27) and (VI.28):

$$s_t = s_{1t} + s_{2t}$$

or

$$s_{t=1\text{ yr}} = \bar{A}h_t + \bar{a}\left(\chi_1 h_t p + \chi_2\frac{\gamma'h_t^2}{2}\right)$$

Given is: $A = 0.02$, $a = 0.001$ m^2/ton, $\gamma' = 1$ ton/m^3, $p = 20$ tons/m^2, $\beta_t = 4$ m $\times$ yr$^{-1/2}$, $t = 1$ year, and $k = 0.02$ m/year.

Substituting numerical values of the quantities in the above equation, we obtain $h_t = \beta_t\sqrt{t} = 4\sqrt{1} = 4$ meters,

$$c_m = \frac{k_m}{\gamma'\bar{a}} = \frac{0.020}{1 \times 0.001} = 20 \text{ m}^2/\text{yr} \qquad r_{\text{th}} = \frac{\beta_t}{2\sqrt{c_m}} = \frac{4}{2\sqrt{20}} \approx 0.5$$

Then from Table 39, $\chi_1 = 0.84$, $\chi_2 = 0.72$, and the settling of the frozen ground layer as it thaws during one year will be

$$s_{t=1\text{ yr}} = 0.02 \times 4 + 0.001\left(0.84 \times 4 \times 20 + 0.72 \times \frac{1 \times 16}{2}\right) = 0.153 \text{ m}$$

i.e., the settling in one year will be 15.3 cm.

It is important to note that in Example 10, which was computed by the author's approximate engineering method, the settling of the same frozen ground layer was found to be 15 cm after one year.

The above data indicate that the two methods discussed above for prediction of the settling of thawing soils as a function of time yield quite comparable results.

In the case of a local load (three-dimensional problem when a load is imposed on the ground by structure foundations), the magnitude and decay of the post-thaw compaction settling of the soil $s_{3t}^{(p)}$ in time is also computed by the equivalent-soil-layer method, assuming a triangular contour of the equivalent compaction-pressure diagram with a base equal to the external load p and a height equal to twice the equivalent soil layer $h_a = 2h_e$.

If we consider only the postcompaction consolidation due to filtration in the thawed ground (as, for example, is acceptable when the soils have small filtering capacity), we have for the settling under the effective load*

$$s_{3t}^{(p)} = \frac{h_a a'_{rm} p}{2} (1 - \chi_1)$$

$$\times \left\{ 1 - \frac{16}{\pi^2} \left[\left(1 - \frac{2}{\pi}\right) e^{-N} + \frac{1}{9} \left(1 + \frac{2}{3\pi}\right) e^{-9N} - \cdots \right] \right\} \qquad \text{(VI.38)}$$

where $N = (\pi^2 c_v / 4h_a^2) t'$;

$$t' = t - t_0 \qquad \text{and} \qquad c_v = \frac{k'_m}{a'_{rm} \gamma_w}$$

a'_{rm} and k'_m are the averaged relative-compressibility and filtration coefficients of the ground after complete thawing, and $h_a = 2h_e$ is the maximum thickness of the active compression zone.

But if the thawed soil is permeable enough (e.g., a fine sand, sandy loam, varved loam, etc.), the filtration consolidation after thaw will be of little significance, and the main settling magnitude would be due to attenuating creep. Then according to Eq. (V.15) we will have

$$\left[s_{3t}^{(p)} \right]' = \frac{h_a a''_{rm} p}{2} (1 - e^{-\delta' t}) \qquad \text{(VI.39)} \cdot$$

a relation that will be valid for the most part in cases of thawed clayey ground with a certain compactness (semisolid, stiff plastic, and others).

But for ground loosened by freezing and for uncompacted ground (sandy, clayey soils with high ice content, etc.), more applicable will be Eq. (V.17),

---

*See, e.g., Eq. (125) in N. A. Tsytovich's book "Mekhanika gruntov" (Kratkiy kurs) (Soil Mechanics [A Short Course]), Izd-vo "Vysshaya Shkola," 1968.

according to which

$$\left[s_{3t}^{(p)} \right]'' = \frac{h_a a_{r\infty} p}{2} \left(\frac{t}{T+t} \right)$$ (VI.40)

where δ and T are parameters of attenuating creep.

7. Settling of Bases on Prethawed Permanently Frozen Ground

In the preceding section, we discussed settling of frozen ground in the thawing process and settling due to further compaction of already thawed soils. But permafrost is sometimes prethawed to improve the structural properties of natural bases in permafrost regions, a practice that, firstly, reduces the total settling of the ground by the amount of the thaw settling and, secondly, produces a more compact state of the ground (under its own weight and sometimes also as a result of the use of special compacting and consolidation methods), so that the soils are less deformable in the thawed state. In the present section, we shall discuss the settling of bases on preliminary thawed permafrost and subsequent subjection to local loads.

To compute the settling of foundations and its time variation, it is necessary to obtain, by experiment, a working compression graph for thawed soils under their own weight alone (without an external load), so that full account can be taken of the looseness that usually appears during freezing of soils, especially with excessive ice segregation.

Experiments conducted in detail have shown that the change in the void ratio of thawed soils, a coefficient that is proportional to the compaction settling, depends nonlinearly on an increase in external pressure, even when it varies in a narrow range (Fig. 120).

Full account must be taken of the nonlinearity and the substantial variations of the void ratios of thawed soils in predicting the settling of foundations on prethawed soil bases.

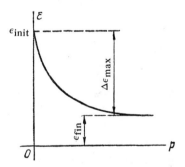

Fig. 120. Nonlinear dependence of void ratio change of thawed soils on external pressure.

The change in the void ratio of a soil thawed only under its own weight can be described by the exponential graph:

$$\Delta \epsilon_i = \Delta \epsilon_{max} \left(1 - e^{-a_c p_i} \right)$$ (VI.41)

where $\Delta \epsilon_{max} = \epsilon_{init} - \epsilon_{fin}$ is the difference between the initial and final (for the entire pressure range) void ratios, and a_c is the nonlinear compression coefficient (the parameter of the exponential curve), has the dimensions cm^2/kg, and can be determined from the rectified compression curve in

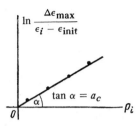

Fig. 121. Determination of nonlinear compression coefficient a_c of thawed soil.

semilogarithmic coordinates ($\ln \Delta\epsilon_{max} / \epsilon_i - \epsilon_{init}$ and p_i) as the slope of the straight line with respect to the pressure axis ($p = \gamma' h$), where $\gamma' h$ is the pressure due to the weight of the ground less the weight of the displaced water (Fig. 121).

In exactly the same way, the filtration coefficient k_f for thawed ground depends on the variations of its porosity coefficient, as expressed by the relation

$$k_f = k_0 \left(1 - \frac{\Delta\epsilon}{\Delta\epsilon_{max}} \right) \qquad \text{(VI.42)}$$

where k_0 is the initial filtration coefficient of the ground.

With nonlinear variation of the void ratio in accordance with (VI.41) and of the filtration coefficient according to (VI.42), the total stabilized settling $s_{t\infty}$ for thawed soils (which were previously loosened by freezing) is determined from a solution of the nonlinear problem of ground compaction under its own weight in accordance with the expression:[16]

$$s_{t\infty} = \bar{A}h_t + \frac{\Delta\epsilon_{max}}{(1 + \epsilon_m)} h_t \left(1 - \frac{1 - e^{-M}}{M} \right) \qquad \text{(VI.43)}$$

where $\bar{A}$ is the thaw coefficient of the ground, ϵ_m is the mean value of the initial void ratio of the thawed ground over the entire compression zone, h_t is the depth of the presumed preliminary thawing of the permafrost, and

$$M = a_c (\gamma' h_t) \qquad \text{(VI.44)}$$

We denote the multiplier in the right-hand side of Eq. (VI.43) by

$$D = 1 - \frac{1 - e^{-M}}{M} \qquad \text{(VI.45)}$$

and then the equation for the full settlement of the thawed ground under its own weight assumes the form

$$s_{t\infty} = \bar{A}h_t + \frac{\Delta\epsilon_{max}}{1 + \epsilon_m} h_t D \qquad \text{(VI.43')}$$

Values of D for various values of M are given in Table 41. Also included in the table are the dimensionless degrees of consolidation $U_\Delta^{(\gamma h)}$ of the thawed ground under its own weight, i.e.,

$$U_\Delta^{(\gamma h)} = \frac{s_t}{s_{t\infty}} \qquad \text{(VI.46)}$$

Then the compaction settling of the thawed ground for any time interval after thawing will be determined by

$$s_t = U_\Delta^{(\gamma h)} s_{t\infty} \qquad \text{(VI.46')}$$

Table 41. $U_\Delta^{(\gamma h)}$ values of consolidation degrees under own weight in preliminary thawed bases

N	Degree of consolidation $U_\Delta^{(\gamma h)}$ as a function of M												
	0.01	0.05	0.1	0.2	0.3	0.4	0.5	0.6	0.7	0.8	0.9	1.0	1.5
0.01	0.0079	0.0078	0.0076	0.0071	0.0065	0.0059	0.0052	0.0045	0.0037	0.0029	0.0020	0.0010	0
0.1	0.0806	0.0795	0.0779	0.0745	0.0708	0.0667	0.0623	0.0575	0.0523	0.0468	0.0408	0.0344	0
0.2	0.1607	0.1592	0.1571	0.1525	0.1474	0.1419	0.1359	0.1294	0.1223	0.1146	0.1064	0.0976	0.0433
0.3	0.2374	0.2359	0.2336	0.2288	0.2235	0.2178	0.2114	0.2046	0.1972	0.1892	0.1805	0.1712	0.1138
0.4	0.3087	0.3072	0.3050	0.3004	0.2954	0.2898	0.2838	0.2772	0.2700	0.2624	0.2540	0.2451	0.1897
0.5	0.3740	0.3726	0.3706	0.3663	0.3616	0.3564	0.3508	0.3447	0.3381	0.3309	0.3232	0.3149	0.2634
0.6	0.4333	0.4320	0.4302	0.4263	0.4220	0.4173	0.4121	0.4065	0.4005	0.3939	0.3867	0.3792	0.3321
0.7	0.4872	0.4860	0.4843	0.4808	0.4768	0.4726	0.4679	0.4628	0.4573	0.4513	0.4449	0.4380	0.3951
0.8	0.5359	0.5349	0.5334	0.5301	0.5266	0.5227	0.5184	0.5138	0.5088	0.5034	0.4976	0.4913	0.4524
0.9	0.5801	0.5791	0.5778	0.5748	0.5716	0.5681	0.5642	0.5601	0.5555	0.5506	0.5454	0.5397	0.5044
1.0	0.6200	0.6192	0.6179	0.6153	0.6124	0.6092	0.6057	0.6019	0.5978	0.5934	0.5886	0.5835	0.5516
2.0	0.8602	0.8599	0.8594	0.8585	0.8574	0.8562	0.8549	0.8536	0.8520	0.8504	0.8486	0.8468	0.8350
2.5	0.9152	0.9150	0.9147	0.9142	0.9135	0.9128	0.9120	0.9112	0.9102	0.9033	0.9082	0.9070	0.8999
3.0	0.9486	0.9484	0.9483	0.9479	0.9475	0.9471	0.9466	0.9461	0.9456	0.9450	0.9443	0.9436	0.9393
∞	1.0	1.0	1.0	1.0	1.0	1.0	1.0	1.0	1.0	1.0	1.0	1.0	1.0
D	0.0050	0.0246	0.0484	0.0936	0.1361	0.1758	0.2131	0.2480	0.2808	0.3117	0.3406	0.3679	0.4821

We find the degree of consolidation $U_\Delta^{(\gamma h)}$ as a function of

$$M = a_c \, (\gamma' h_t)$$

and

$$N = \frac{\pi^2 c_c}{4 h_e^2} \, t$$

with

$$c_c = \frac{k_0 (1 + \epsilon_m)}{\gamma_w \Delta\epsilon_{max} a_c}$$

$$(VI.47)$$

Calculating M and then c_c and specifying values of t, we determine N and use Table 41 to find the degree of consolidation $U_\Delta^{(\gamma h)}$ and the value of D, which we use in (VI.43') and (VI.46') to determine the settling for any time t, i.e., s_t.

Example 12. Determine the total amount by which a soil thawed to a depth $h_t = 10$ m will settle under its own weight if $a_c = 0.05$ cm^2/kg, $\bar{A} = 0.02$, $\Delta\epsilon_{max}/1 + \epsilon_m = 1.5$, $\gamma' = 1.0$ ton/m^3 = 0.001 kg/cm^3.

We perform the calculation using relation (VI.43') with reference to Table 41:

$$\gamma' h_t = 0.001 \times 1{,}000 = 1 \text{ kg/cm}^2 \qquad M = a_c \gamma' h_t = 0.05 \times 1$$

$$= 0.05 \text{ cm}^2/\text{kg}$$

from Table 41 we have $D = 0.0246$; then

$$s_{t\infty} = \bar{A} h_t + \frac{\Delta\epsilon_{max}}{(1 + \epsilon_m)} \, h_t D = 0.02 \times 1{,}000 + 1.5 \times 1{,}000 \times 0.0246$$

$$= 20 + 36.9 = 56.9 \text{ cm}$$

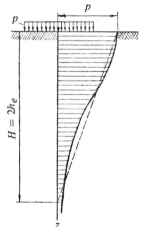

Fig. 122. Calculating diagram of equivalent compaction pressures.

In the case of a local load (pressure from structure foundations) after compaction of the thawed ground under its own weight, we calculate settling by the equivalent-layer method, assuming the compacting-pressure diagram to be triangular with a base equal to the pressure p from the foundations and a height equal to twice the thickness of the equivalent ground layer, i.e., $h_a = 2 h_e$ (Fig. 122).

Here we shall have two theoretical cases: (1) the changes in the void ratio of a thawed ground are large and depend nonlinearly on the external pressure (see Fig. 120) and (2) the compression changes in the porosity are proportional to the external pressure changes (as is usually assumed for unfrozen soils).

1. In the first case, expression (VI.43) with the thaw coefficient $\bar{A}$ set equal to zero and $M' = a_c' p$

Table 42. Degree of consolidation $U_\Delta^{(p)}$

Degree of consolidation $U_\Delta^{(p)}$ as a function of M'

N'	0.01	0.10	0.20	0.30	0.40	0.50	0.60	0.70	0.80	0.90	1.0	2.0
0.01	0.1351	0.1330	0.1304	0.1278	0.1252	0.1227	0.1201	0.1175	0.1149	0.1123	0.1097	0.0826
0.1	0.3724	0.3653	0.3572	0.3487	0.3400	0.3311	0.3218	0.3123	0.3024	0.2924	0.2817	0.1534
0.2	0.4802	0.4717	0.4619	0.4517	0.4411	0.4301	0.4186	0.4067	0.3943	0.3815	0.3680	0.1977
0.3	0.5479	0.5394	0.5295	0.5192	0.5085	0.4973	0.4857	0.4736	0.4609	0.4477	0.4340	0.2567
0.4	0.5982	0.5902	0.5808	0.5711	0.5609	0.5503	0.5393	0.5278	0.5157	0.5032	0.4901	0.3202
0.5	0.6394	0.6320	0.6234	0.6144	0.6050	0.5952	0.5850	0.5744	0.5632	0.5516	0.5395	0.3819
0.6	0.6749	0.6682	0.6603	0.6521	0.6435	0.6346	0.6253	0.6155	0.6054	0.5948	0.5837	0.4395
0.7	0.7063	0.7002	0.6930	0.6856	0.6778	0.6697	0.6612	0.6524	0.6432	0.6335	0.6234	0.4924
0.8	0.7345	0.7289	0.7224	0.7157	0.7086	0.7013	0.6936	0.6856	0.6772	0.6685	0.6593	0.5405
0.9	0.7598	0.7548	0.7489	0.7428	0.7364	0.7298	0.7228	0.7156	0.7080	0.7001	0.6918	0.5841
1.0	0.7827	0.7782	0.7728	0.7673	0.7615	0.7555	0.7492	0.7426	0.7358	0.7286	0.7211	0.6239
1.5	0.8682	0.8655	0.8622	0.8589	0.8554	0.8517	0.8479	0.8439	0.8398	0.8354	0.8308	0.7717
2.0	0.9201	0.9184	0.9164	0.9144	0.9123	0.9101	0.9077	0.9053	0.9028	0.9002	0.8974	0.8615
2.5	0.9515	0.9505	0.9493	0.9481	0.9468	0.9454	0.9440	0.9426	0.9410	0.9394	0.9378	0.9160
3.0	0.9706	0.9700	0.9693	0.9685	0.9677	0.9669	0.9661	0.9652	0.9642	0.9633	0.9630	0.9491
D	0.0050	0.0484	0.0936	0.1361	0.1758	0.2131	0.2480	0.2808	0.3117	0.3406	0.3679	0.5625

substituted for $M = a_c \gamma' h_t$, where p is the external pressure from the foundations of the structure, will hold for the compaction settling:

$$s'_{t\infty} = \frac{\Delta \epsilon_{max}}{1 + \epsilon_m} h_e \left(1 - \frac{1 - e^{-M'}}{M'} \right) \qquad (VI.48)$$

or, denoting $\left\{ 1 - [(1 - e^{-M'})/M'] \right\} = D$, we obtain

$$s'_{t\infty} = \frac{\Delta \epsilon_{max}}{1 + \epsilon_m} h_e D \qquad (VI.48')$$

On the other hand, the degree of consolidation (which we denote in this case by $U_{\triangledown}^{(p)}$) will be different when the compacting pressures vary according to the equivalent triangular diagram, namely:

$$U_{\triangledown}^{(p)} = f(M', N')$$

where

$$N' = \frac{\pi^2 c_c}{4 h_a^2} t \qquad M' = a'_c p$$

To facilitate calculations, values of the degree of consolidation $U_{\triangledown}^{(p)}$ are given in Table 42 as functions of N' and M'.

2. At small external pressures, when the change in the soil void ratio can be assumed to be a linear function of pressure without incurring any major error, the extent and time variation of the settling of foundations on soils that have already been thawed and compacted under their own weight will be calculated from the familiar equations of general soil mechanics.*

*See Chap. III, Ref. 13.

References

1. Tsytovich, N. A.: Osnovy mekhaniki gruntov (Fundamentals of Soil Mechanics), Stroyizdat, 1934.
2. Shusherina, Ye. P.: An Investigation of the Changes in Physicomechanical Ground Properties as a Result of Freezing and Subsequent Thawing, Author's Abstract of Candidate's Dissertation under the sponsorship of N. A. Tsytovich, Moscow, 1955.
3. 1. Tsytovich, N. A., V. G. Grigor'yeva, and Yu. K. Zaretskiy: A Study of the Consolidation of Thawing Ice-Saturated Ground, NIIosnovaniya collection No. 56, Gostroyizdat, 1966; 2. Ponomarev, V. D.: An Experimental Study of the Deformation Properties of Thawing Clayey Soils, Candidate's Dissertation under the sponsorship of N. A. Tsytovich, Industrial and Scientific Research Institute for Engineering Surveys in Construction (PNIIIS), Moscow, 1967.
4. Shusherina, Ye. P.: A Method of Determining the Thawing Coefficients and Compacting Coefficients of Frozen Ground Upon Thawing, Collections Nos. 1 and 2 of "Materialy po laboratornym issledovaniyam merzlykh gruntov pod rukovodstvom N. A. Tsytovicha" (Materials from Laboratory Studies of Frozen Ground Under the Supervision of N. A. Tsytovich), Izd. Akad. Nauk SSSR, 1953 and 1954.

5. Tsytovich, N. A., I. N. Votyakov, and V. D. Ponomarev: Methodological Recommendations for the Investigation of Settling of Thawing Soils, *Izd-vo Akad. Nauk SSSR*, 1961.

6. 1. Lapkin, G. I.: Determination of the Settling of Permanently Frozen Ground Thawing Under Structures, *Byull. Soyuztransproekta*, 1939, No. 12; 2. Lapkin, G. I.: Raschet osadok sooruzheniy na ottaivayushchikh gruntakh po metody kontaktnykh davleniy (Calculation of the Settling of Structures on Thawing Ground by the Contact-Pressure Method), Stroyizdat, 1947.

7. 1. Kiselev, M. F.: A Method of Determining the Relative Compression of Frozen Ground Upon Thawing, Collection "Trudy NIIOSPa," 1952, No. 19; 2. Kiselev, M. F.: Kraschetu osadok fundamentov na ottaivayushchikh gruntakh (Calculation of Foundation Settling on Thawing Soils), Gosstroyizdat, 1957.

8. Ushkalov, V. P.: Determination of Foundation Pressure in a Thawing Base, *In*: "Materialy k osnovam ucheniya o merzlykh zonakh zemnoy kory" (Source Materials for a Theory of the Frozen Zones of the Earth's Crust), No. III, *Izd-vo Akad. Nauk SSSR*, 1956.

9. Tsytovich, N. A.: "Issledovaniye deformatsiy merzlykh gruntov (A Study of the Deformations of Frozen Soils)," vol. 2 of Doctoral thesis, pp. 45, etc., Leningrad, 1940; 2. Tsytovich, N. A.: "Construction Under Permafrost Conditions," abstracts of papers at Conference of USSR Academy of Sciences Council for the Study of Productive Resources (SOPS), USSR Academy of Sciences, 1941; 3. Tsytovich, N. A.: Raschet osadok fundamentov (Calculation of the Foundations Settling), Stroyizdat, 1941.

10. 1. Tsytovich, N. A., Yu. K. Zaretskiy, V. G. Grigor'yeva, *et al.*: Consolidation of Thawing Ground, Doklady k VI Mezhdunarodnomu kongressu po mekhanike gruntov (Papers at Sixth International Congress on Soil Mechanics), Stroyizdat, 1965; see also Ref. 3.

11. Fel'dman, G. M.: A Solution of the One-Dimensional Problem of Consolidation of Thawing Soils with Consideration of Variable Permeability and Compressibility, *In*: "Materialy Vsesoyuznogo soveshchaniya po geokriologii" (Materials of an All-Union Conference on Geocryology), no. 5, Yakutsk, 1966.

12. Malyshev, M. V.: Calculation of the Settling of Foundations on Thawing Ground, "Osnovaniya, Fundamenty i Mekhanika Gruntov," 1956, no. 4.

13. Zaretskiy, Yu. K.: Calculation of Settling of a Thawed Ground, "Osnovaniya, Fundamenty i Mekhanika Gruntov," 1968, no. 3.

14. Terzaghi, K.: Structural Soil Mechanics, Translation from the German, Prof. N. M. Gersevanov (ed.), Gosstroyizdat, 1933.

15. See Ref. 10, this chapter, and N. A. Tsytovich, Yu. K. Zaretskiy, *et al.*: Prediction of Settling of Thawing Ground as a Function of Time, "Materialy V Vsesoyuznogo soveshchaniya po obmenu oputom stroitel'stva v surovykh klimaticheskikh usloviyakh" (Materials of Fifth All-Union Conference for Experience Exchange in Construction Under Harsh Climatic Conditions), No. 5, Krasnoyarsk, 1968.

16. 1. Tsytovich, N. A., Yu. K. Zaretskiy, and M. V. Malyshev: Ukazaniya po raschetu skorosti osadok ottaivayushchikh i ottayavshikh osnovaniy (Instructions for Calculation of the Settling Rates of Thawing and Thawed Bases), *Izd-vo NIIOSPa*, 1967; 2. Tsytovich, N. A., V. G. Berezantsev, *et al.*: Osnovaniya i fundamenty (Bases and Foundations), *Izd-vo "Vysshaya skola,"* 1970.

Part Two

PRACTICAL APPLICATION OF
FROZEN-GROUND MECHANICS

Chapter VII

ENGINEERING-GEOCRYOLOGICAL STUDIES OF PERMAFROST

1. Objectives and Scope of Engineering-Geocryological Studies

Distinctive features of engineering-geocryological investigations of permafrost arise out of the unique properties of frozen ground (which is classed as a temperature- and structurally unstable material) and the cryogenic processes that take place in it, both under natural physicogeological conditions and in interaction with structures built on it.

As the most important element of complex engineering-geological exploration in permafrost regions, engineering-geocryological studies must characterize the natural conditions prevailing in the proposed construction area, predict possible temperature and cryogenic variations in the bases of the structures as a result of the area build-up, and determine the design characteristics of the soils (in their frozen, thawing, and thawed states) that are needed for planning and design of the bases and foundations (and, to a certain extent, of the superstructure) of structures to be built on permafrost.

The natural physicogeological conditions of construction sites in the permafrost area are characterized by:

1) the extent (area and depth) of permafrost containing various interlayers, lenses, and other inclusions of ground ice (and sometimes flow of interstitial water);

2) the cryogenic structure of frozen ground, which varies as a function of composition of the soil and the conditions under which it entered the frozen state;

3) the specific general temperature regime of the soils, with the negative temperature of the permafrost and the variable temperature of the annual freeze-thaw layer, which is determined by recent climatic and paleoclimatic conditions of the construction area;

4) the presence of local criogenic processes (migration of water, frost heave, thermokarsts, solifluction, etc.);

5) the specific physicomechanical properties of ground in the frozen, thawing, and thawed states.

The peculiar properties of permafrost as a base and environment for structures (which are often totally different from those of ordinary unfrozen ground), the predominant among which are: high strength, plasticity in the frozen state, and especially structural instability on thawing, which governs the tendency of the ground to sag, create special difficulties when these soils are utilized and the need to consider properties and processes that are of no practical importance under ordinary conditions. All of this requires more detailed engineering-geocryological exploration and studies, by which we mean a special field (with consideration of the properties of the ground in the frozen and thawing states and the processes that govern these states) that of engineering-geological exploration and study.

The objectives of the engineering-geocryological studies are:

1) a general engineering-geocryological survey of the construction area with determination of the upper and lower boundaries of the permafrost or the adequacy of its thickness for the particular type of structure to be built, with engineering evaluation of the specific strata of permafrost that are characteristic for the region (including layers of underground ice) and the cryogenic processes observed in the region of the study;

2) acquisition of data for prediction of the general and local temperature stability of the permafrost beds in the region under study;

3) engineering evaluation of the stratification of the soils over the entire depth of the active compression zone under the foundations of the structures, with determination of the physicomechanical property indicators of the soils in order to obtain the data needed for calculation of the strength, stability, and deformation of the bases in the frozen and thawing states;

4) generalization on experience gained in construction on nearby sites similar to the site of the contemplated projects in regard to geological structure, and analysis of the experience with the purpose of general prediction of the behavior of structures on the territory to be built up.

It is desirable that these objectives be reflected sufficiently clearly and completely in the engineering-geocryological prospecting report, so that it will be possible to avoid errors in siting and selection of the construction principle and the system of bases and footings.

The scope of engineering-geological prospecting is determined by:

1) the extent to which the construction area has been studied geologically and geocryologically;

2) the complexity of the geological structure of the contemplated construction site;

3) the geocryological features of the construction area;

4) the stage of planning.

Before proceeding with the preparation of an engineering-geocryological prospecting program, literature and departmental materials on the geological structure and geocryological conditions of the prospecting region should be studied.

It must be remembered that the geocryological conditions of a construction site depend on the geocryological zone in which it is located: in the northern, central, or southern regions of the permafrost occurence area.

The region north of the $-5°$ geoisotherm at a depth of 10 meters from the surface on the map of permafrost distribution in the USSR (see Fig. 11) can be regarded as the northern geocryological zone; the southern zone would be the permafrost region south of the $-1°$ geoisotherm at a depth of 10 meters from the surface, i.e., it would consist of areas with permafrost strata temperatures from 0 to $-1°C$; the central permafrost zone would be located on the map between the -1 and $-5°C$ geoisotherms at the 10-meter depth.

The northern geocryological zone is characterized by continuous distribution of permafrost to considerable depths (more than 300 meters, and sometimes in the 500–1000-meter range), with low temperatures at the depth of zero annual amplitudes (below $-5°C$) and a stable temperatures regime in the massif, in most cases with no ground water (except for seacoast areas with salinized ground waters); the southern geocryological zone exhibits, as a rule, an unstable temperature regime in its permafrost strata, which are at temperatures above the limit of substantial (intensive) ice-to-water phase transitions; continuous (in depth) taliks, stratified permafrost beds (with ground layers at positive temperatures in the frozen ground bed), and deep summer thaw and winter freezing (on the order of 2.5 to 4.5 meters, depending on soil composition) with total thicknesses of permafrost ranging from a few meters to 100; the permafrost of the central zone (see Fig. 11) will exhibit intermediate cryogenic properties.

The hard-frozen state of permafrost and the presence of thick ice inclusions are characteristic for the northern zone; for the southern zone, the plastically frozen state is typical, with a preferential stratified-veined texture of the soils, also with fossil ice wedges present in some regions.

All of the above must be taken into account in preparing the engineering-geocryological prospecting program for the particular construction area.

Finally, the stage in planning of the particular structure is an influential factor in the engineering-geocryological prospecting and research program: whether or not we are concerned with a "technical-economic feasibility report (TED)," or a "project specification," or the stage of blueprints preparation.

For the first stage, it is possible in many cases to confine the work to minimal field studies of the soils with full consideration of literature and departmental materials for the region in question. Here, as will become clear below, it is often sufficient to use data from determinations of only the classification indicators of the soils, and of these, as a minimum, their total moisture content W_d, their temperatures at a depth of 10 meters, i.e., $-\theta_0$ (which can sometimes be taken from the map of Fig. 11), and the values of the thaw coefficient $\bar{A}$ as determined by an elementary field test.

The second stage (proposal, or "project specification") must include a complete discussion of the geological structure and geocryological features of the construction site, with mandatory layer-by-layer determination of the thermal and physicomechanical (chiefly deformation and strength) properties of the soils for the entire active compression zone, which is assumed equal to twice the thickness of the equivalent ground layer for the foundations with the largest base areas, but no smaller than the width of the building.

The information on geological structure of the site is refined with the aid of coring and excavation studies and by the use of geophysical survey methods; the thickness of the permafrost stratum, or at least its continuity at a depth of at least 20 meters below the surface, is determined, together with the depth of seasonal thaw (by excavating at the beginning of the fall freeze-up affecting the top layers of soil) for the construction site.

The physical and physicomechanical properties of ground in the frozen and thawing states and sometimes (when the preconstruction-thawing technique is used) in the unfrozen state as well (after thawing of the permafrost) must be determined with exhaustive completeness in the second stage of prospecting.

These studies include, for all representative ground layers of the active compression zone under the foundations, determination of the characteristics, knowledge of which is mandatory for knowledgeable (with appreciation of the up-to-date achievements of science) preparation of the engineering plan (project specification) for structures to be built on the permafrost, with prediction of the general and local temperature instability of the permafrost layer at the site, with calculations of the bases and footings for extreme states, and with justification of the methods by which the earth-moving and foundation work will be done.

The prospecting and investigation done at this stage include the following determinations:

1) the classification indicators of the frozen soils (bulk density, moisture content, specific gravity, and amount of unfrozen water), also the mechanical composition of the soils and, for clayey soils, their consistency;

2) the thermal properties of the permafrost below the bases of the footings (as a minimum, the temperatures of the soils down to a depth of 10 meters, the thermal conductivities and volumetric heat contents of the soils);

3) the mechanical-property indicators of the soils in the frozen state (ultimate long-term compressive strength, parameters of long-term shear strength and stable adfreezing forces, coefficient of total relative compressibility, and creep parameters) and in the thawing state (coefficients of thawing and compacting during thawing, shear-strength parameters in the thawed state).

In the third stage, that of preparing blueprints for the projects, only control cores are taken and control tests performed on the permafrost in excavations (determination of compressibility indicators in the frozen and thawing states by the test-load method, determination of strength characteristics, etc.).

2. Complex Engineering-Geocryological Survey of Territory to be Developed

A composite characteristic of the geocryological conditions prevailing in areas to be developed in the permafrost zone can be obtained only as a result of a competent geocryological ("frost") survey of the terrain to be developed.

A detailed description of the procedures followed in general engineering-geocryological ("frost") exploration would be beyond the scope of the present work. We shall therefore discuss only the fundamental premises of the geocryological survey, referring the reader to the special literature[1] on the problem for detailed study.

According to V. A. Kudryavtsev and B. N. Dostovalov, a geocryological ("frost") survey covers the aggregate of literature and departmental, field, laboratory, and office work done with the purpose of characterizing the geocryological conditions of a particular area in the permafrost zone, setting forth particular and general relationships governing their formation and development as functions of natural *geological-geographic* conditions, and submitting a forecast of possible variations in the geocryological conditions of the region as construction develops, with recommendations for management of these variations.

We should note that the results of the geocryological survey are absolutely indispensible as basic starting materials in the planning of large industrial and residential construction projects and major hydraulic and highway construction.

All engineering geocryological charts should be structured primarily on a geological and geomorphological basis, with reference to charts of quaternary deposits and soil charts.

The following should be shown on the geocryological charts:

1) the distribution of permanently frozen ground, its soils, thickness, stratification conditions, and vertical continuity;

2) the temperatures of the soils (as a minimum, at the base of the layer in which it has an annual variation, at a depth of about 10 meters from the surface);

3) the composition, properties, and genesis of the permafrost and unfrozen ground;

4) the cryogenic structure and ice contents of the permafrost, including the location and thickness of fossil ice interlayers;

5) manifestations of cryogenic processes (heave mounds, icings [Naledy], thermokarsts, solifluction, etc.);

6) data on the seasonal freezing and thawing of site soils (if possible, with natural cover and cleared);

7) the presence and strata position of ground water.

The geocryological survey includes three steps: (1) microdistricting, (2) field studies, and (3) office processing of the survey materials with establishment of the basic geocryological relationships and prediction of the behavior of the permafrost on development of the area by construction.

Step 1 (microdistricting) is initially based for the most part on literature and departmental materials, with subsequent adjustment on the basis of field geocryological studies and compilation of engineering-geocryological charts; here regions characterized by persistence and homogeneity of sets of natural conditions and distinguished from adjacent regions by their geological structure, geomorphological and hydrological conditions, and quaternary-deposit composition, as well as by the nature of their vegetable and topsoil covers, are designated.

Step 2 (field engineering-geocryological studies) must yield materials for preparation of the composite engineering-geocryological chart with establishment of the average working values of the previously enumerated indicators to the physicomechanical properties of the soils in the frozen and thawing states and with evaluation of the manifestations of cryogenic processes occurring on the area under study (heave mounds, icings, thermokarsts, landslides, etc.). Here, together with coring to establish the thicknesses of the individual ground strata and measuring their temperatures and excavating with field testing of the soils, significant importance also attaches to geophysical survey methods for establishing the thicknesses of the permafrost strata and ice inclusions, their stratification conditions, and characteristic composition properties of the frozen ground. Such methods, which are widely used in geocryological surveying, are:

1) electrical prospecting with direct current (vertical electrosounding, electroprofiling, electrometric detection and contouring of underground ice and taliks, etc.);

2) ultrasonic and seismic methods (to establish the stratification conditions of massive crystalline rocks and their fissuring, determination of their deformation properties, etc.);

3) aerial photo-surveying, which is used for preliminary microdistricting of the subject territory and to locate the boundaries of geomorphological relief elements, areas affected by the development of cryogenic processes, and zones with ground deposits differing in composition.

Step 3 (office processing) is an analysis of the experimental data with establishment of general and particular relationships governing the distribution and development of permafrost strata and the shaping of their temperature regime as a function of local physicogeological conditions: climatic, geomorphologic, hydrologic, etc., with special attention to the manifestations of cryogenic processes in the ground.

In generalization of the materials from the engineering-geocryological survey, especially the medium-scale (from 1:500,000 to 1:200,000) and large-scale (from 1:10,000 to 1:50,000) versions and also the detail versions, a general prediction is made of the changes to be expected in geocryological conditions and the general trend of the cryogenic process (its intensification or degradation) on economic development of the territory to be built up, which is very important for practical purposes, and specific forecasts for the extent of thermokarst, icing, frost-heave, etc.

The results of the engineering-geocryological studies are submitted in the format of a geocryological (frost) chart (an example appears in Fig. 123) with an explanatory note giving the data needed for selection of the construction

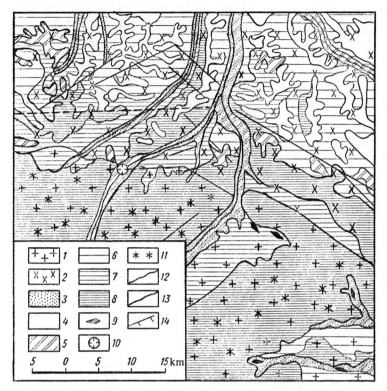

Fig. 123. Geocryological chart of one permafrost area (after K. A. Kondrat'-yeva). (1) Archean crystalline schists and gneisses with ice in fissures down to 100 meters; (2) same, Jurassic sandstones and aleurolites; (3) alluvial sandy-clayey soils cemented by ice; (4) unfrozen rocks; (5) island permafrost up to 50 meters thick with $\theta_{av} \leqslant -1°C$; (6) permafrost up to 100 meters thick at $\theta_{av} \leqslant -1°C$; (7) same, thicknesses 100 to 200 meters, θ_{av} from -1 to $-2°C$; (8) same, > 200 meters thick at $\theta_{av} = -2°C$; (9) icings; (10) thermo-karst; (11) Goletz ice; (12) boundary of permafrost; (13) boundaries of geological formations; (14) tectonic disturbance zones.

principle for structures to be built under conditions of the particular permafrost region (temperature of frozen strata, ice content, lithological composition, etc.) and quantitative indicators to the physicomechanical properties of the frozen ground, which will be needed in planning and designing the bases, foundations, and superstructures to be built under the particular set of conditions.

Let us now turn our attention to the field and laboratory studies of frozen soils conducted in the engineering-geocryological survey of a particular per-mafrost area targeted for construction and absolutely indispensable for the strength and stability calculations to be made for structures built under the particular geocryological conditions, filling in details of the program and methods of the minimal studies without which it would be impossible to design the particular structure rationally or establish a method for its operation that would ensure stability and a long service life under the particular conditions.

3. Field and Laboratory Tests and Studies of Permanently Frozen Ground

The object in field and laboratory testing and study of permafrost is to determine the necessary design data on its physicomechanical properties under natural conditions and, primarily, its thermal, strength, and deformation properties in the frozen and thawing states.

These tests and studies are classified as reconnaissance and basic.

Reconnaissance tests form a basis for preliminary evaluation of the permafrost stratum under the proposed construction site as a base and environment for structures.

As we know (see Chap. I), frozen soils, as four-phase bodies, require knowledge of at least four basic indicators of their physical properties: total moisture content W_d, the bulk density γ of the soil with its structure undisturbed, the specific gravity γ_{sp} of the mineral skeleton, and the unfrozen water contents W_u at various negative temperatures. Structural evaluation of the frozen ground strata also requires determinations of temperature in the permafrost, at least at one point at the depth of zero temperature amplitude, i.e., at a depth of approximately 10 meters, and the relative thermal settling (without load), i.e., the value of the thaw coefficient $\bar{A}$.

Among the possible reconnaissance tests that might be performed on permafrost, the author recommends, as a *minimum minimorum*, determination of three property indicators of the permafrost in the field situation under the specific construction-site conditions:

1) the temperature of the permafrost bed at a depth of 10 m, i.e., $-\theta_0$;

2) the total moisture content of the soils below the footing base level (W_d approximately 1 meter below the upper boundary of the permafrost), and;

3) the relative thermal settling (the thaw coefficient $\bar{A}$ at the level of the footings below grade).

Determination of these quantities requires the most elementary instruments: a thermometer, a ruler, and a balance.

The ground temperature $-\theta_0$ at the 10-meter depth is measured in dry-drilled wells with a "slow" thermometer after sufficient time has elapsed since drilling (several weeks). This temperature is needed for calculation (in accordance with SNiP II-B.6-66) of the maximum design temperature at the footing base level and the equivalent temperature for the foundation heave calculation and other thermal calculations.

The total moisture content W_d of the frozen ground is determined by weighing a sample with the natural moisture content and the same sample after drying to constant weight.

The specific gravity γ_{sp} is taken from the literature data and usually ranges from 2.4 to 2.5 for silty soils with organic matter, while $\gamma_{sp} = 2.65$ for sands and $\gamma_{sp} = 2.73$ for clayey soils.

Then, using, for example, I. N. Votyakov's equation, which has been thoroughly confirmed by experiments, we have the following approximate expression for the bulk density of the frozen soil:

$$\gamma \approx \frac{2.4 \, (1 + W_d)}{2.7 W_d + 0.9} \qquad (VII.1)$$

with an error of no more than 3 percent at $W_d > 5\%$ for pebbly soils, $W_d > 15\%$ for sandy soils, and $W_d > 20\%$ for clayey soils.

Then the approximate expression for the bulk density of the frozen ground skeleton will be

$$\gamma_{sk} \approx \frac{2.4}{2.7 W_d + 0.9} \qquad (VII.2)$$

from which the void ratio is

$$\epsilon = \frac{\gamma_{sp} - \gamma_{sk}}{\gamma_{sk}}$$

It is permissible to determine the unfrozen-water content W_u in permafrost from Table 1 of SNiP II-B.6-66 in reconnaissance tests, as a function of the frozen ground temperature $-\theta°$ and the plastic limits W_L and W_r, determined on soil specimens with undisturbed structure:

$$W_u = k_u W_r \qquad (VII.3)$$

where the coefficient $k_u = f(-\theta°$, $W_p = W_L - W_r)$, i.e., it is determined as a function of the temperature, the type of soil, and the plasticity index W_p, with the coefficient k_u varying from 0.5 to 0.3 for sandy loams, from 0.7 to 0.5 for loams, and from 1.0 to 0.6 for clays as the temperature varies from -0.5 to $-5°C$.

The volumetric ice content is determined from relation (I.5) using the data on the bulk weight γ of the frozen soil, the total moisture content W_d, the specific gravity of the ice $\gamma_{ice} = 0.9$, and the unfrozen-water content W_u:

$$i_{vol} = \frac{\gamma}{\gamma_{ice}} \times \frac{W_d - W_u}{1 + W_d}$$

If $i_{vol} \geqslant 0.5$, the permafrost will belong to the high ice content category with, in most cases, sagging on thawing; but if $i_{vol} \leqslant 2.5$, it will be classed as low ice content permafrost.

Depending on the maximum temperature of the frozen ground $\theta°_{max}$ at footing-base level (determined as a function of $-\theta_0$), on the composition of the permafrost (sands, loams, clays, etc.), and on the volume ice content i_{vol}, normal resistances R^s, kg/cm^2, of the frozen soils to normal pressure in the bases of structures are determined from Table 6 of SNiP II-B.6-66.

The thaw coefficient $\bar{A}$ below footing-base level can be used to estimate the possible amount of thermal settling of the ground (which constitutes, in most cases, a major part of the total settling) and the amount by which the permafrost will sag upon thawing.

We recommend that the thaw coefficient be determined by an elementary field settling test performed on the thawing soil in an excavation, with the soil allowed to settle under its own weight. If the thawing layer of frozen soil is quite shallow (the thawing depth of the soil is determined in the field test with a metal probing bar, while settling is measured with a millimeter rule), for example, less than 0.50 m, the maximum pressure from the soil's own weight will be small (usually less than 0.1 kg/cm^2), and then the second term in Eq. (VI.2) can be neglected for a thaw area of at least 1-2 m^2; the thaw coefficient will then be equal to the ratio of the thermal settling s_{th} of the frozen ground to the depth of thaw h_t:

$$\bar{A} \approx \frac{s_{th}}{h_t} \qquad (VII.4)$$

When the thaw coefficient $\bar{A} > 0.02$, permafrost soils are regarded as sagging soils.

We note that the compaction coefficient of thawing soils $(\bar{a}, \text{cm}^2/\text{kg})$ can be put equal to

$$a \approx \frac{\beta}{E_{min}} \qquad (VII.5)$$

for preliminary calculations without incurring any great error; here β is the lateral pressure coefficient of the ground, which equals 0.8 for sands, 0.7 for sandy loams, 0.5 for loams, and 0.4 for clays, and E_{min} is the minimum value of the total strain modulus of the thawing soils, determined in approximation from Table 13 of SNiP II-B.1-62 for the values of the soil's porosity coefficient ϵ and the moisture content W_r at the plastic limit.

It is clear from the above how important it is to determine $-\theta_0$, W_d, and $\bar{A}$ for the load-bearing stratum of permafrost under natural field conditions on the site for the actual design of the bases and foundations of structures to be built on it, and to use these quantities to find preliminary values for other design characteristics of the permafrost.

The basic field and laboratory tests and studies of permafrost are necessary for selection of the construction principle and for preparation of the basic documents (proposals or "project specification") for planning and design of bases and foundations to be erected on permafrost. They include determination of the following necessary characteristics of the permafrost:

1) the classification indicators of the permanently frozen ground;

2) the thermal properties of the soils in the frozen and thawed states;

3) the mechanical properties of the ground in the frozen, thawing and thawed states.

The *classification indicators* of soils in the active layer and the permafrost bed are: (a) ground temperatures, (b) the granulometric composition of sandy soils and the consistency of clayey soils, and (c) the ice contents of the soils.

Below we shall concentrate our attention on an enumeration of physico-mechanical property indicators of permanently frozen ground whose values must be known for design purposes; concerning the procedure of the determinations, however, we shall comment only as necessary, since many references have already been published on this problem.[2]

To classify permafrost on the basis of temperature state, ice content, and natural compactness, it is, as we noted earlier, in addition to the temperature and granulometric composition of the soils, also necessary to know their basic physical property indicators: γ_{sp}, γ, W_d, and W_u, of which the bulk density γ and total moisture content W_d are determined under field conditions (for natural soil samples), while the others are determined in the laboratory (for specimens with disturbed structure) in exactly the same way as the physical-state characteristics of the thawed ground (density and consistency).

On the basis of temperature state, permanently frozen soils are divided into hard-frozen (with temperatures below the limit of intensive water-to-ice phase transitions) and plastically frozen (at temperatures above that limit); on the basis of ice content (volume ice content i_{vol}), they are either high, or low ice content soils ($i_{vol} > 0.50$ or $i_{vol} < 0.25$), and on the basis of natural compactness in the thawed state they are classified in terms of the porosity ϵ for sandy soils and the consistency B for clayey soils.

Indicators of thermal property of the soils (subscripts f in the frozen state and t in the thawed state) are determined from Table 10 of the SNiP II-B.6-66, namely: the thermal conductivities λ_f and λ_t and the volumetric heat capacities C_f (at a temperature $\theta = -10°C$) and C_t as functions of the type of soil (sand, clay, etc.), their bulk densities γ, total moisture W_d, and content of unfrozen-water W_u (the latter for volumetric heat capacities C_f' at temperatures from -0.5 to $-10°C$).

Also determined in the basic engineering-geocryological field studies are the temperatures in the active layer at various depths and the temperature of the permafrost stratum, at least at a depth of 10 meters, i.e., $-\theta_0$, which is necessary for calculation of the maximum temperature θ_{max} at footing-base level according to the SNiP and the average equivalent temperature θ_e of the permafrost over the height of that part of the foundations or over that length of the piles that is set in the permafrost—to determine the standard shear resistance of the frozen ground at the foundation-to-soil adfreeze surface.

The *mechanical properties* of the ground (strength and deformation properties) are determined in field tests both for the frozen state and during and after thaw (in the thawed state). The procedure for determining the design indicators of stress and strain state of frozen and thawing ground was examined in detail in Part One of this book. Here we shall enumerate the fundamental indicators that must be determined in the field and laboratory studies.

The basic indicators in the frozen state will be:

1) the long-term compressive strength (determined by the ball-plunger method or on a dynamometer instrument) σ_{lt} (see Chap. III, Part One);

2) the ultimate cohesion equivalent c_e, determined by the ball-test method, which is also used to determine the standard design pressure R^s [kg/cm^2] on the ground in the bases of structures;

3) the resultant relative compressibility coefficient a_r^{Σ}, cm^2/kg, of the ground stratum at footing-base level (and deeper if there is an increase in the compressibility of the frozen ground), determined by testing ground blocks in a thermally nonconductive odometer (see Fig. 106) or by the test load method in the excavation;

4) the following are determined in occasional cases: a_r'', the secondary relative compressibility coefficient; δ', the creep parameter of the exponential creep parameter, or the parameter T of the hyperbolic creep factor (see Chap. III, Sec. 5).

The following are determined in the thawing state for soil beds below the footing-base level: the parameters of the fundamental settling equation (VI.3) of frozen ground during thawing, namely: the thaw coefficient $\overline{A}$ and the coefficient of compaction during thaw $\overline{a}$; the consolidation coefficient c_m (from the mean value $\overline{a}_m$ of the compaction coefficient and the filtration coefficient k_m of the thawing soils, from Eq. [VI.31]), which is needed for prediction of the course that will be taken by settling in time.

For particularly critical cases, the thaw coefficient $\overline{A}$ is determined by the test-load method in excavations with layer-by-layer thawing and direct loading of a special hot-steam-heated plunger not smaller than 50 X 50 cm (using the test method elaborated earlier).[3]

On thawing of the ground to a depth equal half the width of the plunger, the thaw settling is determined (without load) using relation (VII.4) to calculate the thaw coefficient

$$\overline{A} = \frac{s_{th}}{h_t}$$

The compaction coefficient $\overline{a}$ during thaw is determined from settling at the first load increment (usually 0.5–1 kg/cm^2 for disperse soils and 1–2 kg/cm^2 for coarse-skeleton soils) from relation (VI.7):

$$\overline{a} = \frac{\Delta s_1}{h_t p_1} \qquad \text{(VII.6)}$$

where Δs_1 is the increase in settling under the action of the first load p_1 after thaw.

In the odometer test of two identical frozen ground blocks, $\bar{A}$ and $\bar{a}$ are calculated from relations (VI.5).

Also determined in these tests is the thermal coefficient

$$\beta_t = \frac{h_t}{\sqrt{t}}$$

where h_t is the thickness of the layer of frozen ground thawed uniformly (plane-parallel) in the thermally nonconductive odometer and t is the time to complete thawing of the soil in the odometer to depth h_t (determined with thermocouples).

The following are determined for thawed ground: the parameters of ultimate shear strength for the state acquired by the ground immediately after thawing, c_{th}, ϕ_{th}, the average filtration coefficient k_m; here the cohesion determined by the ball-plunger method for the thawed ground should correspond to its utlimate long-term strength, i.e., to *long* c_{th}, and the angle of internal friction ϕ_{th} should correspond to the undrained and uncompacted state of the ground determined by the fast shear method.

These last quantities are used to estimate the bearing ability of the thawing ground in the physical state most dangerous for stability of the structures—that in which they have not yet compacted under the load.

In concluding this section, we noted that the individual determinations of the various property indicators of the permafrost in its frozen and thawing states and the field tests are carried out in accordance with a complete test and research program specifically developed for the particular construction site, taking into consideration both the general engineering-geocryological ("frost") conditions of the territory under study as well as the results of reconnaissance prospecting, and also of the design features of the structures to be built and their thermal conditions during use.

In any event, the thermal and especially the mechanical properties of permafrost lying directly under the foundations of the planned structures must be characterized completely for proper determination of the construction principle and the strength and stability calculations of the project specification and technical proposal for the bases and foundations of the contemplated structures.

In the technical report developed from the engineering-geocryological prospecting and testing at the construction site, all of the material obtained is summarized, and an evaluation of the features of various zones of the subject territory is presented from the standpoint of their use as bases and environments for erection of the projected structures.

4. Stationary Observations at Geocryological Stations

In support of construction in permafrost regions, it is necessary to conduct stationary observations of changes in the temperature of the active layer and permafrost stratum under land-development influence, and manifestations of cryogenic processes on the built-up area: the extent and nonuniformity of frost heaving, heave of lightweight structures, the appearance and development of icing processes, sagging, and thermokarsts, landslides and solifluction processes on slopes and embankments, etc.

Also mandatory, in addition to stationary observations of the geocryological conditions of the built-up area, are similar observations of the ground temperature regime in the bases of the structures (and especially of the ground thaw depth under the structures), of the settling of bases and deformations of superstructures, the heave of foundations (if it occurs), and other manifestations of the interaction between the structures and the permafrost stratum.

According to Sec. 1.5 of the SNiP II-B.6-66, "for major construction projects for which the general cost estimate exceeds 10 million rubles, a geocryological station is put in place at the very start of prospecting;" its functions include observations, and their analysis, of the engineering state of the structures built and their interaction with the permafrost bases.

The program of the stationary observations "is established by the (according to SNiP II-B.6-66, Sec. 1.5) planning organization dependent upon the intended use of the buildings or structures, their class and their design features, and also in accordance with soil conditions and the principle chosen for utilization of the soils as bases."

The program of the stationary geocryological "frost" stations must include:

1) observations of the engineering state (settling, strength and stability problems, etc.) of completed structures and variations of the local (under the structures) and general (on the built-up area) geocryological conditions;

2) consideration of construction experience gained both on the newly built-up territory from the outset of building and on nearby sites that were built up earlier;

3) organization and implementation of experimental construction in accordance with the latest and most progressive design proposals from the client agency, with mandatory and thorough-going observations of the temperature regime in the permafrost, settling of bases and foundations, and deformations of superstructures.

We should note that the results of observations made on constructed facilities can be used for later planning only when they are accompanied by data from soil studies at the construction site, most important among which are: the physical properties of the soils and their deformability indicators in the frozen and thawing states. If these data are not available at the start of construction, the program of the stationary geocryological station must include their acquisition.

References

1. 1. Kudryavtsev, V. A.: The Frost Survey as the Basic Type of Frost Investigation, "Merzlotniye issiedovaniya" (Frost Studies), no. 1, *Izd-vo MGU*, 1961; 2. Dostovalov,

B. N. and V. A. Kudryavtsev: Obshcheye merzlotovedeniye (General Geocryology), *Izd-vo MGU*, 1967; 3. Poltev, N. F.: Osnovy merzlotnoy s'emki (Fundamentals of Frost Survey), *Izd-vo MGU*, 1963; 4. Meyster, L. A.: Osnovy geokriologii (Fundamentals of Geocryology), part II, "Metody inzhenerno-geokriologicheskikh issledovaniy" (Engineering-Geocryological Study Methods), chap. XII, *Izd-vo Akad. Nauk SSSR*, 1969; 5. Poltev, N. F. and N. N. Romanovskiy: Surveying and Charting Frozen Soils, Doklady na Mezhdunarodnoy konferentsii po merzlotovedeniyu *(Papers at International Conference on Geocryology)*, N. A. Tsytovich (ed.), *Izd-vo Akad. Nauk SSSR*, 1963.

2. 1. Tsytovich, N. A., I. N. Votyakov, and V. D. Ponomarev: Methodological Recommendations for the Investigation of Settling of Thawing Soils, *Izd-vo Akad. Nauk SSSR*, 1961; 2. Vyalov, S. S., S. E. Gorodetskiy, V. F. Yermakov, *et al.*: "Metodika opredeleniya kharakteristik polzuchesti, dlitel'noy prochnosti i szhimayemosti merzlykh gruntov" (Method for Determination of Creep, Long-Term Strength, and Compressibility Characteristics of Frozen Soils), NIIOSP, Nauka Press, 1966; 3. Nersesova, Z. A., *et al.*: Metodika opredeleniya fiziko-mekhanicheskikh svoystv merzlykh gruntov (Procedure for Determination of the Physicomechanical Properties of Frozen Soils), PNIIST, 1971, and elsewhere.

3. Tsytovich, N. A., I. N. Votyakov, and V. D. Ponomarev: Methodological Recommendations for the Investigation of Settling of Thawing Soils, *Izd-vo Akad. Nauk SSSR*, 1961.

Chapter VIII

THERMAL STABILITY OF PERMAFROST
ON CONSTRUCTION SITES

1. Importance of the Problem and Ways to Its Solution

The problem of the thermal stability of permafrost on areas to be built up is of substantial practical importance for the selection of the construction method—whether to preserve the frozen state of the base soils or to allow for their thawing during the use of the structures.

By the thermal stability of permafrost or, more precisely, the stability of the negative temperature field in permafrost, we shall refer to a temperature state of the ground such that no general temperature increase occurs in the permafrost-stratum on the construction site over the long term (a time comparable to the service lives of the principal structures), i.e., so that no degradation of the permafrost occurs.

The idea of predicting the thermal stability of permafrost and a method for estimation of its general thermal stability on the construction site was proposed by the author back in the early 1930's.[1]

It was assumed that stability of the temperature regime in the permafrost stratum on the site would be guaranteed if the depth of freezing of the ground under the base of the structure, calculated on the basis of heat balance according to actual monthly average ground temperatures at various depths, its humidity and ice content, its thermal conductivity and heat capacity, and consideration of heat losses by the floor of the building, is greater than the depth of the summer thaw determined by the same method.

We note that the above condition was first used in 1931 to calculate the thermal-regime stability of the permafrost under the thermal electric power station built at Yakutsk (YaCES),[2] and that it served as a basis for appropriate measures to preserve the frozen state of the ground bases; the correctness of the premises used here in the thermal calculations has been fully confirmed by trouble-free operation of the plant up to the present time, i.e., more than 40 years, with no unacceptable deformations observed.

Below, based on detailed thermophysical studies (by S. S. Kovner, V. P. Ushkalov, S. V. Tomirdiaro, the thermophysical studies of G. V. Porkhayev, *et al.*, which are especially valuable for practice, and those of L. N. Khrustalev, *et al.*), we shall discuss the problem of both general and local thermal stability in greater detail.

On detailed scrutiny of the thermal-stability problem of the permafrost-strata, we note first of all that any development type in the area will have a significant influence on variations of the permafrost thermal regime as a result of an extremely complex interaction between many factors, chief among which are:

1) Economic development of the area (disturbance of moss, snow, under-growth, and other natural covers, forest cutting, agricultural use, etc.);

2) Building up of the area with heat-releasing buildings and structures, the laying of underground heating and water lines, etc.

Following G. V. Porkhayev and V. K. Shchelokov[3] and L. N. Khrustalev,[4] we can divide the numerous factors that influence the thermal regime of a construction site into three groups: general, local, and specific. The general factors include the components of the external heat and moisture exchange, the local factors, the effects of open ponds, drainage ditches, and various types of buildings and structures (especially with positive indoor temperatures) and underground pipelines (water, heating, etc.); the specific factors include hydro-geological features (the presence of interstratified water, etc.), unusual snow-accumulation conditions, mineralization of ground waters, etc.

All of the above results in an extremely complex interaction between the external factors and the temperature field in the permafrost-stratum; the results of external disturbances may vary quite widely, depending on the general geological-geographic and geocryological features of the region and favorable conditions for the development of certain specific factors. Thus, human activity usually results in weakening (degradation) of the permafrost and lowering of its upper boundary in regions with high-temperature permafrost strata; under other conditions, however, when the permafrost temperatures are low, the tempera-ture may sink farther below freezing and the upper boundary of the permafrost bed may rise. For example, observations in the Vorkuta area showed that the upper boundary of the permafrost soils is sinking and the soil temperatures are rising.

Exactly the same phenomenon is observed at Igarka and in a number of other localities in the southern permafrost zone. The latter is explained firstly by the increase in the total radiation in the built-up areas (towns) as compared with unaffected areas (as a result of pollution of the air over the cities and towns with soot, dust, etc.), secondly by the release of heat by buildings, industrial plants, and transportation, which raises the temperature in the cities by amounts on the order of $+0.5$ to $+1.5°C$ and sometimes even more; thirdly, by the increased turbulence of the air flow through the towns and villages in summer as a result of nonuniform heating of different parts of the buildings and structures; fourthly, by local evaporation of moisture in the built-up areas as compared to open areas and, finally, by the presence of underground heating lines that are not exposed to the weather.

At certain places in the northern and central permafrost zones and under certain conditions (in the absence of ground-water movement and when the upper heat-insulating peat-moss or snow layer is heavily compacted or removed), we observe a rise of the upper permafrost surface and a lowering of the ground temperature at the depth of zero thermal generation (e.g., according to P. A. Solov'yev, the ground temperature at a depth of 10 meters has dropped from − 2 to − 6°C at Yakutsk over the last 300 years), a result of the complex thermal interaction of many factors, including increased thickness and moisture content of the loose upper "cultivated" ground layer, a decrease in the heat-insulating properties of the surface layers (moss, snow, etc.), changes in hydrogeological conditions, etc.

For practice, it is very important to establish both the general direction of the cryogenic process on the area to be built up (whether it will become stronger or degraded with gradual warming and eventual thawing of the permafrost) and the dynamics of the temperature variations in the frozen strata during the lifetime of the structures, since the negative temperature in the frozen ground is a very important factor in determining all of its strength and deformation characteristics, the values of which are used for the strength and stability calculations for the structures to be built on them.

The temperature regime in the ground of the site is influenced not only by the original geocryological conditions and the degree of preconstruction preparation of the area and the subsequent landscaping, but also by the overall density of construction, especially when it is high (above approximately 30–40 percent) and by the presence of local concentrated high-temperature heat sources ("hot" shops of industrial plants that produce large amounts of waste water, public baths, laundries, and especially the laying of unventilated underground water lines).

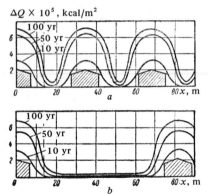

Fig. 124. Increase in ground heat content on a built-up area (according to L. N. Khrustalyev) in kcal/m², after 10, 50, and 100 years for construction densities of: (a) 41%; (b) 17%.

As an example, Fig. 124 shows ground heat content graphs for a built-up area near Vorkuta (with a permafrost layer 80 meters thick and an initial temperature $\theta_f = -2°C$) according to calculations made by L. N. Khrustalyev on the hydraulic integrator designed by Prof. V. S. Luk'yanov. These graphs indicate that at a 17 percent construction density under the particular conditions the heat flux is insignificant, and a steady state can be arrived at in the thawing dish. But if courtyards, streets, and squares are also taken into account in computing the construction density, the actual average integral construction density will be substantially less than 17 percent for cities in the permafrost zone.

In appraising the thermal stability of permafrost soils in construction areas it is necessary to consider separately the general conditions for degradation or growth of the permafrost strata in the particular area and the influence of local factors, chiefly construction and underground pipelines, on the temperature regime in the permafrost.

Solution of this extremely complex thermophysical problem of the temperature field dynamics in the strata at construction sites in the permafrost region can be accomplished by the following methods:

1) Analytically, by drawing up an overall heat and mass exchange balance for the known meteorological, geocryological, and other features of the construction area and the thermophysical properties of the soils;

2) By modeling the thermophysical processes with a stationary temperature field, using the electrohydraulic analogy and Luk'yanov's hydraulic-analogy instrument, and for a nonstationary temperature field on the Luk'yanov device;

3) On the basis of generalization of field observations of the temperature-field changes under structures built under similar geocryological conditions, although this requires long observing times and is not always available.

2. Evaluation of the General Thermal Stability of Permafrost Strata on Construction Sites

Analytical calculations and studies of the temperature regime in permafrost strata by Luk'yanov's hydraulic-analogy method indicate that the principal temperature-field characteristic of the permafrost bed in a built-up area is the variation of ground temperature in the layer with zero annual amplitude $-\theta_0$, i.e., at a depth of approximately 10 meters below the surface.

The value $-\theta_0$, which is characteristic for the temperature regime of the permafrost-stratum is determined, according to studies by G. V. Porkhayev,[5] by considering the soil heat budget components:

$$
\theta_0 = \frac{1}{T}\left\{ \Omega_{a.w} + \frac{1}{\alpha_w}(\Omega_{R.w} - \Omega_{E.w}) + \frac{\lambda_t}{\lambda_f}\left[\Omega_{a.s} + \frac{1}{\alpha_s}(\Omega_{R.s} - \Omega_{E.s})\right]\right.
$$
$$
\left. + 0.47\frac{h'_s}{\lambda'_s}\sqrt{\lambda_t W\left[\Omega_{a.w} + \frac{1}{\alpha_s}(\Omega_{R.s} - \Omega_{E.s})\right]}\right\}\eta
$$

$$(\text{VIII}.1)$$

where

$$
\eta = \sqrt{\frac{1}{1 + 2\sqrt{\frac{\pi}{n^2 a_f T}} + 2\frac{\pi}{n^2 a_f T}}}
$$

$$
n = \frac{1}{\lambda_f\left(\frac{h''_s}{\lambda''_s} - \frac{1}{\alpha_w}\right)}
$$

T is a period of time equal to one year, $\Omega_{a.w}$ and $\Omega_{a.s}$ are the summed monthly average air temperatures during winter and summer, respectively, α_w and α_s are the coefficients of heat transfer from the surface of the snow to the air in winter and from the ground to the air in summer, $\Omega_{R.w}$ and $\Omega_{R.s}$ are the summed monthly-average values of the radiation balance for winter and summer, $\Omega_{E.w}$ and $\Omega_{E.s}$ are the summed monthly average expenditures of heat for evaporation from the snow surface in winter and the soil surface in summer, λ_t and λ_f are the thermal conductivities of the thawed and frozen soils, h'_s and h''_s are the thicknesses of the snow cover during and after freezing of the soil, W is the soil moisture content, a_f is the thermal diffusivity of the frozen soil, and λ'_s and λ''_s are the thermal conductivities of the snow during and after freezing of the ground.

Needless to say, determination of all of the soil-surface heat-budget components—meteorological, radiative, snow-cover thickness, thermal and physical properties of the snow and soil in the thawed and frozen states—encounters substantial difficulties and can be accomplished only to one or another degree of approximation. This complicates the use of relation (VIII.1) in practice, especially when it is considered that it is necessary to use multi-year average data if the results are to be reliable.

Since $-\theta_0$ will appear in later thermal calculations for permafrost bases as the basic temperature indicator for the permafrost-strata of the particular area and almost all SNiP thermal calculations for bases and foundations on permafrost depend on this quantity, it is recommended that if difficulties are encountered in analytical calculation of $-\theta_0$ the temperature of the permafrost at the depth of zero annual amplitude be determined directly, by careful measurement (at a depth of about 10 meters) or that it be projected by geocryological specialists on the basis of appropriate analogs.

L. N. Khrustalyev and his coworkers at the Vorkuta Scientific Research Station of NIIASP investigated the general thermal stability of permafrost-strata on construction areas by the hydraulic-analogy method. We shall present some results of these studies that we consider important for evaluation of the general thermal stability of permafrost soils in construction areas.

Hydraulic-integrator studies of the temperature field in permafrost under structures indicated the following:

1) The influence of heated buildings and structures on the temperature of the permafrost extends to distances on the order of several tens of meters;

2) The soil temperature $-\theta_0$ at the depth of zero annual amplitude and the average integral temperature* θ_{av} of the soil surface are of considerable importance for determining the direction of changes in the permafrost temperature;

3) If the average integral surface temperature is above 0 ($\theta_{av} > 0$), no stationary position of the thaw bowl in the permafrost under the structures will

---

*See Ref. 4, this chapter.

be achieved (even without consideration of underground heat and water lines), under such conditions the permafrost will thaw completely through;

4) If the average integral soil-surface temperature on the construction site is higher than the temperature of the permafrost at the level of zero annual heat fluctuation, then according to Khrustalyev the permafrost will be degraded after the area is developed; if lower, the permafrost layers will become thicker and their temperatures will become lower.

The question of degradation or thickening of a permafrost bed is not, of course, solved by investigating the influence of heat release from buildings only; it will also depend on a number of other general and local heat sources.

3. Influence of Local Factors on the Temperature Field in the Permafrost Strata

When buildings and structures are erected on permafrost, it is very important to establish correctly the fundamental specifications of construction on permafrost and to lay down correct general lines for foundation construction.

The most important factor, and one that may not be overlooked in construction on permafrost, is the thermal effect of the structure on the temperature field and properties of the permafrost, including the formation of the "bowl," which is usually filled with weak, fluidized soils.

Early builders did not take thermal effects into account, since their structures were usually built under normal conditions outside of the permafrost region, where the thermal effects of buildings and structures had no significant effect on the properties of the base soils. But heat effects are of enormous importance when erecting structures on permafrost; as we noted above, importance attaches both to the general habitation of the area and especially to the thermal effects of local factors (erection of heated buildings, structures, laying of underground heat and water lines, etc.).

As we noted earlier, a general lowering of ground temperature is observed on the built-up area in regions with low-temperature permafrost (for example, at Noril'sk, on the Kolyma, etc.), if no leakage from water mains and central heating system pipes takes place; although local thawing of the permafrost may occur under heat sources (heated buildings, warm underground pipelines, etc.), it does not extend far laterally (assuming the absence of convective heat transfer by ground and process waters).

Thus, observations made at the settlement of Myaundzha in the Magadan district (including streets and courtyards)* showed that development of the area (under low-temperature permafrost conditions) was accompanied by a lowering of soil temperature, chiefly due to snow compaction.

---

*This example was taken from the paper by G. V. Porkhayev and V. K. Shchelokov, "Prediction of Changes in the Temperature Regime of Perennially Frozen Rocks on Development of a Site," NIIOSP, 1971.

Figure 125 shows the annual average soil temperatures in Myaundzha as plotted by G. V. Porkhayev and V. K. Shchelokov on I. T. Raynyuk's data.

To clarify the fundamental aspect of the problem, let us consider an elementary scheme with a heated building erected on permafrost having a width many times greater than the depth of the active layer (the layer of annual thaw and freeze of the ground). Then, neglecting lateral heat losses (which is permissible for a large building floor area), we have according to the well-known Fourier equation for the steady-state propagation of heat

$$Q \approx \frac{\theta_r - \theta_f}{R_0} t \qquad \text{(VIII.2)}$$

where Q is the amount of heat released into the ground through the building's floor (kcal/m^2), θ_r is the indoor (positive) temperature in °C, θ_f is the average temperature of the frozen ground in °C, R_0 is the thermal resistance of the building floor in m^2 × hr × deg/kcal, and t is the time in hours.

According to expression (VIII.2), there will be a continuous heat flux from the building into the soil (larger or smaller, depending on the temperature difference $\theta_r - \theta_f$ and the thermal resistance R_0 of the floor), and more heat will continually enter the permafrost-stratum, with the obvious resultant changes in the temperature of the frozen ground and formation of a thaw bowl under the footings of the structure. Only by removal of heat (by means of a subfloor ventilated in the winter or by other methods) can the base ground be kept in the frozen state, as was shown in the very first studies of foundation building under permafrost conditions.[6]

Recent studies (by G. V. Porkhayev, et al., 1970) have shown that the thawing of permafrost in the bases of buildings will have somewhat different character depending on the geographic region in the permafrost area (northern, central, or southern), the dimensions of the footing area of the building, thermal insulation of its floor, etc. (Fig. 126).

However, as certain observations indicate, a thaw bowl will form in all cases (see Fig. 126) except those of very small buildings, when the width of the

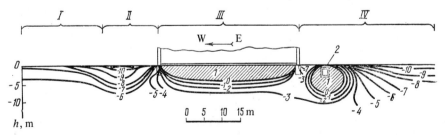

Fig. 125. Temperature field in permafrost strata under heated buildings and streets: (I) areas with undisturbed snow cover; (II) streets with highly compacted snow; (III) heated buildings (House of Culture); (IV) inner courtyard with compacted snow cover; (1) thawed zone; (2) duct of heating line.

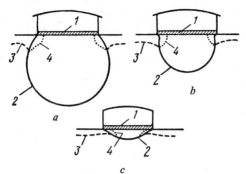

Fig. 126. Thawing of permafrost under buildings: (*a*) in the southern; and (*b*) the northern permafrost zone; (*c*) in the northern zone under small buildings; (1) heat insulation; (2) thaw limit; (3) position of thaw limit at the end of summer; (4) same, at the end of winter.

building is little more than the depth of the winter frost penetration and lateral heat losses into the ground are large; then there may be no thaw bowl although this situation is very seldom observed in practice.

Evaluating local effects on the permafrost temperature field, the heat released by heated buildings and structures, whose influence is considerably stronger than that of any possible variations of ground heat transfer outside the buildings, should be considered first.

Thus, according to calculation by Porkhayev and Shchelokov,[7] the amount of heat entering the ground on 1 m² of its area in the subfloor of a building designed to preserve the frozen state of the base was found to range from 6,600 to 28,000 kcal/m² × yr for various areas in the permafrost region, and to average from 45,000 to 54,000 kcal/m² × yr for structures built to allow thawing of their permafrost base, while the heat-flux variation due to removal of plant and snow cover was only 320 kcal/m² × yr for Skovorodino and 500 kcal/m² × yr for Yakutsk.

We may conclude from the above data that heated structures have a very strong influence on the permafrost thermal regime.

Analytical determination of permafrost temperatures in the bases of structures for various spans of time from the start of construction of formation of the thaw bowl and its stabilization (for thick permafrost-strata) is an extremely complex thermophysical problem in engineering geocryology. The complexity of the calculation stems chiefly from the need to take account of the release of the latent heat of fusion of the pore ice at the boundary between the thawed and frozen zones of the stratum.

Leaving aside the approximate method for solution of this problem that was proposed back in 1932*–based on use of a heat-balance principle for the soils

*See Ref. 1, this chapter.

under the floor and outside the building, and using the actual temperatures and thermal characteristics of the ground and the heat insulation of the floor with separate winter and summer consideration of heat losses by the base of the building—we shall focus our attention on the more rigorous recent calculation methods, taking note of the premises on which they are based and their fields of application.

The one-dimensional problem of thawing of a homogeneous medium, assuming constancy of the heating temperature and a stationary temperature distribution in the thawing medium, but with consideration of the latent heat of thawing was, as we know, first stated and solved by J. Stefan (1890). According to this solution, the depth to which the ground thaws is directly proportional to the square root of time [relation (VI.15)]. Later, an improved solution of the one-dimensional problem of soil thawing was given in the papers of M. M. Krylov, V. A. Kudryavtsev, V. P. Ushkalov, V. S. Luk'yanov, and M. D. Golovko, and others.

The solution for the symmetrical two-dimensional problem for a thawing frozen-ground half-space was first obtained by S. S. Kovner (1933)[8] assuming a stability of the temperature fields in the thawed and frozen zones of the ground and a moving interface between them. Here it was assumed that the ground temperature was zero and that the heating surface had no thermal insulation. According to Kovner's solution, the isotherms in the ground under the belt of heating are arcs of circles that pass through its end points, and the maximum thaw depth occurs on the symmetry axis and is also proportional to the square root of time.

This solution was used in frozen-ground mechanics back in 1937,* but it did not come into extensive use due to the limitations noted.

The two- and three-dimensional problems of permafrost thawing under structures were later developed most fully in the work of G. V. Porkhayev, initially only for the depth of thaw under the center of the heated area, and then also for various other points.[9]

We take note of the semiempirical relation for the thaw depth of permafrost under heated structures that was proposed by Ushkalov,[10] who extended the one-dimensional problem of soil thawing to the three-dimensional case by introducing a correction factor whose values were assumed to be constant and to be dependent only on the width of the building and its length-to-width ratio, which, of course, can be regarded only as an approximate solution. Ushkalov determines the value of the correction factor on the basis of laboratory-experiment results and observations of permafrost thaw depths under structures built in the southern zone of the permafrost region.

Porkhayev showed that the correction factor applied by Ushkalov is a variable, depending on the thaw depth and the dimensions of the thaw zone, and

*See Chap. II, Ref. 31.

that it does not take account of the main heat flux in the direction away from the thaw boundary.

All the above considerations limit the use of Ushkalov's equation to the southern regions of the permafrost region for determination of thaw depths under the centers of buildings.

We take note of a solution of the two-dimensional problem of permafrost thawing under structures that was obtained by S. V. Tomirdiaro and can be regarded as a development of Kovner's problem. The solution is obtained in a simple closed form and makes it possible to determine not only the contour of the thaw bowl in the stationary temperature field of the two-dimensional problem, but also all other isotherms in the heated base.[11]

In his development of Kovner's problem, Tomirdiaro adds the geothermal flux to the heat flux from the building and takes account of heat insulation on the soil surface by increasing the active-layer thickness by a certain amount, equivalent with respect to heat insulation; as pointed out by Porkhayev, this increases slightly the value obtained for the ground thaw limit, since the heat insulation is also extended to the area outside of the building in Tomirdiaro's procedure.

Tomirdiaro's solution takes the following form for the case of a row of buildings situated on the surface:

$$\theta_{xy} = \frac{1}{\pi} \sum_{1}^{n} \left[\left(\frac{\lambda_t}{\lambda_f} \theta_{id} - \theta_0 \right) \left(\arctan \frac{\frac{B_n}{2} - x + l_n}{y + R_{fl}\lambda_t} + \arctan \frac{\frac{B_n}{2} + x - l_n}{y + R_{fl}\lambda_t} \right) \right]$$
$$+ \theta_s + Gy \qquad \text{(VIII.3)}$$

where θ_{xy} is the ground temperature at the point with coordinates x and y, θ_{id} and θ_0 are the indoor temperature and the temperature at the depth of zero heat fluctuation ($\sim$ 10 meters) in the ground, B_n is the width of a single building in the row, l_n is the distance from the coordinate origin (the center of the extreme left-hand building) to the center of each building when several buildings are placed on the surface, R_{fl} is the thermal resistance of the building floor ($R_{fl}\lambda_t$ is the thickness of the ground layer with equivalent heat-insulating value), G is the geothermal gradient (approximately 0.02–0.04 deg/m in the permafrost zone), and θ_s is the annual average temperature of the ground surface (which can be assumed equal to θ_0).

The following approximate expression can be used to determine the highest frozen-ground temperature at any point at depths greater than the thickness of the active layer:

$$\theta_{max} \approx \theta_{xy} + A_i \qquad \text{(VIII.4)}$$

where A_i is the amplitude of the annual temperature fluctuations at the boundary of each ith soil layer, with

$$A_i = \theta_{i-1} e^{-h_i \sqrt{(\pi/a_i^e)T}}$$

Then

$$\theta_{max} \approx \theta_{xy} + \theta_{i-1} e^{-h_i \sqrt{\pi/a_i^e T}} \qquad (VIII.4')$$

where θ_{i-1} is the temperature of the preceding ground layer, h_i is the thickness of the layer, a_i^e is the effective thermal diffusivity (considering the conversion of part of the unfrozen water to ice) for the ith soil layer, and T is the period of the temperature oscillations ($T = 1$ yr = 8,760 hr).

With simple calculations, Eqs. (VIII.3) and VIII.4) can be used to construct isotherms of the stationary temperature field in the ground under heated structures and to determine the highest temperature to which the ground is heated in the bases of the structures.

By way of example, Fig. 127 shows isotherms of the stationary temperature field in permafrost under a single heated ($\theta_{id} = +10°C$) building (width $B = 18$ m) as calculated from relation (VIII.3), and Fig. 128 shows a comparison of the

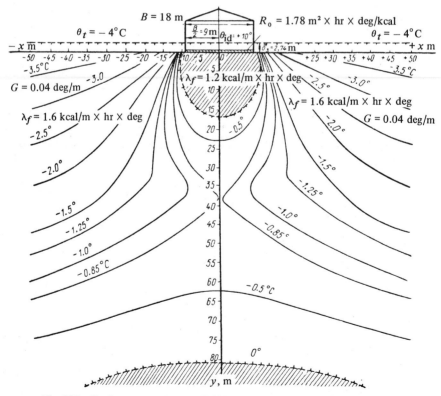

Fig. 127. Stationary temperature field in the ground under a heated building.

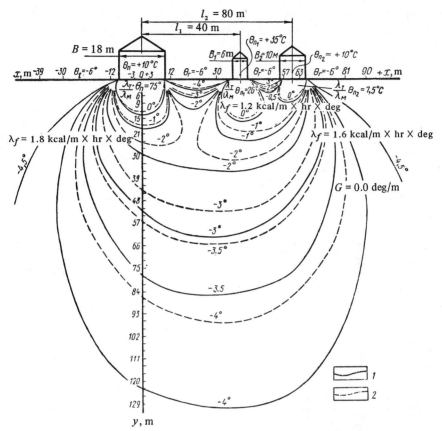

Fig. 128. Comparison of calculated stationary field with the field obtained on the EGDA-9/60 electrohydraulic integrator for a group of three buildings: (1) field isolines obtained on the EGDA; (2) calculated isotherms.

calculated stationary temperature field in the stratum under a group of three heated buildings with insulated floors obtained by Tomirdiaro on the EGDA-9/60 electrical integrator.* These data indicate adequate agreement (in regard to the general outline and size of the influenced zones) between the calculations and the modelling data.

Mention should also be made of V. D. Braun's method for graphical construction of the stationary temperature field for the case of the two-dimensional problem with consideration of the geothermal gradient but without consideration of the difference in the thermophysical properties of the soils in the thawed and frozen states or of the heat insulation of the floor.**

*See Ref. 11, this chapter.
**See Ref. 9, part 2, this chapter.

More rigorous solutions of the thermophysical problem of permafrost thawing under heated structures has been obtained in recent years by Porkhayev (with the aid of the Poisson integral, applied to the specific case of the Dirichlet boundary-value problem) and also by L. N. Khrustalev.[12]

Porkhayev's method has been brought to a form convenient for practical use and is recommended by SNiP II-B.6-66; it is this method that we shall discuss here.

To solve the extremely complex thermophysical problem posed, Porkhayev used a "method of auxiliary temperatures," in which the temperature distribution in each zone (thawed and frozen) is quasi-stationary, and a certain auxiliary temperature defined by*

$$\theta_t = \theta_{0a} + (\theta_s - \theta_{0a}) f(x, y, z) \qquad \text{(VIII.5)}$$

is maintained to preserve the law of migration of the interface (the θ° isotherm); here θ_t is the temperature in the thawed zone, θ_{0a} is the auxiliary temperature at the boundary of the frozen zone, and θ_s is the temperature of the heating surface; $f(x, y, z)$ is a function that depends on the system configuration (it is determined by the Poisson integral).

The heat insulation of the heating surface is taken into account in consideration of two-dimensional and three-dimensional permafrost thawing problems in the bases of structures, but when the shaping of the temperature field is of interest, the auxiliary-temperature method is used, with the calculation based on the annual average ground temperature disregarding the heat flux from the interior of the earth (along the geothermal gradient); but the temperature functions are determined by the Poisson integral.

For the two-dimensional problem

$$\theta(x, z) = \frac{z}{\pi} \int_{-\infty}^{+\infty} \frac{\theta_1(x')}{(x - x')^2 + z^2} dx' \qquad \text{(VIII.6)}$$

Solution of (VIII.6) for a strip of width B (with $-B/2 \leqslant x \leqslant B/2$) within which the temperature is constant at θ_s has the form

$$\theta(x, z) = \theta_0 + (\theta_s - \theta_0) f(x, z) \qquad \text{(VIII.7)}$$

where the configuration function is given by

$$f(x, y) = \frac{1}{\pi} \left(\text{arc tan} \frac{B + 2x}{2z} + \text{arc tan} \frac{B - 2x}{2z} \right) \qquad \text{(VIII.8)}$$

Substituting (VIII.8) into (VIII.7), we obtain

$$\theta(x, z) = \theta_0 + (\theta_s - \theta_0) \frac{1}{\pi} \left(\text{arc tan} \frac{B + 2x}{2z} + \text{arc tan} \frac{B - 2x}{2z} \right) \qquad \text{(VIII.9)}$$

---

*See Ref. 9, part 1, this chapter.

Introducing a factor for the heterogeneity of the thawed-frozen zone (after S. G. Gutman) into (VIII.7), we obtain

$$\theta(x, z) = \theta_0 + \left(\theta_s \frac{\lambda_f}{\lambda_t} - \theta_0\right) f(x, z) \qquad \text{(VIII.7')}$$

Finally, we have

$$\theta(x, z) = \theta_0 + \left(\theta_s \frac{\lambda_f}{\lambda_t} - \theta_0\right) \frac{1}{\pi} \left(\text{arc tan} \frac{B + 2x}{2z} + \text{arc tan} \frac{B - 2x}{2z}\right) \qquad \text{(VIII.9')}$$

In the case of the three-dimensional problem (for a stationary thermal field), the Poisson function is

$$\theta(x, y, z) = \frac{z}{\pi} \int_{-\infty}^{\infty} \int_{-\infty}^{\infty} \frac{\theta_1(x', y') \, dx' dy'}{[(x - x')^2 + (y - y')^2 + z^2]^{3/2}} \qquad \text{(VIII.10)}$$

For a rectangle of width B and length L on whose area the temperature is constant at θ_s, while the soil outside of the contour is at the annual average surface temperature, which can be put equal to the temperature θ_0 at the depth of zero thermal fluctuation in the permafrost stratum, the solution of (VIII.10) with consideration of the heterogeneity of the thawed-frozen medium takes the form

$$\theta(x, y, z) = \theta_0 + \left(\theta_s \frac{\lambda_f}{\lambda_t} - \theta_0\right) f(x, y, z) \qquad \text{(VIII.11)}$$

where the configuration function $f(x, y, z)$ is determined by

$$f(x, y, z) = \frac{1}{2\pi} \left[\text{arc tan} \frac{(x + B/2)(y + L/2)}{z\sqrt{z^2 + (x - B/2)^2 + (y + L/2)^2}} \right.$$

$$- \text{arc tan} \frac{(x - B/2)(y + L/2)}{z\sqrt{z^2 + (x - B/2)^2 + (y + L/2)^2}}$$

$$- \text{arc tan} \frac{(x + B/2)(y - L/2)}{z\sqrt{z^2 + (x + B/2)^2 + (y - L/2)^2}}$$

$$\left. + \text{arc tan} \frac{(x - B/2)(y - L/2)}{z\sqrt{z^2 + (x - B/2)^2 + (y - L/2)^2}} \right] \qquad \text{(VIII.12)}$$

Expression (VIII.7) with (VIII.8) and (VIII.11) with (VIII.12) can be used for direct determination of the temperature at any point of a stationary temperature field in the ground under heated structures.

In the case of a nonstationary temperature field the problem becomes considerably more complicated since it is necessary to base the calculation on the overall heat balance at the interface between the thawed and frozen zones, with consideration of the latent heat of fusion of the ice and the system configuration function.

We turn now to treatment of such problems.

From (VIII.5), putting the temperature at the thawed-frozen interface equal to zero (i.e., $\theta_t = 0$), we obtain an expression for the auxiliary temperature:

$$\theta_{0a} = \theta_s \frac{f(x_i, y, z_i)}{f(x_i, y_i, z_i) - 1} \qquad \text{(VIII.13)}$$

where x_i, y_i, and z_i are the coordinates of some point on the interface between the zones.

Using (VIII.5) for the ground temperature in the thawed zone, we obtain

$$\theta_t = \theta_s \frac{f(x_i, y_i, z_i) - f(x, y, z)}{f(x_i, y_i, z_i) - 1} \qquad \text{(VIII.14)}$$

and for the ground temperature in the frozen zone

$$\theta_f = \theta_0 \frac{f(x_i, y_i, z_i) - f(x, y, z)}{f(x_i, y_i, z_i)} \qquad \text{(VIII.15)}$$

The displacement of the thaw surface is determined from the equation of the heat balance on a unit area of the surface with the coordinates x_i, y_i, z_i:

$$\lambda_t(\theta_t)'_n - \lambda_f(\theta_f)'_n = \zeta \frac{ds}{dt} \qquad \text{(VIII.16)}$$

where $(\theta_t)'_n$ and $(\theta_f)'_n$ are the temperature derivatives on the normal to the unit area, ds is the differential of the flowline passing through the point x_i, y_i, z_i, ζ is the latent heat of ice fusion and t is the heating time.

With the general expressions for the temperature (VIII.14) in the thawed zone and (VIII.15) in the frozen zone, the basic heat-balance equation acquires the following general form:

$$\lambda_t \theta_s \left[\frac{f'(x, y, z)}{f(x, y, z) - 1} \right]_n - \lambda_f \theta_f \left[\frac{f'(x, y, z)}{f(x, y, z)} \right]_n = \zeta \frac{ds}{dt} \qquad \text{(VIII.17)}$$

where $f(x, y, z)$ is the system configuration function and $f'(x, y, z)$ is the derivative of this function.

Substituting the value of the system configuration function [expression (VIII.8) for the two-dimensional problem and (VIII.12) for the three-dimensional problem] and its derivatives along the normals to the elementary area into Eq. (VIII.17) and solving the resulting equation, we determine the temperatures in the thawed and frozen zones and the depth of thaw of the permafrost for the particular problem.

Porkhayev then completed numerical solutions of the problems posed under the following basic premises: The solution is based on the annual average soil temperature; the thermal insulation of the heating surface is taken into account by a layer of soil of equivalent thermal conductivity, but only on the heating area; the heat flux due to the geothermal gradient is left out of account, as well as the heating of the thawed zone of the soil; the three-dimensional character of the problem is taken into account by multiplying the calculated thaw depth by a correction factor equal to the ratio of the thaw depth in the three-dimensional solution of the problem to the thaw depth in the two-dimensional solution.

To facilitate practical use of these extremely complex solutions of the thermophysical problem, Porkhayev reduced them to tabulated solutions and plots of the design functions.

In general form, for example, the soil-thaw depth under a structure to be built on permafrost is determined by the expression

$$h_t = f(\xi_i, k_i)B \qquad \text{(VIII.18)}$$

where the coefficient ξ_i is a function of the following parameters:

$$I = \frac{\lambda_t \theta_s}{qB^2} t \qquad \text{(VIII.19)}$$

$$\alpha = \frac{\lambda_t R_0}{B} \qquad \text{(VIII.20)}$$

$$\beta = \frac{\lambda_f \theta_0}{\lambda_t \theta_s} \qquad \text{(VIII.21)}$$

where R_0 is the resistance of the building floor to heat transfer in $m^2 \times hr \times deg/kcal$ and q is the thaw heat of the frozen ground in $kcal/m^3$.

The value of the coefficient $k_i = f(\beta, I, L/B)$ was tabulated. To determine the influence of the coefficient ξ_i, graphs were plotted for various points of the thaw bowl in the permafrost under the structure.

The procedure for use of the above equations and the tabulated solutions and diagrams for determination of the influence coefficients in calculations of permafrost thaw depths under structures for various time intervals from the beginning of the structures use is set forth in succeeding chapters of this book. These aids will be needed then to predict foundation settling on the thawed soils.

It should be noted that to the ground thaw depth obtained due to development of the locality must be added to the answer obtained for the thaw depth; it can be calculated if the heat flux given by Eq. (VIII.5) is known.

4. Temperature Stability of a Dam Built from Local Materials under Permafrost Conditions

1. Dams built from local materials under conditions of permafrost. The enormous water and energy resources of the permafrost regions and the development of industry in these regions call for the construction of dams for utilization of the water reserves, chiefly those in rivers, although the rivers discharge at highly nonuniform rates over the course of the year.

The remoteness of the regions, the frequent absence of efficient communications, and the need to meet schedules in exploitation of water resources have given rise to the practice of building dams from local materials (chiefly rock-filled types) in the permafrost distribution regions.

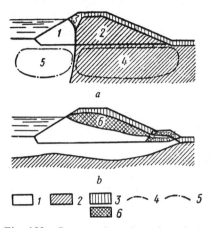

Fig. 129. Cross sections through typical earth dams built on permafrost: (a) dam using frozen ground as material for strength and impermeability to water (nonfiltering dam); (b) filtering dam; (1) zone of permanently thawed soil; (2) zone of permanently frozen soil; (3) zone of alternating thawing and freezing; (4) zone in which natural talik must be frozen; (5) zone in which preliminary thawing is desirable; (6) frost-heave zone with water influx.

In the construction of dams on permafrost, the first question that arises concerns the thermal stability of the soils, since most permafrost soils lose their load-bearing ability on thawing and become highly permeable to water, a property unacceptable in hydraulic-engineering structures.

Following Prof. Ye. V. Bliznyak[13] and Prof. P. A. Bogoslovskiy,[14] we must distinguish two fundamental types of dams built from local materials on permafrost: (1) nonfiltering dams that use the permafrost as extremely strong and water-impermeable materials (Fig. 129a); (2) filtering dams, which are designed with consideration of permafrost thaw in their bases (Fig. 129b).

Dams of the first type (nonfiltering dams) can be designed in either of two versions: (1) to preserve the frozen state of the soils in the base and body of the dam or (2) with consideration of thawing of the permafrost during construction and use of the dam.

In the first case ("frozen variant"), the dam can be built with practically any type of soil, provided that the frozen state is preserved, and with it the strength and water impermeability of the ice-saturated frozen ground. In the second case ("thawed version"), earth and rock-filled and concrete dams can be built only on strong bedrock or on coarse-skeleton ground with low compressibility and mandatory placing of an inpermeable clay core.[15]

As an example, we might cite the 9.5-meter-high flat-profile dam of the metallurgical plant at Petrovsk-Zabaykal'skiy, which was built in the winter of 1792 with frost penetration of the packed soil and survived without incident until 1929, when the temperature regime of the frozen ground was disturbed during a repair of the wooden raceways and the ground thawed in the base and body of the dam, requiring immediate repair and addition of fill. However, since the body of the dam had been laid on incompressible sand and gravel deposits (2 to 5 meters thick) supported by weathered diorite, the dam continues to function to this day in the thawed version after its 1939–1945 rebuilding and repair.

Another example of a quite long-lived dam on permafrost soils is the one on the Dolgaia River at Noril'sk (Fig. 130), a frozen version built in 1943 (height 10 meters, length 130 meters). Artificial chilling with wells drilled at a distance of 2.5 meters, with circulation of a cooling solution of calcium chloride, was used to freeze the body of the dam and the talik under the river bed; the well pipes were then left in reserve for winter cooling by air. It is interesting to note that heating of the downstream bank by the sun was eliminated by building a shade and an ice gallery of the ice-storage type proposed by M. M. Krylov, measures that proved quite effective. The dam is now functioning without auxiliary artificial cooling (although the winter-air-cooling option has been left open).

Thus, even these low earth dams built in the permafrost regions required auxiliary systems to keep them in a stable state.

Examples of higher dams that were built successfully in permafrost regions can also be cited, but all of them were built with artificial freezing of the soils by means of a permanent cooling installation or make use of such an installation periodically (the Irelyakh dam, 20.7 meters high, 115 meters wide at the foot and 8 meters at the crest, with a single row of freezing columns in a loam core) or were built on incompressible bedrock (Vilyuy dam, height 75 meters; Fig. 131, and others).

We should note that there are now no specially high dams built from local structural materials using frozen zones as antifiltration elements in the permafrost region.

This is explained by the following considerations. According to the equations for the rates of changes in the ground temperature field on built-up areas when

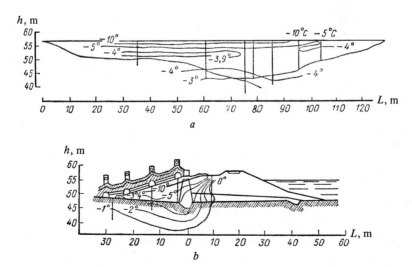

Fig. 130. Dam on Dolgaia River, showing ground isotherms: (*a*) in longitudinal section; (*b*) in transverse section.

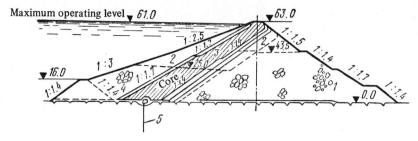

Fig. 131. Vilyuy dam: (1) rock fill; (2) filtering layer; (3) clay core; (4) rock ballast; (5) grout screen.

the integral average surface temperature is higher than the temperature of the permafrost (the temperature at the depth of zero annual amplitude), as is always the case in reservoirs (where the water has always a low, but positive temperature), a thaw bowl will invariably form, and if the width of the reservoir is equal or nearly equal to the depth of the permafrost bed, a continuous talik will form under the reservoir. At smaller reservoir widths, solutions of the two- and three-dimensional thermophysical problems indicate that thawing may also be less than to the bottom of permafrost due to lateral heat losses in the ground. This circumstance, which is highly important for dams built on unconsolidated soil, is not as important for dams built on relatively unfissured bedrock.

It must also be remembered that natural cooling of soils (e.g., on the downstream slope of the dam) will occur only to a depth on the order of 10 meters or slightly larger (if the face is cleared of snow); auxiliary artificial cooling (by galleries, a row of freezing wells, etc.) is therefore required for greater soil thicknesses in the slopes of the dam.

Based on the above, dams of medium height (more than 10 meters) and high dams in particular (heights in the tens of meters), when built according to the cold version (with the ground kept frozen) always require a water-impermeable core with a built-in freezing system appropriately designed to keep the core of the dam in the frozen state.

As for filtering dams built in the permafrost region, they must be constructed only in the warm variant, preferably on bedrock or coarse deterital deposits, and with mandatory consideration of settling of the frozen rocks on thawing.

2. Calculation of the temperature stability of dams to ensure long trouble-free service is paramount when dams are built from local materials on permafrost.

The following must be considered in studying the temperature stability of dams built from local materials on permafrost:

1) The steady-state temperature field as the extreme temperature state of frozen-earth dams, to which the temperature changes in the body and base of the dam caused by disturbance of the natural situation during construction eventually lead;

2) The nonsteady temperature field, prediction of which makes it possible to judge the temperature stability of a dam from the time of its erection and at an arbitrary time during its service life (1, 5, 10, 20 years etc., after construction) after construction is completed.

The problems posed are extremely complex two-dimensional and three-dimensional problems in the thermophysics of freezing and thawing ground (here special difficulties arise for the most part out of the need to consider the changes in the phase composition of the water in the frozen ground at the mobile boundaries between the frozen and thawed zones), the basic trends in solution of which and certain general results of these trends will be considered here. However, it will be necessary to refer the reader to the appropriate literature for more detailed investigation of the problems stated.[16]

3. According to the corresponding calculations, the *stationary temperature field* in a frozen earth dam and in base zones adjacent to it is established only after a very long time, on the order of 50–100 years for low dams (less than 10 meters high) and several hundred years (up to 1,000–2,000) for large dams; thus, the pattern of the stationary temperature fields of frozen dams built on permafrost is only of conceptual nature and useful for establishing the limit to which the temperature variation in the body and base of the frozen dam tend for given boundary conditions during its construction and life.

As we know, the equation of a stationary temperature field is the Laplace equation

$$\nabla^2 \theta = \frac{\partial^2 \theta}{\partial x^2} + \frac{\partial^2 \theta}{\partial y^2} + \frac{\partial^2 \theta}{\partial z^2} = 0 \qquad \text{(VIII.22)}$$

and, in the case of the two-dimensional problem,

$$\nabla^2 \theta = \frac{\partial^2 \theta}{\partial x^2} + \frac{\partial^2 \theta}{\partial y^2} = 0 \qquad \text{(VIII.23)}$$

The solution of the latter equation can easily be obtained with the EGDA-9/60 instrument. As an example, Fig. 132 shows isotherms in a rock-filled dam consisting of rock with ice-filled cavities in the upstream face and an ice core with rock ballast in its downstream face (after a plot by N. V. Ukhova). The isotherms in Fig. 132 show that in order to keep the body of the dam frozen it is necessary to provide constant artificial cooling of the ice core and base of the dam by means of a freezing system; otherwise, thawing and destruction of the base would be inevitable.

Figure 133 shows isotherms of the stationary temperature field in a filtering dam as calculated by P. A. Bogoslovskiy: in this example, it was assumed that the water temperature in the reservoir was $\theta = +3°C$, that the base is perfectly water-impermeable, and that its footing, like the body of the dam above the depression curve, has a constant temperature $\theta = 0°C$. Needless to say, it was

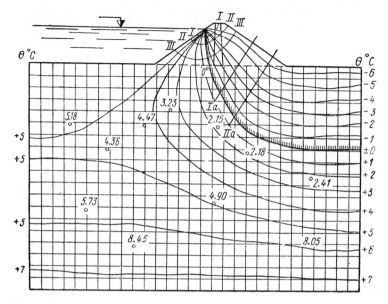

Fig. 132. Temperature variation in body and base of dam 75 years after filling the reservoir.

difficult even in this case to keep the base frozen (without constant artificial support of its temperature at 0°C).

Prediction of the nonstationary temperature fields in frozen dams built from local structural materials (rock fill, earth, etc.) is of tremendous practical importance, since it enables us to estimate not only the variation of the temperature field in the dam body and base at various times after filling of the reservoir starts, but also to plan measures to control these variations in the directions necessary for stability of the dam.

It should be noted that, because of the extreme complexity of the non-stationary temperature-field problem in frozen dams, no rigorous solution has as yet been obtained for the problem; however, an approximate (engineering) method based on a series of assumptions (which have been verified to some degree under laboratory and field conditions) has been developed; it uses the

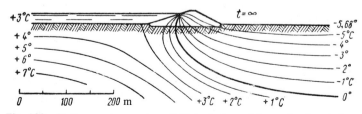

Fig. 133. Stationary temperature field in body and base of a dam built on permafrost (according to calculations and plot by P. A. Bogoslovskiy).

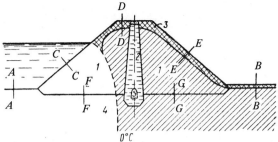

Fig. 134. Design scheme for rock-filled frozen dam on permafrost: (1) rock fill with ice; (2) ice core; (3) dry rock fill; (4) base (permafrost stratum); (5) gallery.

available solutions for the individual thermophysical problems of freezing and thawing.

4. Engineering method developed at the MISI [Moscow Construction Engineering Institute] for forecasting the nonstationary temperature fields of frozen dams built in permafrost regions.[17]

A design scheme for a rock-filled dam with an ice core built under permafrost conditions is shown in Fig. 134.

In this method, the general solution for the temperature stability of a dam built from local materials on a permafrost base is broken down into a number of simpler linear problems; design cross sections are established at locations sufficiently remote from boundary influences so that lateral heat losses can be neglected with accuracy sufficient for practical purposes (for example, sections $A-A$, $B-B$, ..., $G-G$ in Fig. 134). If necessary, two additional one-dimensional problems are used: one in which the zero isotherm is displaced from a region with one set of values of thermophysical characteristics into a region with another set (stratified bed), or one in which the temperature fields formed from various sources are superimposed on one another.

The general temperature field of the dam and base is constructed for each design time value based on the solution of the above problems by curvilinear interpolation method.

For linear problems, either the known solutions of T. Stefan and L. Leybenzon are used or a comparison is made between the thermal balance with consideration of the release of latent heat of ice melting (or water freezing at the boundary between the thawed and frozen zones. If the temperature change occurs without phase changes of the water (when the temperature does not pass through $0°C$ as it varies, and the base is composed of relatively unfissured bedrock, etc.), the known solutions of A. V. Lykov are used.

Thus, the heating temperature of a reservoir bottom composed of bedrock can be determined for sufficient distances from the upstream face (section $A-A$, Fig. 134) using Lykov's solution as for the case of a semibounded laterally thermo-isolated stream:

$$\theta(x, t) = \theta_1 \text{ erfc } \frac{x}{2\sqrt{\pi at}} \int_0^\infty f(\xi) \left[\exp\left(-\frac{(\xi - x)^2}{4at}\right) - \exp\left(-\frac{(\xi + x)^2}{4at}\right) \right] d\xi$$

$$(\text{VIII.24})$$

where $\theta(x, t)$ is the temperature at the point with coordinate x (depth) at time t, θ_1 is the surface temperature of the floor of the reservoir (it is assumed constant), $f(\xi)$ is the initial temperature distribution in the base, a is the thermal diffusivity of the base rock, and erfc $= 1 - $ erf, where erf is the Gaussian error integral.

At a constant temperature at the base ($\theta_0 = $ const), relation (VIII.24) becomes

$$\theta(x, t) = (\theta_1 - \theta_0) \text{ erfc } \frac{x}{2\sqrt{at}} + \theta_0 \qquad (\text{VIII.25})$$

For section B-B, i.e., for the base sufficiently remote from the downstream face, the same relation (VIII.24) is used, but the reservoir bottom surface temperature θ_1 is replaced by the temperature θ_2 that of the base surface on the downstream side.

For section C-C, the thawing of the frozen upper wedge from the heat of the reservoir, with consideration of melting of the ice in the cavities in the rock or soil fill, can be treated as Stefan's one-dimensional thaw problem.

Then the depth of thaw (the penetration depth of the temperature $\theta = 0$) can be determined from Stefan's simplified formula

$$\xi = \sqrt{\frac{2\lambda_t \theta_1 t}{\zeta W_i}} \qquad (\text{VIII.26})$$

where λ_t is the thermal conductivity of the thawed soil or rock fill, ζ is the latent heat of melting of the ice, and W_i is the content by weight of ice in the fill or soil and equals

$$W_i = i_{\text{vol}} \gamma_i = \gamma \frac{W_d - W_u}{1 + W_d}$$

We note that convective heat transfer processes must be taken into account for a certain granulometric composition of the rock fill or soil forming the body of the dam. Studies made at MISI by N. V. Ukhova[18] have shown that when convective heat transfer is taken into account in Eq. (VIII.26) for the thaw depth of rock fill or soil, it is necessary to replace the thermal conductivity λ_t of the thawed soil by the so-called effective thermal conductivity λ_e, which takes account of convective heat transfer; then

$$\lambda_e = \phi \lambda_t \qquad (\text{VIII.27})$$

where ϕ is the coefficient of thermal conductivity increase due to the presence of convection currents and is determined from

$$\phi - 1 = 2.1 \times 10^5 \frac{n^3}{1 - n^2} \times \frac{hd_e^2}{\lambda_t} \qquad \text{(VIII.28)}$$

here n is the porosity of the fill (or the earth in the face), h is the depth of the particular section through the soil from high water level in the reservoir, and d_e is the equivalent diameter of the rock fill or soil (according to Ye. A. Zamarin).

For a given value of the filtration coefficient of the fill or earth, the coefficient ϕ can be determined from the expression

$$\phi - 1 \approx 3.3 \times 10^{-5} \frac{kh}{\lambda_t} \qquad \text{(VIII.29)}$$

where k is the filtration coefficient.

Diagrams were plotted for Eqs. (VIII.28) and (VIII.29) and published in the previously cited work of N. A. Tsytovich, N. V. Ukhova, and S. B. Ukhov.

Convective heat transfer must be taken into account when the coefficient $\phi \geqslant 1.3$.

Cooling of the crest of the dam (section D-D) and the downstream face (section E-E) and part of the upstream face for sections where lateral heat losses can be neglected is determined approximately from Lykov's relation for a semibounded laterally insulated stream:

$$\theta(x, t) = \theta_2 \left(1 - \operatorname{erf} \frac{x}{2\sqrt{at}}\right) \qquad \text{(VIII.30)}$$

where $\theta(x, t)$ is the temperature at a point at distance x along the normal for time t and a is the thermal diffusivity of the dam face material in its frozen state.

Expression (VIII.30) is valid for the case in which the region of the dam body being considered has an initial temperature $\theta = 0°$, and all of the water in the pores of the soil or the cavities of the rock fill is in the form of ice and no phase changes occur in it on further cooling.

The thaw of part of the dam base (section F-F) and heating or cooling of another part (section G-G) under the influence of the temperature field of the base and the variation of temperature in the base are considered on the basis of the thermal-balance equations. Detailed solutions are given for various cases in the aforementioned monograph of Tsytovich, Ukhova, et al.,* which presents practical recommendations for calculation of the temperature fields of frozen heterogeneous dams and their bases and the results of temperature regime studies in models of frozen dams.

We note that a complete finite-difference solution has been obtained by P. A. Bogoslovskiy (two-dimensional problem) for the particular case in which the dam and base consist of the same type of ground and have identical material characteristics.**

*See Ref. 16, part 5, this chapter.
**See Ref. 16, parts 1 and 2, this chapter.

Using the solutions of the aforementioned relatively simple thermophysical problems, we can establish points corresponding to certain temperatures (0°, −1°, −2°, etc.) and use curvilinear interpolation to construct isotherms of the temperature field in the dam and permafrost base from these points. Inferences as to the thermal stability of the earth dam and its base in time can be drawn from the resulting pattern.

To illustrate the method described above for calculation of the thermal stability of dams built from local materials (rock fill) on permafrost, Figs. 135, 136, and 132 present isotherm plots of the temperature field calculated by N. V. Ukhova for a frozen rock-filled dam consisting of three zones: rock fill with ice-filled cavities, an ice core with a gallery for artificial cooling, and rock fill.

The isotherms of Fig. 135 show that there is a thaw danger in the lower part of the core during the initial period of use (5 years), and that the stationary thermal field (see Fig. 132) will not be obtained before 75 years. Comparison of the isotherms in Fig. 132 with the data obtained on the EGDA-9/60 electrohydraulic integrator indicated close agreement. A check in a laboratory experiment on a physical model and comparison with data obtained on the EGDA-9/60 also gave quite comparable results (Fig. 137).

The method set forth above for approximate estimation of the temperature stability of dams built from local materials under permafrost conditions naturally requires further improvements and establishment of the degree of acceptability of the approximate solution (based on the comparison with the rigorous solution, first of all of the two-dimensional problem of the nonstationary temperature field in frozen dams).

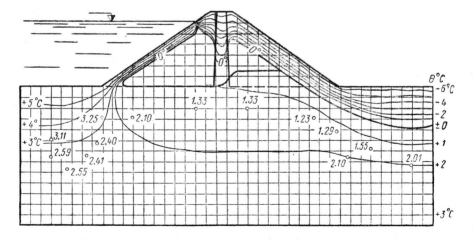

Fig. 135. Example of forecast of temperature variations in body and base of frozen dam five years after filling of reservoir.

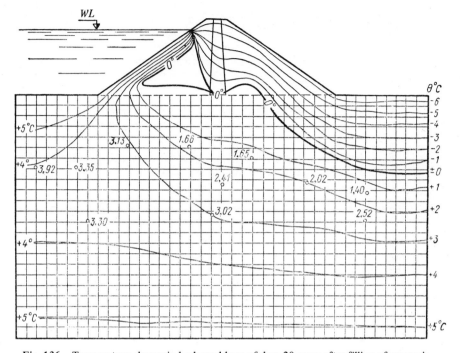

Fig. 136. Temperature change in body and base of dam 20 years after filling of reservoir.

Both the presented and other similar investigation materials* indicate that in order to preserve temperature stability of frozen dams built from local materials to heights greater than 10 meters on permafrost, provision of a water-impermeable core with artificial mechanical refrigeration of the forming taliks is absolutely necessary.

5. The calculation of the freezing of water-impermeable dam elements using mechanical refrigeration should be carried out on the basis of recommendations by Kh. R. Khakimov** or N. G. Trupak.[19]

A not too large freeze pipe radii r (in any event, smaller than two meters) and outside air is used as the refrigerant, the radius R of the frozen ground cylinder can be determined from a relation proposed by A. A. Tsvid:[20]

$$R \approx \sqrt[3]{\frac{3t\theta_s r \lambda_f}{q}} + 0.5r \qquad \text{(VIII.31)}$$

where t is the time of operation of the cooling system in hours, θ_s is the average temperature of the pipe surface, λ_f is the thermal conductivity of the frozen ground in kcal/m × hr × deg, and q is the amount of heat needed to freeze 1 m³ of soil in kcal/m³.

*See Ref. 16, this chapter.
**See Chap. IV, Ref. 35.

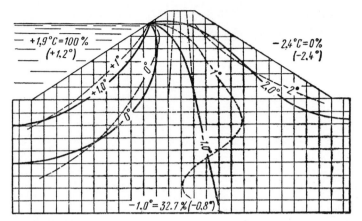

Fig. 137. Comparison of standard temperature field plotted on EGDA-9/60 (solid line) with that obtained in the MISI laboratory on a physical model in experiment No. 3 (dash line) for $t = 40$ years.

Using the theory of similarity and its methods, combined with the solution of many particular problems on the Luk'yanov hydraulic integrator, R. M. Kamenskiy[21] obtained empirical relationships for determination of the operating time of the freezing system up to the point at which a solid frozen ground wall of definite dimensions has formed, under consideration of the mutual thermal effects between two adjacent freezing columns. These relationships can be used in the design of antifiltration elements in dams built from local materials.

The time of freezing-system operation until the frozen ground cylinders fuse will be

$$\text{Fo} = 0.87 \exp\left(-1.66\text{Bi}^{0.114}\right) \text{Ko}\left[\frac{\theta_a}{\theta_i}\right]^{-0.133} \left(\frac{h}{r_1}\right)^{-2.45} \quad \text{(VIII.32)}$$

and the time to formation of a frozen ground wall with a minimum width of $2h$

$$\text{Fo} = 1.13 \exp\left(-0.897\text{Bi}^{0.295}\right) \text{Ko}\left[\frac{\theta_a}{\theta_i}\right]^{-0.15} \left(\frac{h}{r_1}\right)^{2.4}$$

$$+ 2.8\left(\frac{\text{Bi}}{3.2}\right)^3 - \text{Ko}^{\left(\sqrt{\frac{h}{r_1}} - 0.08\frac{h}{r_1}\right)} \quad \text{(VIII.33)}$$

In these equations

$\text{Fo} = \lambda_f t / C_f r_1^2$ is the Fourier criterion
$\text{Bi} = \alpha d_1 / \lambda_f$ is the Biot criterion
$\text{Ko} = q / C_f(\theta_a)$ is the Kossovich criterion

λ_f and C_f are the thermal conductivity and volume heat capacity of the frozen ground, t is the operating time of the freezing system, r_1 and d_1 are the outside radius and the diameter of the freezing column, h is the spacing of the columns, α is the coefficient of heat transfer from the air to the pipe walls in the annular jacket of the column, θ_a is the air temperature in the column averaged over the design period, θ_i is the initial ground temperature, and q is the expenditure of heat for phase change in the water in 1 m³ of soil.

R. M. Kamenskiy's paper includes nomograms for calculations using relations (VIII.32) and (VIII.33) and an example of the calculation for one of the design versions of the dam.

6. Strength of the ice core. In the design of frozen ground dams to be built under permafrost conditions, it is highly important to estimate the strength of the ice core and ensure that the ice will not be forced out at the top under the pressures from the supporting rock-fill prisms (Fig. 138).

Assuming that the rock fill in the prisms is at stable equilibrium, and considering the plane symmetrical problem of the static equilibrium of heavy supporting prisms, the value of the maximum pressure (active E_a) of the rock fill on the walls of the ice core was determined as follows:[22]

$$E_a = \frac{\gamma z^2}{2} m \qquad \text{(VIII.34)}$$

where

$$m = \frac{\sin(j - i)\cos(\alpha_{max} - i)\sin(\alpha_{max} - \phi)}{\cos^2 i \, \sin(\psi - \alpha_{max} - \phi)\cos(\alpha_{max} - j)} \qquad \text{(VIII.35)}$$

and z is the depth from the top of the rock fill.

The maximum angle of inclination of the slip plane to the horizon, α_{max}, is determined by graphical construction of the functional relation $E_a = f(\alpha)$, assuming that the other inclination angles (i between the face of the ice core and the vertical and j between the vertical and the face of the rock fill; $\psi = 90° - \phi - i$, see Fig. 138) are constant.

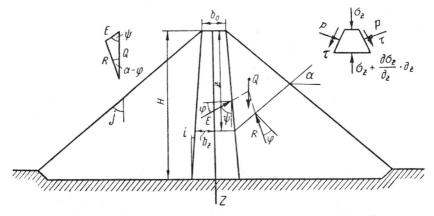

Fig. 138. Calculation scheme to determine active pressure of rock fill on ice core.

The vertical pressure distribution is determined by the expression

$$p_a = \frac{\partial E_a}{\partial z} = \gamma z m \qquad (VIII.36)$$

Then the total pressure on the lateral faces of a symmetrically loaded ice core, with consideration of the buoyant effect of the water, is

$$p = (\gamma' m + \gamma_a)z \qquad (VIII.36')$$

where γ' is the bulk weight of the rock fill reduced by the buoyant effect of the water.

Then, following the Prandtl' problem of plastic extrusion of material pressed between two planes, and assuming a constant volume of ice in the core, constant stresses both at the surface of the ice and in sections of the core, and small shear stresses, we obtain the condition under which an upward plastic extrusion of the ice will not occur:

$$2\sigma_s \ln \frac{b_0 + 2H \tan i}{b_0} \; 1 + \frac{1}{2} \cotan i \; < (\gamma' \cos \psi m + \gamma_a \sin i) \, H^2$$

$$(VIII.37)$$

where b_0 is the thickness of the ice core across the top, H is the height of the rock-filled dam, and σ_s, the long-term compressive strength of the ice, is its yield point at the given negative temperature.

We should note that σ_s is very small, and this often makes it difficult to satisfy condition (VIII.37).

The stability of the core against extrusion can be increased by reinforcing the ice with inclusions of rigid materials (stones, gravelly soils, etc.) and by lowering its negative temperature further. While the rigid restraining structure (the edges of the rock fill) will increase the resistance to extrusion of the ice, it will not, of course, eliminate it, and it will evidently be necessary in the design of future dams to take account of the magnitude and time increase of the viscoplastic flow of the ice in dam ice cores; this will require a corresponding development of the forecasting theory. In a number of cases of frozen dams built from local materials for long service lives, it will be necessary to abandon the ice-core principle because of its inevitable deformability and low extrusion resistance, and to replace it with a conventional water-impermeable clay core placed in compacted layers and provision of a system for artificial refrigeration of the soils.

All of the above enables us to recommend, for almost all cases of dam building on permafrost, that dams higher than 10 meters be built according to the cold (frozen) variant with the use of refrigeration of the ground (especially in the core of the dam) by means of a row of wells: in some cases (as in the dams on the Dolgaia and Irelyakh Rivers) with forced circulation of natural cold air and in others by the use of refrigeration to freeze the soils.

References

1. Tsytovich, N. A.: "Lektsii po raschetu fundamentov v usloviyakh vechnoy merzloty" (Lectures on the Design of Foundations Built under Permafrost Conditions), Izd. Leningradskogo instituta sooruzheniy, 1933; see also Chap. II, Ref. 31.
2. Tsytovich, N. A., N. I. Saltykov, V. F. Zhukov, et al.: "Fundamenty elektrostantsii na vechnoy merzlote" (Power Station Foundations on Permafrost), USSR Academy of Sciences Press, 1967.
3. Porkhayev, G. V., G. M. Fel'dman, V. K. Shchelokov, et al.: "Teplofizika promerzayushchikh i protaivayushchikh gruntov" (Thermophysics of Freezing and Thawing Soils), G. V. Porkhayev (ed.), Chap. V, "Changes in Ground Temperature Regime on Development of a Site," Nauka Press, 1964.
4. "Teoriya i praktika metzlotovedeniya v stroitel'stve" (Permafrostological Theory and Practice in Construction), "Influence of Construction on the Temperature Regime of Perennially Frozen Ground," Chap. II, L. N. Khrustalev, L. A. Bratsev, and V. F. Zhukov (eds.), Nauka Press, 1965.
5. Porkhayev, G. V., et al.: "Teplofizika promerzayushchikh i protaivayushchikh gruntov" (Thermophysics of Freezing and Thawing Soils), Chaps. III and V, Nauka Press, 1964.
6. 1. Tsytovich, N. A.: Design of Foundations for Structures to be Built on Permafrost, Trudy Gipromeza, no. 2, 1928; 2. Tsytovich, N. A.: Selection of Foundation Types under Permafrost Conditions. "Stroitel'naya Promyshlennost'," 1930, Nos. 6, 7.
7. Porkhayev, G. V., and V. K. Shchelokov: Influence of Construction on Temperature and Humidity Regime of Perennially Frozen Ground, in collection "Materialy k osnovam ucheniya o merzloy zone zemnoy kory" (Source Materials for a Theory of the Frozen Zone of the Earth's Crust), no. VII, USSR Academy of Sciences Press, 1961.
8. Kovner, S. S.: A Problem of Heat Conduction, Geofizika, vol. III, no. I, 1933.
9. 1. Porkhayev, G. V.: Temperature fields in the Bases of Structures, in collection "Doklady na Mezhdunarodnoy konferentsii po merzlotovedeniyu" (Papers at International Conference on Geocryology), N. A. Tsytovich (ed.), USSR Academy of Sciences Press, 1963; 2. Porkhayev, G. V.: "Teplovoye vzaimodeystviye zdaniy i sooruzheniy s vechnomerzlymi gruntami" (Thermal Interaction of Buildings and Structures with Permafrost), Nauka Press, 1970.
10. Ushkalov, V. P.: "Glubina i skorost' ottaivaniya merzlogo osnovaniya" (Depth and Thaw Rate of the Frozen Base), Gosstroyizdat, 1962.
11. Tomirdiaro, S. V.: "Teplovyye raschety osnovaniy v rayonakh vechnoy merzloty" (Thermal Calcuations for Bases in Permafrost Regions), Izd-vo SVKNII, Magadan, 1963.
12. Khrustalev, L. N.: A Procedure for Calculation of the Temperature Fields in Soils on Built-Up Areas, in collection: "Teoriya i praktika merzlotovedeniya v stroitel'stve" (Permafrostological Theory and Practice in Construction), Nauka Press, 1965.
13. Bliznyak, Ye. V.: Design and Construction of Dams Under Permafrost Conditions, "Gidrotekhnicheskoye stroitel'stvo," 1937, no. 9.
14. Bogoslovskiy, P. A.: Construction of Earth Dams in Regions with Perennially Frozen Ground, Trudy Gor'kovskogo ICI, no. 29, 1958.
15. Bogoslovskiy, P. A., A. V. Stotsenko, et al.: Dams in the Permafrost Region, "Doklady na mezhdunarodnoy konferentsii po merzlotovedeniyu" (Papers at the International Conference on Permafrostology), USSR Academy of Sciences Press, 1963.
16. 1. Bogoslovskiy, P. A.: Calculation of Multiyear Temperature Variations in Earth Dams Based on Frozen Ground Strata, Trudy Gor'kovskogo ICI, no. 27, 1957; 2. "Nauchnyye doklady vyssshey shkoly "Stroitel'stvo," 1958, no. 1; "Izvestiya vysshikh uchebnykh zavedeniy," 1958, no. 5, 1963, 11-12 and elsewhere; 3. Moiseyev, I. S.: Calculation of the Temperature Regime in Earth Dams in the Permafrost Region, in collection "Trudy MISI, no. 29, 1959; 4. Grandilevskiy, V. N.: Use of a Finite-Difference Method to Solve Three-Dimensional Problems of Nonstationary Heat

Conduction, *Trudy Gor'kovskogo ICI*, no. 37, 1961; 5. Tsytovich, N. A., N. V. Ukhova, and S. B. Ukhov: "Prognoz temperaturnoy ustoychivosti plotin iz mestnykh materialov na vechnomerzlykh osnovannyakh" (Temperature Stability of Dams Built from Local Materials on Permafrost Bases), Stroyizdat Press, 1972.

17. 1. Ukhova, N. V.: A Study of the Nonstationary Temperature Regime in Frozen Dams Built from Local Materials on Permafrost Bases, dissertation under the sponsorship of Prof. N. A. Tsytovich and Docent V. A. Veselov, MISI, 1967; 2. see also Ref. 16, no. 5, this chapter.

18. Ukhova, N. V.: Consideration of Convective Heat Transfer During Thawing of Water-Saturated Ground. "Materialy VIII Vsesoyuznogo mezhvedomstvennogo soveshchaniya po geokriologii (merzlotovedeniyu)" (Materials of 8th All-Union Interdepartmental Conference on Geocryology), no. 4, Yakutsk, 1966.

19. Trupak, N. G.: Spetsial'nyye sposoby provedeniya gornykh vyrabotok (Special Mine Excavation Methods), Ugletekhizdat, 1951.

20. Stotsenko, A. V.: Special Problems in Major Hydraulic-Engineering Construction in the Permafrost Zone, "Materialy VII Mezhvedomstvennogo soveshchaniya po merzlotovedeniyu" (Materials of the 7th Interdepartmental Conference on Geocryology), USSR Academy of Sciences Press, 1959.

21. Kamenskiy, R. M.: Thermal-Engineering Calculations for a Frozen Ground Antifiltration Shield of a Dam with Consideration of Reciprocal Effects Between Columns, "Gidrotekhnicheskoye stroitel'stvo," 1971, no. 4.

22. Tsytovich, N. A., A. L. Kryzhanovskiy, and G. Ter-Martirosyan: Approximate Estimation of the Strength of the Ice Core of a Rock-Filled Dam, "Trudy Koordinatsionnogo soveshchaniya po gidrotechnike" (Transactions of Coordination Conference on Hydraulic Engineering), no. XXIII, VNIIG, 1965.

Chapter IX

BASES AND FOUNDATIONS ON PERMAFROST WITH PRESERVATION OR INTENSIFICATION OF THEIR FROZEN STATE

1. Fundamental Premises for Selection of Construction Principle and Foundation-Design Method

In building structures under permafrost conditions, the builder must first of all solve the problem (which is almost never confronted under ordinary conditions) of selecting a construction principle, i.e., a basic direction that must be adhered to at all times during the design, construction, and use of the facility if it is to be sufficiently strong and stable and not to experience inadmissible deformations.

The builder must resolve the cardinal question as to whether it is necessary to preserve the frozen state of the base soils or whether it is possible to allow them to thaw during the lifetime of the structure. The entire subsequent design, construction, and use of the structure will depend on the basic construction principle chosen at this point. We should note that experience in the construction of various facilities under permafrost conditions has shown that building adjacent structures on different principles (which, generally speaking, often have conflicting effects on the bases of the structures) always results in totally inadmissible deformations of the structures. This calls for even greater care in selection of the construction principle and, as a rule, it should be the same for the entire built-up area.

Selection of the construction principle should be based on evaluation of the geocryological conditions on the building site (on the basis of detailed field and laboratory studies of the temperatures in the permafrost strata, their ice contents, thaw settling, etc.) and the characteristics of the structures to be built (their thermal regime, the dimensions of the construction site, design solutions, etc.).

In selection of the construction principle, full use should be made of data from the engineering-geocryological surveys (described in Chap. VII) and, as a minimum, without which it is totally impossible to justify the choice of either construction principle, the following characteristics of the permafrost: temperature $-\theta_0$ at a depth of zero annual amplitudes, total moisture content W_d of the frozen soils, and the value of the thaw coefficient (thermal settling coefficient) $\bar{A}$.

The latter data make it possible to: (1) assign the permafrost stratum to one or another construction cryozone (northern subarctic, central, or southern) on the basis of its temperature $-\theta_0$, a factor of substantial importance for selection of the construction principle; (2) evaluate the physical state of the frozen ground (also on the basis of temperature, as hard-frozen or plastic-frozen, according to Sec. 4, Chap. I, or from SNiP II-B.6-66); (3) calculate the ice content of the frozen ground from data on W_d and the amount of unfrozen water (from the moisture contents W_p and W_t, using Table 1 of the same SNiP) or to determine it directly (roughly) by making measurements on ice interlayers, which makes it possible to classify the permafrost as high ice content ($I_{vol} \geqslant 0.50$) or low ice content ground ($I_{vol} \leqslant 0.25$) and, finally, (4) establish (from the coefficient of $\bar{A}$) whether the frozen soils are subsiding (if $\bar{A} > 0.02$) or nonsubsiding ($\bar{A} < 0.02$) types.

All of the above, even with the minimum of permafrost property indicators mentioned, permit (with consideration of the features of the projected structures) a sounder approach to selection of the construction principle for the permafrost in the region under consideration, with adjustment (in preparation of the design specification) of the solution adopted on the basis of the more detailed engineering-geocryological survey data and tests of the permafrost directly at the site.

Proper selection of the construction principle, which, with appropriate technical measures, ensures reliability and stability of the physical state in the base throughout construction and operation of the facilities, creates suitable conditions for construction on permanently frozen ground.

Thus it must be established first of all which construction principle should be chosen under the particular geocryological (frost) conditions: principle I (according to SNiP II-B.6-66) "utilization of permafrost in the frozen state throughout the entire service lives of the buildings and structures," or principle II (also according to SNiP) "use of the ground in the thawed and thawing states," and secondly, which method should be used for the design and calculation of the bases and foundations to ensure strength and stability of structures built on either principle.

Several methods are currently in use for the design and calculation of bases and foundations and the erection of structures on permafrost: the method of preserving the frozen state of soil bases; the constructive method (consideration of the thaw settling of the ground during design of the foundations for maximum permissible strains); and the method of preconstruction thawing and reinforcing of the bases.

The method of preserving the frozen state of ground bases is convenient and simple to use in the northern (subarctic) and central permafrost zones (where the frozen strata are quite thick and often hard-frozen) and when the structures to be built do not release substantial amounts of heat and do not occupy large areas in plan view (for example, metallurgical shops, or concentration plants, etc.), and in all other cases when the constructive method (consideration of thaw settling) cannot be used.

Thus, for example, in the presence of very icy ($i_{vol} \geqslant 0.50$) permafrost with thaw subsiding ($\bar{A} > 0.02$) soils and soils in the viscoplastic state (at negative temperatures $-\theta_{\max}$ at footing level above the boundary $-\theta_{\text{ph.t}}$ of rapid water-to-ice phase transitions), the constructive method cannot be used conveniently and is totally unworkable in many cases. On the other hand, placing the foundations with preservation of the frozen state at the bases, which completely prevents thawing of the permafrost at the bases of the structures, permits the use of all types of frozen ground as bases for structures and anchoring of the foundations in a nonthawing frozen-ground stratum— practically the only reliable measure against buckling of reinforced-concrete column-, pile-, and other types of foundations.

Experience in constructing and the theory of the problem indicate that the best foundation system in this case is the system of foundations with a basement that can be ventilated in the winter (or year-round). The method of building structures on permafrost with winter-ventilated basements was stumbled on by the builders after a whole series of unsuccessful trials. Thus, for example, certain older buildings (the two-story Bishop's residence at Yakutsk and the engineer stores at Chita), which have ventilated basements, have been preserved without significant deformations for long periods of time.[1] But no inferences were drawn from this experience, and no engineering calculations or equations for the dimensions of the basements and the foundations themselves for permafrost conditions appeared until the end of the 1920's, when the author* first developed the calculation fundamentals for foundations of structures to be built on permafrost by the method of preservation of the frozen state in the bases in response to inquiries from the field while working at Gipromez [State Institute for the Design of Metallurgical Plants]. It should be noted that, for example in the paper by Prof. V. Statsenko (1916),[2] one of the four foundation types that he described as having been used for permafrost was a design with vaulted passages for winter basement ventilation, that had proven more successful compared to the other three, which were totally unsatisfactory; but even in this case the appropriate conclusions were not drawn at the time from this construction experience.

The theoretical design and structural basis for the principle of using permafrost as bases for buildings and structures by preserving their frozen state was, as we noted earlier, laid down in the late 1920's in connection with the design and construction of the Petrovsk-Zabaykal'skiy Metallurgical Plant (with blast furnaces, rolling mill, etc.) and the Yakutsk central thermal power station (YaCES).** This method is now generally accepted and universally used, since it permits optimum use (with preservation of the frozen state of the ground) of the high construction properties of all frozen soils, and many industrial structures and whole cities (Noril'sk, and others) have been built successfully by this method.

The constructive method (with allowance for thaw settling of the ground) was developed scientifically some time after the proposal of a method for

*See Chap. II, Ref. 51.
**See Chap. VIII, Refs. 1 and 2.

predicting the settling of foundations on thawing soils (N. A. Tsytovich, 1939; G. I. Lapkin, 1939, and elsewhere). On the basis of settling calculations (N. I. Saltykov, 1946-1952),[3] gave the fundamentals of static design for foundations to be built on permafrost by the constructive method (with consideration of the settling of the thawing bases).

The constructive method, e.g., the method of adaptation of the foundation and superstructure to nonuniform settling of the thawing soil bases, can be used in cases when the temperature regime of the permafrost at the construction site is unstable (for example, when the temperature of the permafrost beds is close to $0°$ or not higher than -0.5 to $-1°C$) and, on thawing, the soils do not settle too severely and their settling is less than the maximum tolerable for the particular type of structure.

The latter can be established only if data on compressibility of the thawing ground (the quantities $\bar{A}$ and $\bar{a}$) and on the dimensions of the foundation footings (after preliminary trial and error) are available, so that it will be possible to calculate the settling of the ground as a result of thawing of the load imposed by the foundations.

Calculations indicate that, as a rule, building by the constructive method can be done only on gravel, crushed rock, sandy, and other skeletal soils that compact due to thawing and under load and are not squeezed out from under the footings, something that can occur when structures are built on high ice content high-temperature clayey soils.

It is mandatory when the constructive method is used to predict the extent of settling of the foundations and the course that it takes in time during gradual thawing of the ground under the foundations of heated structures and (in the case of clayey soils) the secondary compaction settling of the soils after they have been thawed completely through the entire depth of the active compression zone under the foundations.

If the predicted settling of the thawing and thawed soils is found to exceed the limits for the particular type of structure (according to SNiP II-B.6-62) it is necessary to provide preconstruction preparation of the base (preliminary thawing of the soils to the depth necessary according to the calculations, compacting, strengthening, and sometimes even reinforcement of the thawed soils, and other similar measures).

The preconstruction thaw method and the improvement of the bases is used in cases when it is necessary to reduce the future settling of thawing and thawed soils (by preliminary gravity compaction or by the use of certain technical measures) and when it is necessary to reduce differential settling when the bases of the structures contain soils of highly nonuniform compressibility in the frozen and thawed states (for example, spots with rocks of low compressibility, e.g., bedrock, or highly compressible soils, e.g., layers of peat, or multiple ice wedges, etc. not far below the surface).

Use of preconstruction thawing is also found to be advantageous when practically incompressible rocks are found in a continuous bed not far below the surface (5-10 meters) and pile columns are not to be used for some reason, and when the structure being designed has spots with concentrated heat sources that introduce substantial nonuniformity into the base-thaw process; in this case, if ground conditions permit, it is economically advantageous first to thaw and compact a certain volume of the frozen ground with the object of reducing the nonuniformity of the settling. In some cases, for example when warm and hot process waste waters are discharged and the structure has a large base area, it may be found advisable (again, if ground conditions permit) to prethaw a certain part of the area to a certain depth in the permafrost bed, although this should be done only in accordance with a special plan based on consideration of the property indicators of the soils in the thawing and thawed states that accounts for their ice content, load-bearing capacity, and deformability in the thawed state. Here partial thawing of the ground in the bases of the structures is permitted only with mandatory subsequent allowance for .the difference between the possible settling depths of the individual base areas (when the structures are to be built on the second principle) or with subsequent artificial freezing of the thawed volumes of ground (if the structure is to be erected on the principle in which the frozen state of the ground bases is preserved).

It must also be remembered here that, according to the general premise stated earlier, which follows from the condition for temperature-field stability in the base soils, both the entire structure and the entire construction site must be designed for high-temperature permafrost on one of the two basic construction principles for construction work in the permafrost zone. Mixing of principles either for adjacent buildings and structures on the same construction site and especially for a single structure, even one that occupies a large area, is totally inadmissible in such cases. The only exception is the low-temperature northern part of the permafrost region.

Thus, the method of designing the foundations and erecting the structures is established on one of the two principles based on the geocrylogical conditions at the construction site, the particular permafrost region, the geological structure of the terrain, the physical and mechanical properties of the ground in the frozen, thawing, and thawed states, the temperature inside the buildings, and the design features of the structures (their sensitivity to differential settlement).

2. Methods of Preserving the Frozen State in the Base

First of all, we note once again the fundamental aspect of this extremely important problem of preserving the frozen state in the bases of buildings and structures built on permafrost. At a consistently positive temperature $+\theta_r$ inside the rooms of the buildings or structures (Fig. 139) and a constantly negative permafrost temperature $-\theta_f$, as we noted above, the amount of heat released by the floor of the building will, according to the Fourier equation for the steady state [expression (VIII.2)], be directly proportional to the temperature difference $(\theta_r - \theta_f)$ and inversely proportional to the thermal resistance R_0 of the

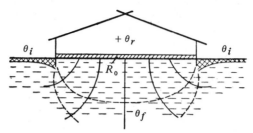

Fig. 139. Diagram of heat fluxes in building with floors resting on soil under permafrost conditions.

floor, and will increase continuously with increasing time t, i.e., the heat flux from the building into the permafrost will be continuous, and this will result in the formation of a thaw bowl unacceptable in the foundation-design method under consideration here.

If the thermal resistance R_0 of the floor is increased, the heat flux is reduced, but it cannot be eliminated by heat insulation alone, since it is practically impossible to provide heat insulation with a thermal resistance larger than 3–5 $m^2 \times hr \times deg/kcal$ from ordinary structural materials owing to the excessive bulkiness (thickness) of the insulating material.

Thus, the thaw bowl will invariably form as indicated by thermal-engineering calculations, and in regions with high-temperature permafrost it may not even reach a stationary state as it does in the subarctic (northern) region, i.e., the thaw depth will continue to increase.

In the latter case, stability of the buildings can be insured, for example by the use of deep-laid foundations (pile, column, supporting-shell, and other types), i.e., foundations are placed below the maximum depth of the steady-state thaw bowl. An effective method of preventing formation of a thaw bowl and fully preserving the frozen state of the base soils is diversion of the heat released by the building or structure away from the ground. The engineering solution of the problem (with a certain safety margin) would be the complete removal of the heat released by the building or structure over its entire basement area, i.e., the use of a basement that is ventilated in the winter or the year-round, which ensures preservation of the frozen state in the base soils and may even strengthen it. Heat can also be removed through special pipes (channels) with natural or forced ventilation and by providing coarse-pored ventilated rock fill under the entire heated area at the base of the building. However, the winter (or continuously ventilated) basement has come into widest use in practice because of its simplicity and total reliability.

The height of the ventilation openings (ducts) or continuous slits in the basement walls can be calculated for complete removal of the heat released by the building floor into the base ground by the ventilation systems.

The first attempt at analytical determination of the necessary size of the openings in the winter-ventilated basement to ensure preservation of the frozen state in the base soils of buildings and structures was made by the author back in 1932 in the design of the YaCES to be built on the banks of the river Lena in Yakutsk;* to obtain a safety margin, the calculation was based on the total absence of wind.

Practical experience has fully confirmed the basic propositions used in the calculation, since the frozen state of the base soils has not only been preserved under the YaCES building to this day, but has even been strengthened (the frozen-ground boundary has risen 0.8–1.2 meters above its original position and is still continuing to rise).

A later study by N. I. Saltykov and N. N. Saltykova[4] was devoted to improvement of calculations for ventilated basements to preserve the frozen condition of the ground under the footing; based on the heat budget, they developed a procedure that takes account of the heat flux through the deck over the basement and through the foundation socle of the building during seasonal freeze and thaw of the ground, performing a monthly calculation by simultaneous solution of the heat-balance equation and the equation of the seasonal ground freezing and thawing under the building.

N. I. Saltykov introduced the concept of the ventilation modulus (the ratio of the area of the ventilation openings in the foundation walls of the building to the total area of the building floor) and prepared a table of approximate values of areas adequate to preserve the frozen state in the base soils under buildings of various widths with various thermal resistances in floors for the three geocryological zones: northern (subarctic), central, and southern.[5]

In the section of G. V. Porkhayev's work** devoted to study of basements, it was established that the basic factor determining air turnover in the ventilated basement is the wind pressure, which was considered in detail in the development of a thermal engineering method that can be used to design a building to have a specified temperature regime in the ventilated basement sufficient to ensure preservation and strengthening of the frozen state of the ground in the base.[6]

We note briefly the essentials of the method proposed by the author for calculation of the winter-ventilated basement that ensures preservation of the frozen state in the base and has played a major role in the introduction of winter-ventilated basements into construction practice, although it is now chiefly of methodological interest, since Porkhayev and other specialists subsequently developed improved calculation methods.

In the calculation for the winter-ventilated basement, the area of the ventilation openings in the socle wall of the building (ducts) that would be adequate to remove all of the heat incoming from the basement ceiling by natural thermal exchange was determined under the following assumptions:

*See Chap. VIII, Refs. 1 and 2.
**See Chap. II, Ref. 31.

1) All of the heat released by the floor of the building is removed by the ventilated basement;

2) Lateral heat losses from the floor are left out of account (to obtain a safety margin);

3) The influence of the wind pressure on the ventilation of the basement is not considered.

The amount of heat released by the building floor (basement ceiling) is, according to Eq. (VIII.2),

$$Q = \frac{\theta_r - \theta_{av}}{R_0} t \qquad (l_1)$$

where θ_r is the indoor temperature, θ_{av} is the average air temperature in the basement, R_0 is the thermal resistance of the building floor (basement ceiling), and t is the time interval considered. It can be assumed that

$$\theta_{av} \approx \frac{\theta_{b.c} + \theta_o}{2} \qquad \text{where} \qquad \theta_{b.c} = \frac{(\theta_r - \theta_o)\alpha_0}{R_0} + \theta_o \qquad (l_2)$$

where $\theta_{b.c}$ is the temperature of the basement ceiling, θ_o is the temperature of the outside air (taken from the meteorological data for the coldest month), and α_0 is the coefficient of heat transfer from the basement ceiling to the outside air (usually $\alpha \approx 0.05$).

The volume L of air that must be removed from the ceiling by thermal load (without consideration of the wind and using an air-exchange coefficient of two) will be determined by the expression*

$$L = \frac{2Q (1 + \alpha\theta_{av})}{0.31 (\theta_{b.c} - \theta_o)} F \text{ m}^3/\text{hr} \qquad (l_3)$$

where $\alpha = 1/273$ is the coefficient of volumetric expansion of air F is the floor area, and 0.31 is the heat capacity of air.

Using Torricelli's formula for the outflow of gases from orifices to express the thermal load in the ventilation openings in the foundation wall, and putting it equal to the air load produced by the difference between the densities $\gamma^a_{\theta_o}$ and $\gamma^a_{\theta_{av}}$ of air at the outside temperature and at the average air temperature in the basement, respectively, and considering a neutral air flow zone through the middle (at half-height) of the ventilation openings, we obtain

$$\frac{v^2 (\gamma^a_{\theta_{av}})}{2g\beta^2} = \frac{h_b}{2} (\gamma^a_{\theta_o} - \gamma^a_{\theta_{av}}) \qquad (l_4)$$

where v is the air flow velocity in the ventilation openings that is adequate for passage of the entire discharge of heated air, β is the contraction coefficient of the stream in the opening and can usually be put equal to 0.65, g is the

*See Chap. VIII, Refs. 1 and 2, and Chap. IV, Ref. 4.

acceleration of gravity, and h_b is the height of the ventilation openings in the basement.

We note that if v in relation (l_4) is expressed in m/sec, $\gamma_{\theta \text{ av}}$ in kg/m³, and g in m/sec², the dimensions of the load will be kg/m².

On the other hand, the air flow velocity in the ventilation openings for passage of the entire volume of air L m³/hr (1 m/hr = 1/3600 m/sec) with a total area F_0 of all ventilation holes will be

$$v = \frac{L}{F_0} \tag{l_5}$$

We introduce the basement ventilation coefficient

$$M = \frac{F_0}{F} \tag{IX.1}$$

Then, multiplying and dividing the right-hand side of (l_5) by F and applying expression (IX.1), we obtain

$$M = \frac{L}{vF} \tag{l_6}$$

Substituting the value of the velocity from expression (l_4) and L from (l_3) into (l_6), we obtain

$$M \geqslant \frac{2Q \left(1 + \alpha\theta_{\text{av}}\right)}{0.31\beta \left(\theta_{\text{b.c}} - \theta_o\right) \sqrt{h_b g \left[\dfrac{\gamma_{\theta o}^a}{\gamma_{\text{av}}^a} - 1\right]}} \tag{IX.2}$$

After a simple transformation, putting $\beta \approx 0.65$ and, consequently, $\beta \times 0.31 \approx 0.2$, we finally obtain

$$M = \frac{10Q \left(1 + \alpha\theta_{\text{av}}\right)}{\left(\theta_{\text{b.c}} - \theta_o\right) \sqrt{h_b g \left[\dfrac{\gamma_{\theta o}^a - \gamma_{\text{av}}^a}{\gamma_{\theta \text{ av}}^a}\right]}} \tag{IX.3}$$

Having determined the basement ventilation modulus M from (IX.3), we find the total area of all ventilation openings in the foundation wall for winter ventilation needed to preserve the frozen state in the base from the equation

$$F_0 = MF \tag{IX.4}$$

or, knowing the basement height h_b, we determine the total width B of all ventilation openings:

$$B = \frac{MF}{h_b} \tag{IX.5}$$

We note that use of the above computation method for the dimensions of passages in ventilated basements to preserve the frozen state of the base soils has justified itself fully in practice, as exemplified by the YaCES building. A study made by the Yakutsk Scientific Research Station of the air-temperature regime and air movement in the ventilated basement of the YaCES showed that the removal of heat from the basement space is accomplished largely by the wind pressure, and that the velocity of the air flow in the basement depends entirely on the presence of wind.

As we noted earlier, thermal engineering calculation of ventilated basements (to preserve the frozen state in base soils) was developed by N. I. Saltykov and especially by G. V. Porkhayev,* who gave a complete analytic problem solution for preserving the frozen state in the bases of buildings erected on permafrost, both by means of ventilated basements and by the use of cooling tubes and ventilated air-permeable coarse-skeletal fill.

Without dwelling here on the complicated thermophysical calculations (which are based on drawing of the heat balance and consideration of the thermal and wind loads of the air in the basement), since they would be beyond the scope of the present book, we note only Porkhayev's extremely simple solution[7] for determination of the average basement ventilation modulus, which is often used in practice, and cite the results of Saltykov's calculations** of the basement ventilation moduli needed for the various geocryological zones to keep the base soils frozen; in conclusion, we shall set forth construction methods that are used in practice for winter-ventilated basements.

When it is only necessary to preserve (and not to lower) a negative temperature in the base (by means of vents or slits open in all directions in the basement of the building) and it is unnecessary to determine the maximum ground temperature (for its load-bearing capacity estimates) at the foundation-bottom level, the basement ventilation modulus M can, according to Porkhayev, be determined by the following simple expression, assuming the depth of the seasonal thaw under the building to be equal to the thawing depth outside of it and assuming that the wind pressure is uniformly distributed over the year:

$$M = k_d \frac{(\theta_n - \theta_{av}) + N_p}{300 R_0 v_{av} (\theta_{av} + \theta_o)} \tag{IX.6}$$

where k_d is a construction density coefficient that depends on the distance l between buildings and the height h of the buildings ($k_d = 1$ for $l > 5h$; $k_d = 1.2$ for $l = 4h$, and $k_d = 1.5$ for $l = 3h$); θ_r, θ_{av}, and θ_o are the indoor temperature, the annual average basement temperature and the annual average outside-air temperature, in °C; v_{av} is the annual average wind velocity in m/sec, R_0 is the thermal resistance of the deck over the basement in $m^2 \times hr \times deg/kcal$, and N_p is the heat release from water and sewage piping into the basements, determined

*See Chap. VIII, Ref. 9.
**See Ref. 5, this chapter.

by the equation

$$N_p = \frac{R_0}{FT} \sum_{i=1}^{i=n} \frac{l_{pi}}{R_{pi}} (\theta_{pi} - \theta_{av}) t_{pi} \qquad \text{(IX.7)}$$

where n is the number of heat-releasing pipes, l_{pi} is the length of the internal pipeline in meters, θ_{pi} is the temperature of the heat carrier in $°C$, R_{pi} is the thermal resistance of the heat insulation on the pipe in $m^2 \times hr \times deg/kcal$, F is the floor area in m^2, T is the length of the year ($T = 8{,}760$ hr), and t_{pi} is the operating time of the pipe during the year.

According to the "Attachment to SNiP II-B.6-66," the annual average air temperature θ_{av} in a basement that is ventilated year-round is assumed equal to the annual average temperature in the ground at the level of zero heat fluctuation, i.e., $\theta_{av} \approx \theta_0$. But if the base of the structure is composed of plastically frozen ground (at temperatures from $0°$ to a temperature not lower than the limit for intensive phase transitions of pore ice to water), then, according to Sec. 3.24 of SNiP.II-B.6-66, the base-ground temperature must be lowered; here it is assumed that:

$$\text{at } \theta_0 \geqslant -0.5°C \ldots \theta_{av} = 4\theta_0$$
$$\text{at } \theta_0 < -0.5°C \ldots \theta_{av} = 3\theta_0$$

We note that for buildings with increased heat release, it is necessary to allow separately for the summer and winter heat budgets and the aerodynamic coefficients of the air flow around the building as they depend on the shape of the building and its orientation in respect to the points of the compass; this is done in a special thermophysical calculation.

In practice, the total height of the ventilated basements is specified on the basis of the use conditions of the structures and on the basis of SN 353-66: no less than 0.5 meter for buildings up to and including 18 meters wide, and at least 1 meter for buildings wider than 18 meters. However, construction experience gained at Noril'sk has shown that the height of the ventilated basement should be at least 1.2 meters in all cases, and at least 1.8 meters for wide buildings and when many water and other pipelines are present in the basement.

To determine ventilated-basement vent sizes adequate to preserve the frozen state of the base soils, reference may be made to Table 43, which lists values of the ventilation modulus compiled by Saltikov.

Using the Luk'yanov hydraulic integrator to study the temperature field in bases laid on permafrost, M. D. Golovko and V. K. Shchelokov[8] showed that foundation columns that pass through the basement space have practically no effect on the general pattern of the ground-temperature field under buildings and structures. However, we should note that these experiments were apparently performed only with small foundation models whose widths were smaller than the depth of the frozen ground base below the surface. Indeed, in such cases, positive annual-average soil temperatures will not reach the footings of the

Table 43. Values of basement ventilation modulus M
(Without consideration of wind pressure)

Building width, m	Thermal resistance R_0 of floor, m^2 hr-deg/kcal	Indoor tempera-ture θ_r, °C	Ventilation modulus M for permafrost zones		
			Northern (subarctic)	Central	Southern
15	1	15	0.0015	0.005	0.025
	1	30	0.0050	0.015	0.030
	2	15	0.0010	0.003	0.012
	2	30	0.0020	0.007	0.020
	3	15	0.0005	0.002	0.007
	3	30	0.0010	0.004	–
30	1	15	0.0025	0.008	Open
	1	30	0.0075	0.025	Open
	2	15	0.0015	0.006	0.015
	2	30	0.0035	0.010	0.030
	3	15	0.0007	0.003	0.007
	3	30	0.0022	0.005	0.010
50	1	15	0.003	0.010	Open
	1	30	0.010	0.030	Open
	2	15	0.002	0.007	0.020
	2	30	0.005	0.015	Open
	3	15	0.001	0.004	0.10
	3	30	0.003	0.007	Open

columns (Fig. 141) at moderate indoor temperatures, as was also indicated by our experiments in the 1930's in which thermocouples were used to measure temperatures in physical models of concrete foundations[9] (Fig. 140), and by observations in the experimental building of the Gipromez scientific station at Petrovsk-Zabaykal'skiy.*

But if the foundation width is equal to the thaw depth of the ground under the building, or greater, local disturbances of the temperature field in the frozen ground will undoubtedly take place, and a thaw through under the footings may occur.

For wide foundations, therefore, it is necessary to consider their effect on the temperature field in the bases of structures built on permafrost and to add to the total amount of heat lost by the floor of the building (the basement ceiling) the amount of heat that penetrates into the ground through the foundations of the structure, or to take measures to remove the heat transmitted into the ground by the foundations (by provision of vents, ventilation ducts, trench masonry, etc.).

*See Chap. II, Ref. 31.

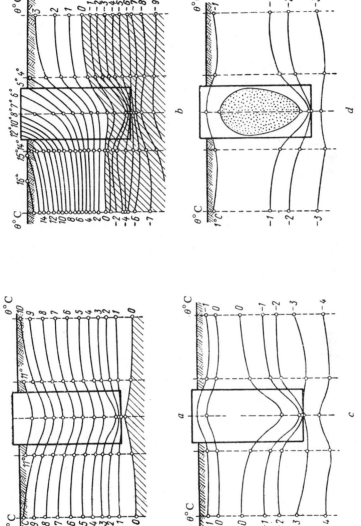

Fig. 140. Ground isotherms and concrete-foundation models. (*a*) For uniform heating from above; (*b*) for unsymmetrical thawing of the upper layer of frozen ground; (*c*) and (*d*) for freezing of the upper ground layer.

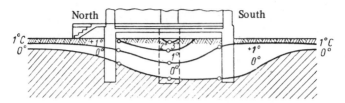

Fig. 141. Annual average ground and foundation temperatures under the experimental building at the Petrovsk-Zabaykal'skiy scientific station (as of 1933).

This makes it necessary to extend the thermal-engineering calculation to include the stability of the temperature regime in the frozen ground under individual foundations that occupy substantial areas (when their widths exceed the thaw depth), using the same method as in calculating the temperature stability of the bases of an entire building, except that the lateral heat losses through foundations within the buildings perimeters can be neglected.

To conclude the present section, let us cite certain examples of field observations of frozen-ground boundary motion in the bases of buildings erected in the permafrost region. Thus, Fig. 142 shows a sectional drawing of the quarry stone strip foundation of the old library building at Yakutsk after installation of a ventilated cellar.[10]

We see from this example that the frozen-ground level has risen considerably (by nearly four meters) and that the whole foundation is now supported by sufficiently strong frozen ground, bringing an end to substantial deformations of the building that had occurred earlier.

Figure 143 presents a second example of observation results and measurements of the frozen-ground boundary under buildings in Yakutsk. The building whose foundation is shown in section in Fig. 143 was built in 1948–1949 on 40 X 40-cm reinforced-concrete pillars sunk 4 meters into the ground, with an all around ventilated basement ranging in height from 65 to 80 cm; the spaces

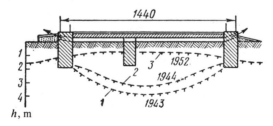

Fig. 142. Gradual refreezing of the thaw bowl formed under an old building in Yakutsk after installation of a winter ventilated cellar. (1) Established thaw boundary, formed before construction of the cellar; (2) rise of the frozen ground boundary one year after cellar installation; (3) same, after 8 years.

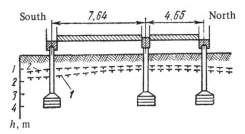

Fig. 143. Decrease, 1 and 2, of seasonal thaw depth in soils under school building in Yakutsk in the presence of a winter-ventilated basement.

between floor joists ranged from 20 to 40 cm. The ground is clayey (down to 1.3–1.6 meters) and is underlaid by fine silty sand with $W_d = 23$–25 percent. After about three years, the frozen-ground level had risen 40–46 cm. The building is fully stable.

Thus, we conclude, on the basis of everything said in the present section, that a basement open on all sides year round (or only in the winter) or a closed basement that has ventilating passages (with a total area correctly calculated) fully guarantees preservation of the frozen state of the ground under structures built on permafrost.

Other methods also exist for preservation of the frozen state of base soils: keeping the first story of the building cold (unheated) (using it to accommodate heating, gas, water, and other plumbing), use of special cooling pipes and channels underground or in fill—a practice that also requires the corresponding thermal-engineering calculations and has been used only in industrial construction.

At the present time, the use of the ventilated basements to keep the ground frozen is one of the fundamental methods for construction of stable structures in permafrost regions.

3. Measures to Strengthen the Frozen State of Base Soils

By strengthening the frozen state of base soils we understand measures that lower the permafrost temperature throughout the entire thickness of the potential warming zone to a level at which the ground will be hard-frozen rather than plastically frozen, i.e., such that the temperature is kept lower than the boundary of intensive (substantial) water-to-ice phase transitions throughout the entire service life of the building or structure.

When the frozen state is permitted to remain a plastically frozen, high-temperature permafrost (at temperatures from 0 to $-1°C$), even a slight rise in the ground temperature of the construction site may be sharply detrimental to the mechanical properties of the frozen ground, and may even cause it to thaw in certain cases. To avoid this situation, G. A. Borisov, M. V. Kim, and G. A.

Pchelkin proposed a method for advance strengthening of the frozen state of the ground in the base (before erection of the structure) by natural or artificial chilling of the ground to a certain depth. They tested it at Noril'sk; it was subsequently developed further by G. N. Maksimov and also by A. A. Konovalov.

According to Maximov,[11] the basis of the method is the creation, under the structure, of a primary cooled zone of optimum dimensions, which is later preserved and even enlarged by the appropriate use of a basement with good ventilation in winter or by the use of other cooling devices that stabilize the frozen state of the ground and prevent its temperature from rising to a certain limit in the design zone.

The simplest procedure for creation of a cooled ground zone (one with a temperature lower than the natural temperature at the particular locality) is surface cooling during one or more winter seasons by systematic clearing of the winter accumulation of snow from the construction site before erection of the structures begins.

An industrial method for cooling the bases (strengthening their frozen state, i.e., lowering the temperature and converting part of the unfrozen water into ice) with the aid of drilled wells into which specially equipped piles are lowered was developed by G. N. Maksimov; the basics for thermal engineering calculations of air cooling by forced ventilation was worked out by A. A. Konovalov.[12]

Based on a thermal budget drawn up for a cooled block of soil with specified dimensions (usually to a depth equal to half the width of the building), the thermal properties of the ground, and the number and dimensions of the cooling wells, Maksimov proposed an approximate equation for determining the radius of ground cooling by circulation of cold outside air in specially equipped wells, which might, for example, be drilled for the sinking of piles.

Construction experience has shown that the hard-frozen state of the base soils, once achieved, is successfully maintained by winter ventilation of basements. Needless to say, the frozen state of the cooled block of soil must be monitored by systematic measurement of its temperature at various points.

Maksimov also proposed a procedure for cooling blocks of frozen ground (of length L and depth h) (Fig. 144a) by means of cold air circulated in drilled wells (Fig. 144b); here the cooling radius R_1 around the well is determined from an equation he derived from the heat-balance equation of the cooled soil block.

We should note that other equations have also been proposed (by Kh. R. Khakimov, A. A. Tsvid, and others) for calculating the ground cooling by forced circulation of refrigerants (cold air, cooled brine, etc.) in wells; they are valid for artificial ground freezing and for freezing of the cores of earth dams built on permafrost.

Analysis of experimental data, for example curves of the time to cool the block of soil vs. the negative temperature of the air circulated in the wells (Fig. 144c), as plotted according to Maksimov's scheme for cooling the ground from $\theta_c = -1°C$ to a required temperature $\theta_v = -2°C$, and other similar data,

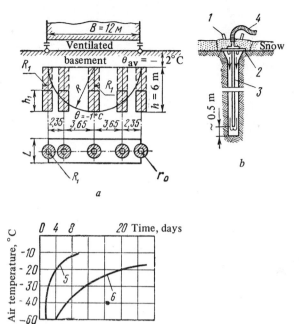

Fig. 144. Diagram illustrating calculation of ground cooling time in base of building with circulation of cold air in drilled wells. (*a*) Working diagram; (*b*) equipment of freezing well; (*c*) time to cool a given volume of ground plotted against negative temperature of outside air (curve 5 for sands and curve 6 for clayey soils); (1) intake pipes; (2) cover; (3) circulation pipe; (4) hose.

indicates that cooling of frozen ground with cold air is effective only at low air temperatures (below $-20°C$), and that a substantially longer cooling time is required for clayey soils (because of their high unfrozen-water content) as compared to sandy soils.

Strengthening of the frozen state of base soils under foundations built on the principle of preserving the frozen state is a factor that improves the performance of the frozen bases and is especially effective in the zone of high-temperature permafrost when winter-ventilated basements are used; if possible, all measures should be taken consistently (and first of all the simplest) to strengthen the frozen state of bases.

But it must be remembered that in planning the cooling of a ground massif to a given depth under a winter-ventilated basement, it is necessary to calculate not only the time to cool the given volume of frozen ground to a given temperature (by circulation of cold air or another refrigerant), but also the temperature redistribution (from the appropriate expressions arriving from analysis of the heat budget) during the summer in the cooled ground massif, and also the temperature stability of the cooled ground throughout the entire service life of the structure.

Enhancing the frozen state of permanently frozen ground by cooling it (with air, liquid refrigerants, brines, etc.) requires a much smaller expenditure of energy (because it is necessary to freeze only a small part of the water present in the frozen ground, i.e., its unfrozen water) as compared to thawing of frozen ground (especially high ice content ground), and is economically highly advantageous. However, Maksimov[13] showed that air cooling is most effective in the zone north of the approximate position of the $-3°C$ geoisotherm at the depth of zero heat fluctuation in the ground; but for geocryological zones south of this isotherm, additional installations are required.

We note that the Krasnoyarsk PromstroyNIIproyekt has now developed and recommended the construction of hollow piles and drilled wells for use when forced circulation of cold air is resorted to for the purpose of strengthening the frozen state of ground massifs and has published recommendations for the design of air-cooling systems for permafrost bases and a procedure for measurement of temperatures in the cooled ground massifs.

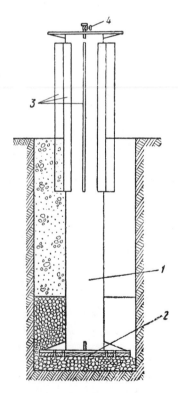

In the 1960's, automatic systems for cooling permafrost and strengthening its frozen state came into use almost simultaneously in the USSR (system designed by S. I. Gapeyev, Soviet Patent No. 163,541) and the USA[14] (the so-called thermopiles by E. L. Long), chiefly in the southern and central geocryological zones.

The Long Thermopile (Fig. 145) is a steel pipe (1) with a bottom cap made from steel plate 75 mm thick resting on wooden planking (2) on gravel fill. The cooling liquid condenses on the part of the pipe above ground, which is made in the form of a special finned radiator (3) for better cooling, with a valve (4) for filling the pile with the liquid.

Propane, which performs effectively from 0 to $-16°C$, is used as the working fluid. The maximum gas pressure in the pile is 4 kg/cm². Long made temperature observations in the frozen ground around a thermopile over more than a year, and showed that the proposed design works well in providing substantial propane cooling (cold air is used in winter) of

Fig. 145. Ground cross section with a Long Thermopile.

plastically frozen ground (in the southern geocryological zone).

The Gapeyev automatic system is used successfully for cooling high-temperature permafrost when it is necessary to stop local thawing of or to intensify the frozen state of plastically frozen ground. Figure 146 presents a diagram of the Gapeyev system. It is made of metal pipes with different diameters (1 and 2) and a welded-on expander pipe (3). The expander terminates in a threaded fitting (4) and attached cover (5). Kerosene is used as the circulating heat carrier in this system; as a result of this arrangement of the pipes, the kerosene descends from the top of the pipe in winter, displacing less dense kerosene and thereby facilitating heat exchange. The rate of kerosene exchange for a temperature gradient of 40°C is 4 cm/sec, but this drops to 1 cm/sec at a 10°C temperature difference. During the summer, the circulation of the kerosene automatically stops. A system using 30 liters of the heat carrier (kerosene) is inexpensive, and the system is reliable in operation.

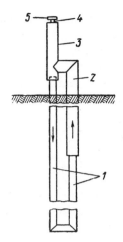

Fig. 146. Scheme of automatic device for cooling permafrost. S. I. Gapeyev's system.

4. Calculations for Foundations to be Built with Preservation of the Frozen State of the Base

Data needed for calculation. In proceeding to the calculation and design of foundations for structures to be built on permafrost on the principle of preserving the frozen state of the bases, it is necessary first of all to have a clear conception of the features of the calculation and of foundation design for the conditions at hand. These conditions are mainly governed by the absolute necessity of considering additional characteristics of the frozen base soils that are not usually required in the design of foundations built on unfrozen ground, such as: the depth h_t of the deepest summer thaw penetration under winter-ventilated basements of the buildings and structures; the depth of the foundations below the deepest summer thaw; the maximum temperature at footing level (for determination of the design strengths) and the temperature of the permafrost layers at the level of zero annual heat fluctuation (required by the SNiP for thermal engineering calculations), as well as the thermal, mechanical, and rheological properties of permafrost of the particular granulometric composition, ice saturation (ice content), and negative temperature.

We recall that the following quantities must be known for calculation and design of foundations in the light of scientific data from mechanics of frozen ground: the basic physical properties of the frozen soils (γ_{sp}, γ, W_d, W_u) and, from the available tables (SNiP II-B.6-66), the corresponding thermal properties (λ_t, λ_f, C_t, C_f, C_f'), which enable us to perform thermophysical calculations of temperature, thawing and freezing depth, etc. (for example, the maximum depth of the summer thaw, the highest soil temperature at footing level, which is needed for specification of the design strengths, etc.), and the long-term strength indicators of the frozen soils (ult σ_{lt}, c_{lt}, τ_{lt}) to calculate the bearing capacities of the bases and the deformability indices of the frozen ground (a_r^Σ, δ and δ_1 or T and $a_{r\infty}$ or a_r' and a_r'', etc.).

According to SNiP II-B.6-66, with consideration of the amount of unfrozen water at the given negative temperature

$$C_f' = \frac{1}{W_d} \left[C_f (W_d - W_u) + C_t W_u \right] \text{ kcal/m}^3 \times \text{deg} \qquad \text{(IX.8)}$$

where C_t and C_f are the volumetric heat capacities of the ground in the thawed and hard-frozen (at $\theta = -10°C$) states.

The design of foundations for structures to be built on permafrost *with preservation of frozen state* in the base includes the following calculations:

1) Calculation of the temperature regime stability in the permafrost base under the projected structures;

2) Analytical determination of the foundation depth below grade (if a ventilated basement is to be used) and calculation of the highest temperature in the permafrost at footing level;

3) Design of the foundation on the basis of the bearing capacity of the permafrost and the maximum settling of it in the plastic frozen (high-temperature) state.

The stability of the frost temperature regime in the ground below the projected structures is ensured throughout the service life of the structures by the use of a winter-ventilated basement with ventilation-opening dimensions determined by thermal-engineering calculations (see Secs. 2 and 3 of this chapter), or by use of special ventilation tubes in the fill to remove the heat released by the building, or other similar devices. Then the question arises as to the necessary foundation depth below grade as it depends on the thaw depth of the ground under the basement.

Foundation depth below grade is determined from the following equation for structures to be built with preservation of the frozen state in the base:

$$h_\phi = h_t + h_f \qquad \text{(IX.9)}$$

where h_t is the depth of the summer thaw under a ventilated during the winter or year-round cellar to preserve the frozen state of the base soils and h_f is the depth at which the foundations are placed in the permafrost (below the bottom of the seasonally thawing layer).

The depth of the summer thaw in a ventilated basement can be determined analytically by considering the heat balance for the soils in the summer based on the annual average air temperature, the duration of the above-zero temperatures period, the thermal properties of the ground in the frozen and thawed states with consideration of water-to-ice phase transition, etc., all of which requires knowledge of many characteristics and a special thermal-engineering calculation.

However, according to SNiP II-B.6-66, Sec. 4.2, it is permissible to determine the theoretical thaw depth for buildings with cold basements by applying a correction factor to the seasonal thaw depth h_t^n given in the norms (the notation is ours.—N.Ts.) by the equation

$$h_t = m_t h_t^n \tag{IX.10}$$

where m_t is the heat-effect coefficient of the buildings, which is assigned the following values for structures with winter-ventilated basements: $m_t = 1.2$ for outside walls with asphalt or shingle siding, $m_t = 1.0$ for the same walls but without the asphalt siding, and $m_t = 0.8$ for inside walls.

As for the seasonal thaw depth norm h_t^n, it is determined analytically by the improved solution of the one-dimensional problem [Eq. (42), "Appendix to SNiP II-B.6-66"], which also requires knowledge of many characteristics, but it is permissible to determine the seasonal thawing depth norm from the averaged results of measurements made over many years and from charts of thaw-depth isolines (see Figs. 17 and 18 in Chap. I) with application of the correction factor for ground moisture content:

$$h_t^n = k_W h_{t,c} \tag{IX.11}$$

where $h_{t,c}$ is the thaw depth determined from the isoline maps for the summer ground thaw of the ground and k_W is the correction factor, whose values (according to the SNiP for sandy and clayey soils) are given by diagrams at the lower left corner of each seasonal thawing depth map (see Figs. 17 and 18).

Practical data can be used to determine foundation implacement depths in permafrost beds [h_f in relation (IX.9)] for column foundations with small footing areas (if the width of the area $b < h_t$) and pile foundations that have little influence on the temperature regime in the permafrost after they have been placed (with subsequent refreezing of the partially disturbed and thawed ground mass).

Thus, according to Table 2 of SNiP II-B.6-66, the minimum implacement depth of the structures foundations to be built with preservation of the frozen state of the ground bases can be stated as: 1 meter below the maximum summer thaw for column foundations, i.e., $h_c = h_t + 1$ m, and 2 meters below it for pile foundations, i.e., $h_p \geqslant h_t + 2$ m.

However, it must be noted that for foundations with large footing areas (width $b \geqslant h_t$), and especially when the temperature at the upper face of the foundations is higher than that in the area immediately surrounding them, a special thermal engineering calculation is required to determine the depth to which positive temperatures penetrate through the body of the foundation into the ground (it is similar to the calculation made for structures to be built on thawing bases; see Chap. X) and deepening of the foundations by a greater amount, i.e., it is necessary to have $h_f > h_t$ (where h_t is the maximum possible thaw depth under the footings of these larger area foundations and higher surface temperature).

Besides the thermophysical conditions, it is also necessary to consider the design features of the projected structures in determining the depth of the foundations below grade and primarily the presence of underground pipelines.

It is necessary to know the maximum permafrost temperature underneath the footing in order to determine the design strengths of the frozen ground (long-term compressive and shear strengths, etc.).

Using a relation given by V. V. Dokuchayev[15] between the maximum temperature θ_{max} and the annual average temperature of the permafrost, which is assumed constant at θ_0,

$$\theta_{max} = \alpha_\theta \theta_0 \tag{IX.12}$$

where

$$\alpha_\theta = 1 - \exp\left(-\frac{h_f}{5.3}\sqrt{\frac{C_f}{\lambda_f}}\right) \tag{IX.13}$$

and also assuming that the temperature amplitudes in the ground damp according to the same law as in the one-dimensional problem, G. V. Porkhayev* obtained the following expression for the highest temperature of the frozen ground:

$$\theta_{max} = \alpha_\theta \theta_0 - (\theta_0 - \theta_0')\left[f(x, y, z) - f(x, y, h_t) \exp\left(-\frac{h_f}{5.3}\sqrt{\frac{C_f}{\lambda_f}}\right)\right] \tag{IX.14}$$

where θ_0' is the annual average ground temperature at the bottom of the seasonal thaw layer under the structure and $f(x, y, z)$ is the system configuration function (which is determined by the Poisson integral).

For the two-dimensional problem ($y = 0$), the system configuration function is given by

$$f(x, z) = \frac{1}{\pi}\left(\text{arc tan }\frac{B/2 + x}{z} + \text{arc tan }\frac{B/2 - x}{2}\right) \tag{IX.15}$$

Then, according to Porkhayev,** the maximum ground temperature under the center of the structure (at $x = 0$) will be

$$\theta_{max} = \alpha_\theta \theta_0 - \left(\frac{\theta_0 - \theta_0'}{\pi}\right)\left[\text{arc tan }\frac{B}{2(h_f + h_t)} - \text{arc tan }\frac{B}{2h_t} \exp\left(-\frac{h_f}{5.3}\sqrt{\frac{C_f}{\lambda_f}}\right)\right] \tag{IX.16}$$

Certain transformations can be applied to Eq. (IX.16) to bring it to a simple form

---

*See Chap. VIII, Ref. 0.
**See Ref. 7, this chapter.

$$\theta_{max} = \alpha_\theta k_{c\theta} \theta_0 \tag{IX.17}$$

where $k_{c\theta}$ is the coefficient of thermal effect.

The magnitude of the coefficient of thermal effect by a building or structure on the ground temperature under the center of a structure is

$$k_{c\theta} = 1 + \frac{2}{\pi} \times \frac{1}{\alpha_\theta} \left(\frac{\theta'_0}{\theta_0} - 1 \right) \left[\text{arc tan} \frac{B}{2(h_f + h_t)} - \text{arc tan} \frac{B}{2h_t} \exp \left(-\frac{h_f}{5.3} \sqrt{\frac{C'_f}{\lambda_f}} \right) \right] \tag{IX.18}$$

According to SNiP II-B.6-66, Sec. 5.8, it is admissible to determine the maximum temperature of permafrost at a working depth (i.e., at the foundation depth below grade or at the lower end of a pile) from the relation

$$\theta_{max} = \alpha_\theta k_\theta \theta_0 \tag{IX.19}$$

Here reference is made to tabulated values of the coefficient α_θ, and the following average values are assumed for the heat-effect coefficient of buildings and structures as it affects the annual average ground temperature under them:

For ground at outside walls .... $k_\theta = 1.0$
Same, at inside walls .... $k_\theta = 0.8$
At inside walls when sewage lines are laid through the basement.. $k_\theta = 0.6$

The values of the dimensionless coefficients α_θ and α_e (where α_e is the coefficient for the averaged temperature on the surface on a foundation column or pile for the length in contact with permafrost) are determined from Table 44 as functions of

$$h_f \sqrt{\frac{C'_f}{\lambda_f}}$$

where C'_f is the volume heat capacity (determined as a function of unfrozen water at the particular negative temperature and the total moisture content of the ground).

In the design of capital buildings and structures to be erected on permafrost, it is, of course, necessary to determine the highest temperature reached (heating temperature) at the footing level by means of a detailed thermal engineering calculation, e.g., by Porkhayev's method or using Tomirdiaro's simple equations, which were given in Chap. VIII [Eqs. (VIII.3) and (VIII.4)], and incorporate a certain margin of safety.

Table 44. Values of the coefficients α_θ and α_e

$h_f \sqrt{C'_f/\lambda_f}$	0	25	50	75	100	125
α_θ	0	0.42	0.63	0.77	0.87	0.94
α_e	0	0.27	0.43	0.54	0.62	0.69

We present an expression for determination of permafrost temperature at any depth z below thaw depth (according to V. V. Dokuchayev),[16] based on the Fourier equation for a periodic steady-state heat flux:

$$\theta_z = \theta_0 \left[1 - e^{-z\sqrt{\pi/a_f t_{yr}}} \cos\left(\frac{2\pi t}{t_{yr}} - z\sqrt{\frac{\pi}{a_f t_{yr}}}\right)\right] \qquad \text{(IX.20)}$$

where θ_0 is the annual average temperature of the permafrost (and can be put approximately equal to the average temperature at the depth of zero annual temperature amplitude), a_f is the thermal diffusivity of the frozen soils in m^2/hr, t_{yr} is a time equal to one year, and t is the time elapsed since the deepest summer thaw of the soil.

The soil temperature will be highest at the depth z when the cosine in (IX.20) assumes a value of unity:

$$\theta_{z\,max} = \theta_0 \left(1 - e^{-z\sqrt{\pi/a_f t_{yr}}}\right) \qquad \text{(IX.21)}$$

Equation (IX.21) gives results that differ slightly (by less than 6 percent) from the values obtained on the hydraulic integrator and from field observations.

Foundation designs for bearing strength of underlying permafrost strata are based on the following data (which must be known beforehand):

1) The depth to which the foundations must be sunk into permafrost, $h_c = h_t + h_f$, and the type of foundation;
2) The design load N on the base in the least favorable combination;
3) The values of the ultimate long-term compressive strength $ult\sigma_{cm}$ of the permafrost base and the ultimate long-term shear strength $ult\tau_{lt}$ against the material of the foundations that correspond to the maximum temperature $-\theta_{max}$ in the permafrost at the calculated footing depth and the equivalent negative temperature $-\theta_e$ along the length of a pile.

In the case of plastically frozen (high-temperature) permafrost, it is also necessary to design the foundations for maximum permissible deformations (settling in the frozen state) of the bases; this requires knowledge of the deformability characteristics of the plastically frozen ground.

Determination of the placement depth of the foundations into the permafrost was discussed above; the selection of the type of foundation is based on the following considerations.

Various types of foundations are used to preserve the frozen state in the base soils: concrete column foundations, reinforced-concrete and wooden (sometimes rough stone) foundations, single-row and double-row pile foundations, deep footings, and sometimes continuous concrete or quarry-stone foundations.

The foundation types now most extensively used in building structures to utilize the ground in its frozen state are reinforced-concrete (separate) foundation columns and especially pile foundations, chiefly of reinforced concrete because of their greater economy, superior resistance to frost heaving, and greater ease of fabrication, which permits extensive use of mechanization in construction operations and a limited volume of earthmoving operations, which are highly time-consuming in frozen ground.

The design load N on the foundation base includes the weight of the superstructures and the useful loads transmitted onto the foundations through the superstructures in their least favorable combination, the weight of the foundation itself (from basement-ceiling level to the bottom of the footing), and the weight of the backfill in the excavation, which is transmitted onto the shoulders of the footings (Fig. 147).

The ultimate long-term strengths of permafrost at footing level depend on the composition of the frozen soils, their ice content (volumetric $i_{vol} = \gamma/\gamma_i \times (W_d - W_u)/(1 + W_d)$ or the inclusion ice content I_i, determined by direct measurements on continuous in area ice interlayers) and the negative temperature $- \theta$. Determination of the ultimate long-term compressive strength ult σ_{lt}^{cm} of frozen ground was discussed in Chaps. III and IV of the present book. We recall that the ultimate long-term compressive strength of a particular frozen ground at a given negative temperature can be determined most simply by direct experiments: on a dynamometer instrument [Eq. (III.13′)] or on a ball-test instrument [Eq. (III.11′)]; preference should be given here to Tsytovich's ball-test method for clayey permafrost, since it determines most simply and automatically not only the ultimate long-term compressive strength ult σ_{lt}^{cm}, but also the initial critical load init p_{cr} from the equivalent cohesion, which will be necessary to establish the validity limits of the equations for calculation of foundation settling in plastic frozen (high-temperature) ground.

The maximum temperature $- \theta_{max}$ of the permafrost at footing depth is established on the basis of direct temperature measurements at the given depth in the permafrost during the onset of freezing in the upper part of the active

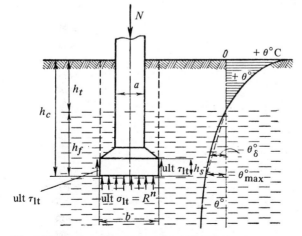

Fig. 147. Design diagram for column foundation to preserve frozen state of base soils.

layer with consideration of the changes in the ground-temperature regime during use of the buildings or structures, or else it is calculated from the analytical relations given above from the known average air temperatures (outdoor and indoor), the duration of the season with positive air temperatures, data on the thermal and physical properties of the soils in the summer-thaw layer and in the upper layer of the permafrost bed, etc. However, it is admissible to determine the maximum soil temperature $-\theta_{max}$ at footing depth from the simplified equation (10) in SNiP II-B.6-66 or, in our notation, from Eq. (IX.19) on the basis of the temperature $-\theta_0$ at the depth of zero ground-temperature amplitudes (~ 10 meters) with consideration of the penetration coefficient α_θ and the coefficient k_θ of the thermal effect of the building or structure as recommended by the SNiP.

In foundation design, the long-term compressive and shear strengths ult σ_{lt} and ult τ_{lt} of the permafrost are equated to the normalized resistance of the permafrost to normal pressures R^n and to the normalized resistance of the frozen ground to shear along the adfreezing surface, R_{sh}^n.

Knowing the temperature $-\theta_{max}$ of the permafrost at footing level, the composition of the soils (coarse fragments, sandy, clayey, with high ice-inclusion contents, etc.), and their ice content (i_{vol} or I_i), Tables 6 and 5 of SNiP II-B.6-66 are consulted to determine the normalized strengths R^n and R_{sh}^n of permafrost, which are needed in the bearing-capacity calculation for the foundations.

We recall here that if difficulties are encountered in determining the ice content I_i due to ice inclusions from the cumbersome in application expression (4) of the SNiP, the ice contents of permafrost can be estimated when Tables 5 and 6 are used to determine R_{sh}^n and R^n from the total volumetric ice content i_{vol}, the limits $0.2 \leqslant I_i \leqslant 0.4$ must be replaced by $0.25 \leqslant i_{vol} \leqslant 0.50$.

With the above data, we can calculate the bearing capacity for foundation columns or piles under structures to be built by the principle of preserving the frozen state of the soils.

For column foundations, we recommend that the calculation begin with a preliminary selection of the foundation dimensions, specifying the height h_s of the bottom step of the shoe (see Fig. 147) and considering the external load N on the column, the column's own weight (from basement-ceiling level to footing depth) and the weight of the fill on its shoulders. For preliminary selection of the column's footing area, we can recommend the following extremely simple relation, which derives from the condition for equilibrium of all forces applied to the foundation (here the adfreezing forces according to the SNiP are taken into account only for the lateral surface of the bottom step of the column):

$$F \geqslant \frac{N - R_{sh}^n F_{sh}}{R^n - \gamma_c h_c k_{fill}} \tag{IX.22}$$

where F_{sh} is the shear area, which is assumed equal to the lateral surface area of the bottom step of the shoe, i.e., $F_{sh} = u h_s$ (u is the perimeter of the shoe and

h_s is its height; in preliminary selection of the foundation dimensions, it can be assumed that $h_s = 0.25$ m and $u = 4b$, where b is the width of the square footing area); γ_c is the specific gravity of the foundation masonry, in tons/m^3, and k_{fill} is a fill factor that takes account of the fact that part of the prism with base F will be filled with masonry (concrete) and part of it with packed earth of lower specific gravity, so that it is necessary to use a value $k_{\text{fill}} = 0.7$–0.8.

Specifying the area F_{sh} for the shear resistance to be taken into account, we determine the required footing area F and can then establish the overall dimensions of the projected foundation.

Knowing the dimensions of the foundation and the physico-mechanical properties of the frozen ground, we determine the bearing capacity of the foundation.

The bearing capacity of a foundation on permafrost is determined by the shear strength of the part of the foundation frozen into the permafrost (as we noted above, only the shear resistance of the bottom step of the column is considered), and by the resistance of permafrost to the normal pressure over a footing of width b (see Fig. 147).

The bearing capacity of a permafrost-stratum in its interaction with foundation columns is determined for all types of column and pile foundations from the equation

$$N_{\text{adm}} = k_1 m_1 \sum_{i=1}^{n} R^n_{\text{sh}i} F_{\text{sh}i} + k_2 m_2 R^n F \qquad (IX.23)$$

where k_1 and k_2 are the homogeneity coefficients of the permafrost, which are assumed equal to $k_1 = k_2 = 0.8$ in accordance with the SNiP, this value being multiplied by a coefficient $(1 - I_k)$ for driven and drilled-and-driven piles; m_1 and m_2 are working-conditions factors determined from Tables 3 and 4 of SNiP II-B.6-66, the values of m_1 varying from 0.9 to 1.1, while $m_2 = 1.0$ for all types of foundations (except pile foundations) and varies from $m_2 = 1.0$ to $m_2 = 2.5$ for pile foundations depending on the ice content I_i, the larger value applying for pile foundations in coarse clastic detritus ground without significant ice inclusions (at $I_i < 0.03$ or $i_{\text{vol}} < 0.05$). $\sum\limits_{i=1}^{n}$ must be extended over all layers from the first to the nth, surface of adfreeze contact between the column or pile and the ground.

The normalized resistance of permafrost to normal pressure (R^n) and its shear strength (R^n_{sh}) at the adfreezing surface are determined, as we noted above, as functions of the composition of the permafrost, its ice content i_{vol}, and its temperature $-\theta°$; here the resistance to normal pressure is determined at $-\theta_{\text{max}}$, which is calculated from one of the equations given above, for example (IX.19), and the normalized shear strength at a temperature slightly higher ($-\theta_s$, see the right-hand drawing in Fig. 147), which can be determined fairly accurately and with a certain safety margin by assuming linear variation of the permafrost temperature from the depth h_t of the summer thaw to footing level h_f (dashed on the right-hand drawing of Fig. 147, which shows curves of ground temperature vs. depth during the period of the deepest thaw).

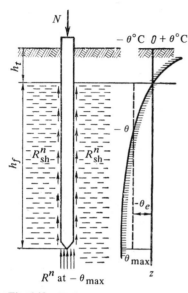

Fig. 148. Design diagram for determination of bearing capacity of piles in permafrost.

Then the temperature at the top of the column shoe $-\theta_s$ can be assumed to be

$$\theta_s \approx \theta_{max}\frac{h_f - h_s}{h_f} \qquad (IX.24)$$

It is this temperature or the average temperature

$$av\ \theta_s = \frac{\theta_{max} + \theta_s}{2}$$

that is used to determine the long-term (normalized) shear strength R_{sh}^n of the frozen ground.

The bearing capacity of pile foundations (Fig. 148) is determined from the same equation (IX.23), but with different (slightly higher) values of the normalized working-conditions coefficient, (see the SNiP). For pile foundations placed into hard-frozen ground (with permafrost temperatures $-\theta_0$ below $-2°C$ at the 10-meter depth), the norms permit determination of the load-bearing capacity from the average equivalent value of the normalized shear strength of the frozen ground at the lateral adfreezing surface, $R_{sh.e}^n$, determined from the equivalent temperature $-\theta_e$:

$$\theta_e = \alpha_e k_\theta \theta_0 \qquad (IX.25)$$

where α_e is the dimensionless coefficient (function of $h_f \sqrt{C_f'/\lambda_f}$) determined from Table 44.

Then relation (IX.23) assumes the following simpler form for pile foundations:

$$N_{adm} = k_1 m_1 R_{sh.e}^n F_{sh} + k_2 m_2 R^n F \qquad (IX.23')$$

We note that according to Sec. 5.12 of SNiP II-B.6-66 the maximum lateral pressure on the permafrost for eccentrically loaded column foundations may not exceed 1.2 times the average normalized pressure on the ground for a symmetrical load, i.e., $1.2 k_2 m_2 R^n$.

Design calculations for static stability of foundations (including pile foundations) and their bases, required in case of steady horizontal stresses, as well as foundations surrounded by downward sloping surfaces or bases composed of fissured bedrock, are conducted using analytical relations (V. V. Sokolovskiy and V. G. Berezantsev) from the Ultimate Stressed State theory of soils.* For perennially frozen ground the ultimate design shear (slip) strength characteristics are accepted as follows: The angle of internal friction ϕ_{1t} and cohesion c_{1t}, are determined from direct experiments at a negative temperature corresponding to its average value for the potential slip failure surface.

The design of bases and foundations for maximum deformation (settling) of perennially frozen ground is based on the following fundamental relations:

$$\text{av } s_{des} \leqslant \text{av } s_{lim} \tag{IX.26}$$

$$\Delta s_{des} \leqslant \Delta s_{lim} \tag{IX.27}$$

where s_{des} and s_{lim} are the design and ultimate settling of the permafrost bases and Δs_{des} and Δs_{lim} are the differences in design and ultimate settling of the permafrost bases.

The left-hand sides of inequalities (IX.26) and (IX.27) are determined by calculation with equations derived in Chap. VI, and the right-hand sides are taken from experimental data that generalize the results of direct observations of settling and deformation of structures built on plastically frozen ground under preservation of the frozen state: here it is permissible (in the absence of necessary data) to take av s_{des} and Δs_{des} from Tables 10 and 11 of SNiP II-B.1-62, i.e., the same values as for structures built on unfrozen soils.

Calculation of foundations to be built on permafrost for the maximum permissible deformation (settling) of their bases will be essential only in cases when the structures are to be built on plastically frozen (high-temperature) soils with preservation of negative ground temperatures and under prolonged action of loads such that the settling may be of a magnitude requiring consideration in their design.

To calculate the settling on high-temperature plastically frozen ground, it is necessary to know the average value of the relative compressibility coefficient of the permafrost a_{rm}^{Σ} (for example, the data given in Tables 32 and 33 can be used for rough calculations) and the creep-deformation parameters δ and δ' or T and $a_{r\infty}$, as well as the dimensions of the footing area (on the basis of a preliminary selection) and the specific load p imposed by the foundations on the base.

The equations derived earlier for the compaction settling of the ground will be valid for determination of the summary compaction settling of plastically frozen ground [see, e.g., (V.24)], but under the condition that the pressure imposed on the frozen ground by external forces does not exceed the initial critical pressure init p_{cr} determined from Eq. (IV.10'), i.e., the value of

$$\text{init } p_{cr} = \pi c_{1t} + \gamma h_c$$

Then the average settling of foundations on plastically frozen ground with

*See Chap. IV, Ref. 32, chap. III, sec. 12.

negative temperature held constant will be determined by expression (V.24):

$$\text{av } s_\infty = h_e a_{rm}^\Sigma p$$

If the external pressure on a plastically frozen ground exceeds init p_{cr}, i.e., $p > \text{init } p_{cr}$, creep settling will occur, and will fade at pressures less than the ultimate long-term strength, i.e., at $p < \text{ult } \sigma_{lt}$. The attenuating creep settling of plastically frozen ground can be determined with the equation derived earlier e.g., (V.28) or (V.29) if the parameters of the creep factor are known.

For an exponential creep factor according to (V.29)

$$s_{ct} = h_e a_{r\infty}^\Sigma p \, (1 - e^{-\delta' t})$$

and for a hyperbolic factor according to (V.28)

$$s'_{ct} = h_e a_{r\infty}^\Sigma p \, \frac{t}{T + t}$$

These expressions can be used to calculate the average total settling of the foundations on plastically frozen ground and the course that it takes in time.

5. Control of Stability and Strength of Foundations Under Frost-Heave Forces

As set forth in Chap. II, Sec. 6 of this book, substantial frost-heave forces appear and act on the foundations of structures upon freezing of moist ground in the active layer around them, owing to the volume increase of the freezing soils and their adfreezing to the foundation materials.

When the seasonal freeze-and-thaw layer contains clayey, fine-sandy, and especially silty soils or coarse-grain soils containing up to 30 percent or more of particles smaller than 0.1 mm, SNiP II-B.6-66 makes mandatory the control of the foundations for determination of the effects of frost-heave forces.

A design scheme for frost-heave resistant foundations was first proposed back in 1928,* and subsequent studies by various authors (M. N. Gol'dshteyn, S. S. Vyalov, B. I. Dalmatov, V. O. Orlov) fully confirmed its fidelity to the observed phenomena. Thus, summarizing the results of his experiments and a theoretical analysis of the heaving process, Gol'dshteyn** concludes that "the forces that heave the structures are determined by the shear resistance of the frozen bed against the foundation," whence it follows that "the design scheme originally proposed by Prof. N. A. Tsytovich is fully justified, and misgivings as to its applicability are explained not by errors in the scheme, but by incorrect estimation of the adfreezing stress under natural conditions by the various investigators."

During recent decades, views as to the nature and magnitude of the frost-heave forces and their effects on structures have been refined substantially

*See Chap. II, Ref. 51.
**See Chap. II, Ref. 28.

by comparison with the original conceptions, both on the basis of field experiments (and primarily the experiment of N. I. Bykov, see Chap. II, Sec. 6) and by subsequent laboratory studies of the long-term shear strength of frozen ground and the stable adfreeze strength between the soils and foundation materials with consideration of stress relaxation and the rate of load increase.

The design frost-heave forces and the ultimate long-term adfreeze strength between soils and foundation materials are now taken from the results of direct field tests and laboratory studies of the frozen ground.

As was shown in Chap. II, a number of factors exert a strong influence on frost-heave forces: the composition of the soils, their moisture contents and water-influx conditions, negative temperature, rate of freezing, etc., and this sometimes makes it impossible to estimate the heaving forces and the influence of the various factors upon them from the results of field determinations. For this reason, the so-called stable adfreezing strength τ_{st} is often taken in accordance with relation (II.16) as the value of the soil frost-heaving forces τ_h in the interaction of the soils with the foundations of structures, i.e.,

$$\tau_h = \tau_{st}$$

A diagram of the forces acting on a column foundation the bottom section of which is load in a permafrost bed, while its upper section is subject to a frost-heave force τ_h appears in Fig. 149.

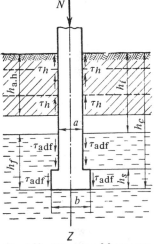

Fig. 149. Diagram of forces acting on column foundation during heaving of soils in seasonal freeze-thaw layer.

We introduce the following notation: τ_h^{av} is the average value of the frost-heaving forces over the entire depth of the active (in the sense of heaving) part of the freeze-thaw layer $h_{a.h}$; lt τ_{adf}^{av} is the average ultimate long-term adfreeze strength between the ground and the foundation material over the depth h_f of the foundation in permafrost; h_i is the thickness of any of the layers differing in composition, ice content, and average temperature.

The values of τ_h and lt τ_{adf} should be chosen with consideration of the ground properties and their negative temperature; their weighted-average values can be introduced into the design equations:

$$\tau_h^{av} = \frac{\tau_{h1}h_1 + \tau_{h2}h_2 + \dots}{h_1 + h_2 + \dots} \tag{IX.28}$$

and

$$\text{lt } \tau_{adf}^{av} = \frac{\tau_{adf1}h_1 + \tau_{adf2}h_2 + \dots}{h_1 + h_2 + \dots} \tag{IX.28'}$$

The height of the layer $h_{a.h}$—the layer active in soil heaving—is determined experimentally to establish the depth distribution of tangential heave forces of the freezing ground, and if the annual freeze-thaw layer is underlaid by permafrost without water influx, it can, considering certain dehydration of the lower third of the freezing layer, be assumed to be two-thirds to three-fourths of the summer-thaw depth h_t (e.g., in accordance with the observations of B. I. Dalmatov):

$$h_{a.h} \approx \frac{2}{3} h_t \quad \text{to} \quad \frac{3}{4} h_t \qquad (IX.29)$$

Further, using also u_1 for the average perimeter of the foundation over the length $h_{a.h}$, u_2 for the corresponding perimeter over length h_f (Fig. 149), and N_1 for the permanent vertical load including the weight of the foundation and the ground resting on its shoulders, we write the equation for the equilibrium of all forces acting on a foundation column subject to ground frost-heave in the seasonal frost layer:

$$\Sigma Z = 0$$

or

$$N_1 + \text{lt } \tau_{adf}^{av} h_f u_2 - \tau_h^{av} h_{a.h} u_1 = 0 \qquad (IX.30)$$

from which

$$\text{lt } \tau_{adf}^{av} h_f u_2 = \tau_h^{av} h_{a.h} u_1 - N_1 \qquad (IX.30')$$

For stability of the foundation under the action of frost-heave forces, it is necessary that the left side of Eq. (IX.30′) be larger than the right-hand side:

$$\text{lt } \tau_{adf}^{av} h_f u_2 \geqslant \tau_h^{av} h_{a.h} u_1 - N_1 \qquad (IX.31)$$

From (IX.31) we obtain the foundation placement depth into the permafrost in order to ensure stability:

$$h_f \geqslant \frac{\tau_h^{av} h_{a.h} - N_1}{\text{lt } \tau_{adf}^{av} u_2} \qquad (IX.32)$$

The author considers it unnecessary to introduce the homogeneity coefficients, working-conditions, and overload conditions into (IX.32) in view of the uncertainty surrounding these conditions; it is necessary to obtain only a design value of h_f 10–15 percent larger than the preliminary design placement depth into permafrost.

The prevailing norms (SNiP II-B.6-66) strengthen inequality (IX.31) by introducing a series of homogeneity coefficients, working-conditions, and overloads (load-underestimation) coefficients:

If these coefficients are introduced into inequality (IX.31) according to the SNiP, we have

$$km \text{ lt } \tau_{adf}^{av} h_f u_2 \geqslant n\tau_h^{av} h_{a.h} u_1 - n_1 N_1 \qquad (IX.33)$$

where km is the product of the homogeneity coefficient by the working-conditions coefficient, assumed to be $km = 0.9$; n is the overload (safety) heave-force coefficient, which equals 1.2 if the seasonally frozen layer merges with permafrost and 1.4 if it does not; n_1 is the overload coefficient (more precisely, the load-underestimation coefficient), which equals 0.9.

We note that in the absence of experimental data on the long-term adfreezing strength lt τ_{adf}, it may be replaced by the normal shear strength of the frozen ground at the adfreezing surface, i.e., by R_{sh}^n, and the heave forces can be assumed (in our opinion, only in the absence of water influx from the outside and in the presence of good surface-water drainage) to average, in accordance with the SNiP, $\tau_h^{av} = 0.8$ kg/cm^2 for regions with ground temperatures $\theta_0 = -3°C$ and above at a depth of 10 meters and $\tau_h^{av} = 0.6$ kg/cm^2 for regions with temperatures below $-3°C$ at that depth.

But it must be remembered that in the presence of some influx of ground water into silty soils (loams, sandy loams, and clays), the heave forces may become substantially larger (on the order of 1.5–3 kg/cm^2 according to the data of Table 8 in Chap. II), and this must be taken into account in special cases.

We note also that if the foundation is anchored (if a shoe is provided), the perimeter of the shoe (the bottom step) of the foundation is taken as the working perimeter u_2, and if the footing rests in unfrozen (thawed) soil it is necessary to replace the retaining adfreezing strength lt τ_{adf}^{av} in (IX.33) by the frictional resistance of the unfrozen soil against the foundation, which is approximately 2–2.5 ton/m^2 for clayey soils and 3 to 6 ton/m^2 for sandy soils (depending on particle size).

The frost-heave stability of the foundations is checked not only for the load corresponding to the completion of construction, but also for uncompleted construction (suspension of work over the winter) based on the actual weight of the installed parts of the structures, and if stability is inadequate, measures must be taken to prevent frost penetration into the ground.

In calculating the retaining force from anchors designed to increase the resistance of the foundations to frost heaving [if relation (IX.33) is not satisfied], it is permissible to take account of the normal pressures on the upper surface of the anchor (foundation shoe) that arises from the action of frost-heave forces and is directed downward.

The additional (restraining) pressure on the top surfaces of the anchors is taken into account by introducing a reduction coefficient of the effective heave forces $k_h \leqslant 1$ into the first term in the right-hand side of (IX.33), i.e., the expression $n\tau_h^{av}h_{a.h}u_1$ must be replaced by $nk_h\tau_h^{av}h_{a.h}u_1$.

For anchored foundations with seasonal frost penetrating permafrost and sufficiently deep foundation placement in the case of taliks (with $h_t + h_m - h_s > (5-10)a$ or $> 5-10b$, see Fig. 149), the reduction coefficient for the upward heaving force should be assumed equal to unity, i.e., $k_h = 1.0$; in other cases k_h is determined by a special calculation that takes account of the distribution of additional pressures on the upper surface of the foundation shoe (anchor) due to heave of the upper ground layers.[17]

Foundations are designed for tension under frost-heave considering the forces determined by the right side of the fundamental stability equation (IX.33) for foundations under the action of heave forces.

Denoting the tensile force by P and the load on the foundation, including the weight of the part of the foundation above the cross section considered, by N_2 we have

$$P = n\tau_h^{av}h_{a.h}u_1 - n_1N_2 \qquad (IX.34)$$

Analysis of (IX.33) and (IX.34) indicates that it is necessary to build foundations on permafrost from materials with good resistance to tensile forces, with consideration of the action of frost-heave forces in the annual frost layer on these materials. Such materials are reinforced concrete, wood, and to some extent (with small admissible tensile stresses) concrete; on the other hand, materials that do not withstand tension, such as crushed-stone, stone, and brick masonry, cannot be used for foundations to be built on permafrost. If foundations are built from crushed stone on permafrost, cracks will inevitably appear in them, which filled with water and subsequent freezing damage the foundation masonry, causing unacceptable deformations in the structures.

6.   Measures To Reduce Frost-Heave Forces On Foundations

In Chap. II, which was devoted to the description of cryogenic processes in freezing moist soils and to estimation of the frost-heaving forces, we noted two basic trends in the application of antiheaving measures: (1) antiheaving melioration of the ground (salinization with various salt mixtures; a combined method with simultaneous salinization and compacting, etc.), used in winter earth work (placement of water-impermeable clay cores in dams built from local materials, correction of local heaving in embankments, etc.); (2) antiheaving stabilization of soils, which ensures stability of the foundations when frost-heave forces act on them.

The first trend was discussed in Sec. 7 of Chap. II, where salinization was mentioned as the basic soil-amelioration measure, but one that cannot be used when the structures are built to preserve the frozen state of the base ground. It is therefore appropriate to examine other antiheaving measures that might help reduce the forces of ground frost heave and its adfreezing to the foundations of structures and thereby ensure stability of the structures when they are acted upon by frost-heave forces.

The general measures to reduce ground frost-heave forces acting on the foundations of structures will be:

1) Ground drainage, since the volume increase of freezing ground is proportional to its moisture content, which can be lowered by providing a common drainage system, diverting surface waters by means of wide water-impermeable screenings, and (if possible) lowering the water table to eliminate influx toward the freezing front, which increases ground heave substantially;

2) Heating of the ground around the foundations, which strongly affects and in some cases even eliminates migration of water in the freezing ground directly around the foundations and sharply reduces ground heave during freezing. The ground can be heated by two methods: (a) by the use of heat-insulating screenings around the foundations, which tend to make the water migrate away from the foundations toward the periphery of the fill, which freezes more rapidly; (b) artificial heating by means of steam pipes, which can not only reduce, but even completely eliminate ground frost heave and buckling of the foundations, but requires constant resource expenditure during the entire winter;

3) Increasing counteracting against frost-heave forces on the foundations, including: (a) increasing the load on the foundations and the specific pressures on their lateral surfaces; (b) backfilling the foundation trenches with unfrost susceptible materials, and (c) strengthening the anchoring of the foundations.

Loads on foundations can be increased by the use of foundation columns with minimal cross section, which requires the use of materials with good resistance to compression and tension. Practical observations indicate that this measure is highly effective and substantially reduces and, under certain conditions, even eliminates the heave of the foundations. Thus, for example, the abutments of bridges with long spans that carry heavy loads, as well as foundation columns of multistory buildings, practically never buckle.

Antiheaving backfill materials include (a) dry gravel and pebbles protected by wooden planking from silting up and communicating with a drainage system (Fig. 150a); (b) a vertically layered reversed filter (Fig. 150b); (c) materials with low freezing temperatures (a layer of clay treated for water repellancy or a special asphalt compound (Fig. 150c)); another measure that reduces heaving to some extent is tapering the faces of the foundation columns and cladding them with steel after casting the concrete.

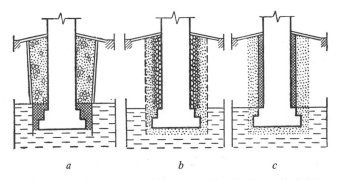

a *b* *c*

Fig. 150. Antiheaving backfill around foundations. (*a*) Pebble backfill protected from silting in by wooden boards with diversion of water (drainage); (*b*) vertically layered backfill (reverse-drainage type); (*c*) backfill of nonadfreezing material.

The anchoring of foundations can be strengthened by:

1) Backfilling the part of the foundation in the frozen-ground bed with fine moistened sand, which, on freezing, will develop the largest adfreezing forces;

2) Increasing the width of the foundation pad, i.e., using an anchor, although this makes the foundation work more expensive, since it requires increasing the dimensions of the trenches in the permafrost. This is the reason for the more extensive use of pile foundations, which can be placed into permafrost without excavation work.

In plastically frozen ground (according to the experience of Vorkutstroy), reinforced-concrete piles are easily driven from above ground with heavy vibrohammers, and in hard-frozen soils (according to Noril'skstroy), piles are driven into wells drilled in advance by rotary drills with subsequent filling of the space between the cylindrical bore of the well and the prismatic pile with a chilled "soil mortar" (moistened drilling mud with fine sand added to it).

7. Examples of Practical Solutions

Here we shall examine the most typical examples of buildings and structures erected on permanently frozen ground, most of them designed to preserve the frozen state in the bases.

As we noted above, when buildings and structures are constructed without sufficient attention to the effects of permafrost on their strength and stability (or when the allowance made for these factors is fundamentally incorrect), e.g., when poorly heat-insulated factory floors are relied on to preserve the frozen state of the ground, when the foundations are sunk deeper than those of structures built under ordinary conditions, etc., a thaw bowl forms under the buildings or structures, and the latter are subject to unacceptable deformations.

Thus, Fig. 151 gives a section of a single-story shop building built at Vorkuta in 1939 (it was described by N. I. Saltikov[18] and P. D. Bondarev[19]), indicating the positions of the thawing boundary of the frozen ground over eight years of use of the building.

Although the building was provided with numerous vertical settling joints, the design strengths of the base soils were assumed to differ depending upon the properties of the soils (1.5 kg/cm^2 for the southern part of the building, 1 kg/cm^2 for the northern part, and only 0.9 kg/cm^2 for the center), and very thick reinforced-concrete foundations were used (columns with 2.5 × 2.5-meter footing areas for the northern part, and continuous reinforced-concrete strips with footing width of 1.4 meters for the central part, in the form of massive beams with rigid parallel flanges), substantial thawing of the permafrost under the building was observed (the thaw boundaries are shown on Fig. 151), and by 1947 the northern and southern ends of the building had settled by 20 cm with a 5-8 cm differential, causing cracking and separation of the settling joints to 10 cm. Thus, use of the massive and costly foundation columns did not protect the

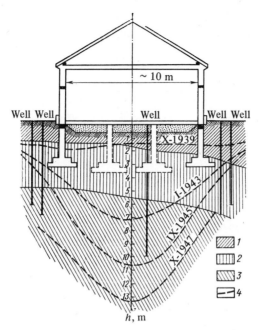

Fig. 151. Example of deep thawing of permafrost under a building in Vorkuta. (1) Cover loam; (2) loams of upper moraine; (3) loams of lower moraine; (4) upper permafrost boundary.

building from inadmissible nonuniform deformations (although the building is still being used in its deformed shape), which could have been prevented altogether if the building had been constructed with a winter-ventilated basement.

How fast permafrost may thaw under structures can be judged from another example, cited in Bondaryev's report, of the boiler station built in Vorkuta in 1946. This building was erected on high ice content moraine loams encountered at a depth of 1.6–1.7 meters under one part of the building and at a depth of 6.2 meters under the rest of it (Fig. 152). Solid concrete-block foundations were designed for the boilers (without

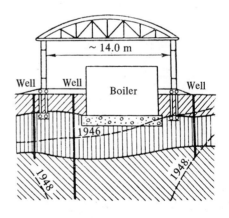

Fig. 152. Sectional drawing of building and base soils of boiler station in Vorkuta. (1) Cover loams; (2) upper-moraine loams; (3) lower-moraine loams; (4) sand-and-gravel fill; (5) upper permafrost boundary.

ventilation openings), and this caused rapid thaw of the permafrost, which had, after 2 years, reached a depth of 12 meters under the boiler footings and caused inadmissible cracks to appear in the walls of the building, with deflections of up to 20 cm.

This example demonstrates the incorrectness of the principle followed in designing the foundations and constructing the boiler building, which did not ensure uniformity in the settling of the foundations, which led to major deformations of the building.

Even faster permafrost thaw was observed under a foundry-shop structure,* the temperature field in the base of which appears in Fig. 153 (according to the Noril'sk Scientific Research Station). Placement of a heat-insulating layer of slag fill 50-cm thick under a layer of molding sand (thickness 1 meter) did not, of course, appreciably reduce the continuous heat flux from the shop into the ground. The structure of the shop building, including its walls and floor, was severely deformed.

Figure 154[20] shows sectional drawings of buildings of fundamentally correct construction to preserve the frozen state in the base soils—with year-round ventilated basements (and especially in winter) (Fig. 154, *a* and *c*) or using cooling ventilation channels below ground level (Fig. 154*b*).

Figure 154*a* is a sectional drawing of the building for a Hoffmann brick kiln that was built at the village of Pokrovskoye (Yakutiya) using brick trench masonry (with air passages) under the bottoms of the kilns with a high ventilated basement open on all sides and pile foundations for the walls of the building and the heavy firing ovens. Despite the fact that the permafrost at the site was high ice content clayey ground and it was necessary to freeze in the piles after driving them into steam-drilled wells and they were kept from floating in the liquefied

*See Ref. 16, this chapter.

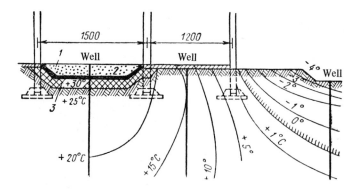

Fig. 153. Thawing of permanently frozen ground under foundry shop (showing temperature isolines). (1) Sand; (2) insulation; (3) excavation boundary.

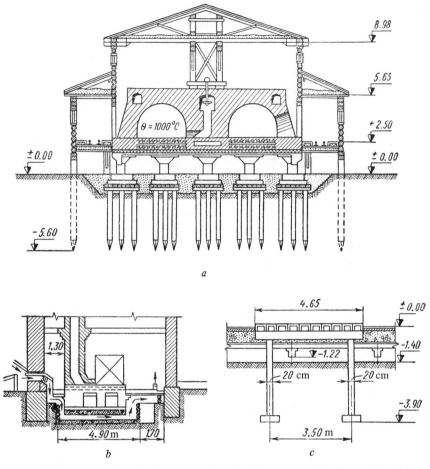

Fig. 154. Examples of foundations of structures built to preserve the frozen state of their base soils. (*a*) Hoffmann kiln for brick firing; (*b*) half-underground boiler station; (*c*) foundation under a steam boiler.

steamed ground by horizontal beams, the building has now functioned for many years without inadmissible deformations.

Figure 154*b* illustrates the successful use of ventilation passages under the floor of a boiler plant (after a proposal by P. I, Mel'nikov) at Yakutsk with the object of preserving the frozen state in the base; although the passages were underground, a steady air draft has prevented disturbance of the frozen state in the soils.

Figure 154*c* is a sectional drawing of a boiler building at the settlement of Amderma, which was built on foundation columns with a ventilated basement and has now been in operation for a long time without unacceptable deformation of the building.

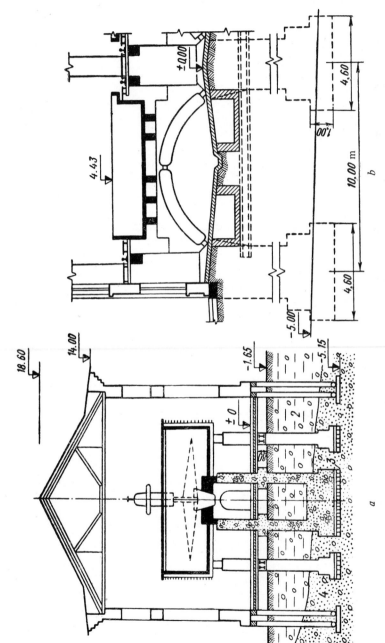

Fig. 155.   Examples of foundations under industrial equipment. (a) Under a condensing vat; (b) under an electrolyte bath.

Figure 155* shows examples of foundations for heavy industrial equipment. Figure 155a shows a column foundation for a concentration-plant condensing vat, with the columns sunk deep below the thaw bowl and resting on dense pebble fill, with the use of a ventilated basement for the entire building and the foundation columns under the equipment; Fig. 155b shows a sectional drawing of a building with hydrolysis-shop equipment installed on a three-hinged arch that rests on the massive supports of the building. In the author's opinion, however, stability of the foundations of the hydrolysis-shop building could have been ensured only by laying the heavy supports on bedrock.

A well-designed open basement in a building erected under harsh climatic conditions (at Churchill, Canada; latitude $58°47'$, annual average air temperature $-7.3°C$) on foundation columns set on gravel fill (Fig. 156)** placed on permafrost with subsequent freezing of the fill and blanketing with a bed of dry peat protected with gravel has preserved the frozen state in the bases well for more than 10 years of building use. Helpful factors here included both proficient execution of the construction work (with prefreezing of the gravel fill) and the careful design of the high ventilated basement with its ceiling lined with reflecting surfaces of aluminum foil on a 10-cm layer of cork, which is, of course, very expensive.

*Figure 155 was taken from the book of Yu. Ya. Velli, et al. (See Ref. 20, this chapter.)
**See Ref. 16, this chapter.

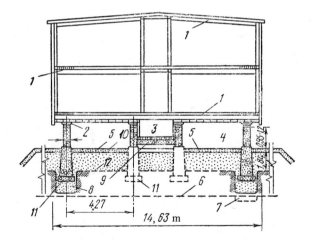

Fig. 156. Sectional drawing of a building built in Canada to preserve the frozen state in the base. (1) Steel girder floor decks; (2) wooden heat insulation; (3) service tunnel; (4) open basement; (5) layer of moss covered with gravel; (6) permafrost boundary; (7) pit cut in frozen soil for gravel fill; (8) packed gravel covered with concrete; (9) concrete; (10) gravel fill; (11) concrete pad; (12) original ground level.

Erecting structures on permafrost by the method of preservation of the frozen state, it is necessary to give special attention to the prevention of thermal regime disturbances in the bases by various types of water-, gas-, sewer-, and other pipelines.

Improperly designed building entrances of various pipeline systems, which may act as local ground-heating sources, can disturb the thermal regime in the base soils in particularly troublesome fashion. To prevent this, it is recommended that all entrances and exits of heating lines and water mains be of the overhead type, suspended from the basement ceiling and run into the buildings in special ducts laid on ventilated slabs and entering the ground or coming out of it at a distance of several meters from the perimeter of the building. Heating and water lines are laid in special galleries outside of the building (Fig. 157), in which the presence of the heating pipes prevents freezing of the water in the water lines and accommodation of both in the same gallery improves inspection and repair conditions; also in use is suspension from scaffolding, masts, walls and other structural elements.[22]

Figure 158 shows sectional drawings of two-level reinforced-concrete galleries for underground pipelines as proposed by K. Borisov and G. Pchelkin on the basis of construction experience at Noril'sk; they were ventilated by taking cold

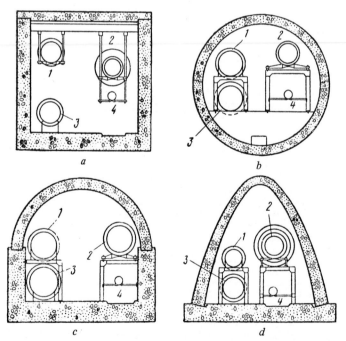

Fig. 157. Underground (utilidoors) gallery designs for pipelines in the Alaskan permafrost. (a) Box-section; (b) circular; (c) half-round; (d) parabolic; (1) water main; (2) steam line; (3) sewer line; (4) condensate.

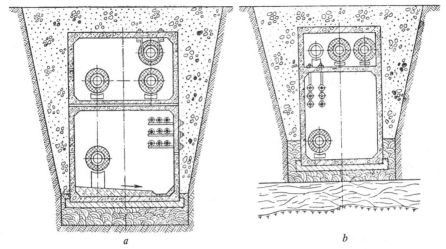

Fig. 158. Two-level reinforced-concrete sewer mains at Noril'sk. (*a*) with drainage trough; (*b*) without trough.

air from the basements of the building and expelling it through special heated shafts.

Problems of laying heating, water-supply, and sewer lines have a number of aspects all their own, and we refer the reader to the specialized literature for discussions of these problems.[23]

References

1. Lukin, G. I.: A History of the Development of Foundation Design in Yakutiya, in collection: "Fundamenty sooruzheniy na merzlykh gruntakh Yakutii" (Foundations of Structures on the Frozen Ground of Yakutiya), Nauka Press, 1968.
2. Stotsenko, V.: "Chasti zdaniy" (Building Elements), Petrograd, 1916.
3. 1. Saltykov, N. I.: Foundations Designs by the Constructive Method, "Merzloto-vedeniye," 1946, no. 1; 2. Saltykov, N. I.: "Teoreticheskie osnovy proyektirovaniya fundamentov na ottaivayushchem osnovaniy" (Fundamentals of the Design of Foundations on Thawing Bases), USSR Academy of Sciences Press, 1952.
4. Saltykov, N. I., and N. N. Saltykova: Thermal-Engineering Design of Ventilated Basements, in collection: "Issledovaniye vechnoy merzloty v Yakutskoy respublike" (An Investigation of Permafrost in the Republic of Yakutiya), no. 2, USSR Academy of Sciences Press, 1950.
5. "Osnovy geokriologiy" (Fundamentals of Geocryology), part II, "Inzhenernaya geokrio-logiya" (Engineering Geocryology), section written by N. I. Saltikov, "Thermal-Engineering Calculations for Above-Ground Cooling Systems," USSR Academy of Sciences Press, 1959.
6. Porkhayev, G. V.: Design of Ventilated Basements for Buildings Built on the Principle of Permafrost Preservation, "Trudy Instituta merzlotovedeniya Ak SSSR", vol. XI, USSR Academy of Sciences Press, 1952.
7. Porkhayev, G. V.: Calculation of the Temperature Regime in the Bases of Buildings and Structures with Year-Round Ventilated Basements, in collection "Fundamenty

sooruzheniy na merzlykh gruntakh v Yakutii" (Structures Foundations on Frozen Ground in Yakutiya), Nauka Press, 1968, and also Posobiya k SNiP II-B.6-66, A Guide to the Construction Norms and Rules II-B.6-66, Gosstroyizdat Press, 1969.

8. Shchelokov, V. K.: "Ratsional'noye ustroystvo fundamentov zdaniy v rayonakh rasprostraneniya vechnomerzlykh gruntov pri sokhraneniy merzlogo sostoyaniya osnovaniya" (Rational Design of Building Foundations in the Permafrost Regime to Preserve the Frozen State of the Bases), Izd. TsNIIS Minstroya, 1959.

9. Tsytovich, N. A.: "O raspredelenii tepla v modelyakh fundamentov, postavlennykh na merzlyy grunt" (Heat Distribution in Foundation Models Placed on Frozen Ground), Bull. No. 25 Vsesoyuznogo instituta sooruzheniy (Bulletin No. 25 of the All-Union Institute of Structures), Izd. "Kubuch," 1932.

10. Lukin, G. I.: Observations of the Temperature Regime in the Bases of Buildings at Yakutsk, in collection "Fundamenty sooruzheniy na merzlykh gruntakh v Yakutii" (Foundations of Structures on Frozen Ground in Yakutiya), Nauka Press, 1968.

11. Maksimov, G. N.: Cooling of High-Temperature Permafrost in the Placement of Pile Foundations, "Osnovaniya, fundamenty i mekhanika merzlykh gruntov," 1968, no. 1. Also, A Method of Stable Air Cooling of Plastically Frozen Base Soils for Large-Panel Buildings, "Doklady i soobshcheniya vsesoyuznogo naychno-tekhnicheskogo soveshchaniya" (Proceedings and Communications of the All-Union Scientific Technical Conference), Krasnoyarsk, 1969.

12. Konovalov, A. A.: Solution of Certain Thermal Problems in the Control of Ground Temperature Regime, "Trudy VI soveshchaniya-seminara po obmenu opytom stroitel'stva v surovykh klimaticheskikh usloviyakh" (Transactions of Sixth Conference-Seminar for Exchange of Experience in Construction under Harsh Climatic Conditions), Krasnoyarsk, 1970.

13. Maksimov, G. N.: Artificial Air Cooling in the Placement of Pile Foundations on Permafrost, NIIOSP Collection No. 55, Stroyizdat Press, 1964.

14. Proceedings of the International Permafrost Conference, Lafayette, November 1963, U.S. National Academy of Sciences, 1965.

15. Velli, Yu. Ya., V. V. Dokuchayev, and N. F. Fedorov: "Zdaniya i sooruzheniya na Kraynem Severe" (spravochnoye posobiye) (Buildings and Structures in the Far North; A Reference Guide), Gosstroyizdat Press, 1963.

16. Dokuchayev, V. A.: "Osnovaniya i fundamenty na vechnomerzlykh gruntakh" (Bases and Foundations on Permanently Frozen Ground), Gosstroyizdat Press, 1963.

17. 1. See Ref. 16, above; 2. Vyalov, S. S., and G. V. Porkhayev (eds.): "Posobiye po proektirovaniyu osnovaniy i fundamentov zdaniy i sooruzheniy na vechnomerzlykh gruntakh (primenitel'no k SNiP II-B.6-66)" [A Guide to the Design of Bases and Foundations for Buildings and Structures on Permanently Frozen Ground (For Use with SNiP II-B.6-66)], Gosstroyizdat Press, 1969.

18. Saltykov, N. I.: Building Foundations in the Region of the Bol'shezemel'skaya Tundra, "Trudy Instituta merzlotovedeniya," vol. IV, USSR Academy of Sciences Press, 1947.

19. Bondarev, P. D.: "Deformatsii zdaniy v rayone Vorkuty ikh prichiny i metody predotvrashcheniya" (Deformation of Buildings in the Vorkuta Region; Causes and Methods of Prevention), USSR Academy of Sciences Press, 1957.

20. Velli, Yu. Ya., et al.: "Zdaniya i sooruzheniya na Kraynem Severe" (Buildings and Structures in the Far North), Gosstroyizdat Press, 1963.

21. Vyalov, S. S., P. I. Mel'nikov, et al.: "Merzlotovedeniye i opyt stroitel'stva na vechnomerzlykh gruntakh v SShA i Kanade" (American and Canadian Geocryology and Experience in Construction on Permafrost Soils), Stroyizdat Press, 1968.

22. "Rekomendatsii po proyektirovaniyu sanitarno-tekhnicheskikh setey v rayonakh rasprostraneniya vechnomerzlykh gruntov" (Recommendations for the Design of Sewer Networks in the Permafrost Regions), Izd. Krasnoyarskogo PromstroyNIIproekta, 1970.

23. 1. Velli, Yu. Ya., V. V. Dokuchayev, and N. F. Fedorov: "Zdaniya i sooruzheniya na Kraynem Severe" (Buildings and Structures in the Far North), Section written by N. F. Fedorov, "Sewage Lines and Constructions," Gosstroyizdat Press, 1963; 2. Saltykov, N. I.: "Kanalizatsiya v usloviyakh vechnoy merzloty" (Sewerage under Permafrost Conditions), USSR Academy of Sciences Press, 1944; 3. Dodin, V. Z.: "Sooruzheniye kanalov podzemnykh kommunikatsiy" (Construction of Galleries for Underground Pipelines), Stroyizdat Press, 1965.

Chapter X

BASES AND FOUNDATIONS
ON THAWING GROUND

1. Conditions Under Which Construction On Thawing Ground is Permissible

We should note at the outset that up to now most structures have been built with thawing of the permafrost bases, often unintentionally, so that neither the magnitude or nonuniformity of the thaw settling of the permafrost bases was taken into account.

Disregard of permafrost bases thawing, or allowing them to thaw without taking account of the consequences, inevitably results in inadmissible deformation of buildings and structures put up under permafrost conditions. These distortions interfere with use of the structures or result in their destruction.

Foundation design with consideration of the thaw settling of permafrost bases in accordance with SNiP II-B.6-66 under Principle II—use of permafrost in the thawing state—only with full consideration of settling of the thawing bases under the condition that the design settling of the thawing bases will not exceed the allowable settling due to progressive thawing of permafrost base soils during construction and operation of the buildings and structures.

To produce competent designs for the foundations of the structures and the structures themselves, with consideration of the thaw settling of the bases, it is absolutely necessary to have data on the physical and mechanical properties of the base soils in the frozen and thawed states, the determination of which was considered in detail in Part One of this book and in Chap. VII, Sec. 3 of Part Two. We list them once again.

1. Thermal properties

λ_t and λ_f are the thermal conductivities in the thawed and frozen states (determined from Table 10 of the SNiP as functions of the composition of the soils, their total moisture content W_d, and their bulk density γ), in kcal/(m)(hr)(deg); R_0, the thermal resistance of the floor in $(m^2)(hr)(deg)/$ kcal, θ_r and θ_0, the indoor temperature and the temperature of the frozen ground at zero-amplitude depth ($\sim$10 m), in $°C$; $q = \zeta(W_d - W_u)\gamma_{sk}$, the heat of thawing of the frozen soil (ζ is the heat of fusion of ice which equals 80,000 kcal/ton; W_d and W_n are the total moisture content and the unfrozen-water moisture content expressed as fractions of unity; and γ_{sk} is the specific weight of the frozen-ground skeleton in tons/m^3), in kcal/m^3.

2. Strength properties

c_{thd} is the cohesion of the thawed soil and ϕ_{thd} is the angle of internal friction of the thawed soil.

3. Deformation properties

$\bar{A}$ is the relative thawing coefficient and $\bar{a}$ is the relative coefficient of thaw consolidation.

In addition, to make a tentative selection of the foundation cross section to take the load from the structure (see Sec. 3 of this chapter), it is necessary to know the dimensions of the foundations-footing area and the load transmitted from the structure to the base soils. However, as we noted earlier (Sec. 1 of Chap. IX), the erection of structures on Principle II, i.e., allowing thawing of the bases, cannot be recommended for any ground, but only for those that have adequate bearing capacity (those that are compacted but not extruded out under load from the structure's foundations) and do not settle beyond the limit stated by the SNiP. Such soils (apart from bedrock) are usually large-fragment, gravelly, and coarse sandy types and, under certain conditions (adequate strength and low ice content), disperse soils.

N. I. Saltykov[1] states the conditions for erection of structures on thawing bases (Principle II) as follows:

1) The average settling of the thawing base should not exceed 25 mm per meter of thaw depth at pressures up to 2 kg/cm², i.e., $s_{av} \leqslant 25$ mm/m at $p \leqslant 2$ kg/cm²;

2) The settling difference between adjacent foundations must not exceed one-fourth of the average settling, i.e., $s \leqslant \pm s_{av}/4$; it is also desirable that the ratio of the thaw coefficient $\bar{A}$ to the thaw-compacting coefficient $\bar{a}$ not exceed three, i.e.,

$$\bar{A}/\bar{a} \leqslant 3$$

These recommendations are, of course, only preliminary ones, since the permissible settling of the bases and the permissible tilt of the foundations will depend not only on the properties of the thawing bases, but also on the dimensions of the foundations, their rigidity, the width of their footing area, the loads on the base, and the sensitivity of the superstructures to nonuniform settling of the bases.

SNiP II-B.1-62 Tables 10 and 11 states norms for the average permissible settling of foundation bases, the difference between the settlings of building columns, maximum deflections of load bearing walls, and tilt of continuous foundations, allowing the following ranges of variation: from $0.005l$ to $0.013l$ for differential settling (where l is the distance between the centerlines of the foundation), from 8 cm (for buildings with unreinforced large-block and brick walls) to 15 cm (buildings with walls stiffened by reinforced-concrete belts) and even up to 30 cm (for monolithic cast foundations for blast furnaces, smokestacks, and other similar structures) for the average settling.

We should note that according to Sec. 5.22 of SNiP II-B.6-66 the average foundation settling limits on thawing soils may be set at larger strain limit values for buildings and structures especially adapted (see SN353-66) for increased nonuniform base deformations (e.g., with provision of rigid reinforced belts, settling joints, or subdivision of the buildings or structures into individual compartments that can settle without affecting the neighboring parts of the building).

In addition, the earlier norms (SN 91-60) also stated values of the maximum settling rates; settling rates from 4 to 8 cm per year were permitted for buildings and structures that were relatively rigid and sensitive to nonuniform settling (rigid reinforced-concrete frames, etc.), up to 10 cm per year for nonrigid structures (buildings and structures with sectional steel bearing elements) and up to 12 cm per year for (statically determinable) wooden structures.

These maximum settling rate values must be taken into account in the event of rapid permafrost thaw in the bases of structures, when the settling rates of the thawing bases may exceed them.

It is quite obvious from all of the above that the principal criterion for the correctness of a foundation design for thawing ground will be consideration of foundation settling, which must be below the limit for the particular type of structure. This makes it necessary to give special attention to prediction of the settling of frozen ground during thaw under the structure.

Thus future calculations and designs for foundations to be built on thawing ground must include the following:

1) Determination of the thaw depth of the permafrost under the structure for various time intervals and at the steady-state limit;

2) Determination of the design strength (bearing capacity) of the thawing base soils;

3) Determination of the reactions of the foundation base, with consideration of nonuniform settling of the thawing bases;

4) Determination of the dimensions and design of the foundations;

5) A check of the foundations for frost heaving.

We note that general relationships for determination of the thaw depths of permafrost under structures were considered in Sec. 3 of Chap. VIII, and that checking frost heave of the foundations makes use of the same formulas and data as were given in Sec. 5 of Chap. IX.

2. Bearing Capacities of Thawing and Thawed Ground Bases

In considering the bearing capacities of thawing ground bases, it is necessary, as in other cases considered earlier, to bear two critical loads in mind: the initial critical load init p_{cr}, which has a value close to the proportional limit, i.e., the stress-strain relation for the particular soil can be regarded as linear up to this load in the calculations, and the ultimate load ult p_{cr}, which corresponds to the bearing capacity limit of the soil bases.

Both quantities are determined from the shear-strength characteristics of the soils for the particular physical state involved, i.e., immediately after thawing.

With accuracy sufficient for practical purposes, the shear resistance of thawing soils can be taken in conformity to the Coulomb-Mohr strength theory, i.e., from the equation

$$\text{ult } \tau = c_{thd} + \tan \phi_{thd} p$$

where p is the external pressure and the parameters ϕ_{thd} and c_{thd} must be determined by the method of undrained unconsolidated soil tests, i.e., at the density and moisture content corresponding to the physical state of the ground at the instant of thawing; this can be done by using the so-called fast (nonstandard) shear test.

Permitting in the particular case a certain propagation of local shear zones outside of the critical planes passed across the faces of the footing to the depth $z_{max} = b \tan \phi_{thd}$ (b is the width of the footing, Fig. 159), as was proposed by N. N. Maslov for unfrozen soils[2] which enables us to take indirect account of the influence of ground properties upon the configuration of the shear zone and to use a somewhat larger load on the soil, which can be regarded as a quite admissible specific pressure on the thawing base, we can determine the initial critical load by the improved equation by N. P. Puzyrevskiy:

$$\text{init } p_{cr} = \frac{\pi(\gamma h_f + \gamma b \tan \phi_{thd} + c_{thd} \text{ co tan } \phi)}{\text{co tan } \phi_{thd} + \phi_{thd} - \dfrac{\pi}{2}} + \gamma h_f \qquad (\text{X.1})$$

where ϕ_{thd} and c_{thd} are the shear parameters for the thawing ground with no change in its density and moisture content.

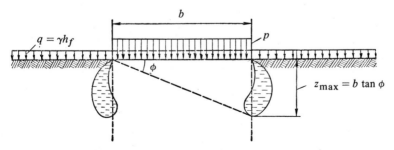

Fig. 159. Configuration of local shear zones under the foundation outside planes passed across the faces of a foundation.

In the design of foundations, the value of init p_{cr} can be taken as a perfectly safe pressure on the ground near the limit of proportionality between pressure on it and the settling of the soil.

To determine the ultimate load on the soil, which corresponds to total exhaustion of its bearing capacity in the thawing state, we may make use of Eqs. (IV.15) and (IV.16) and Table 28, which gives V. G. Berezantsev's soil bearing capacity coefficients A_p, B_p, C_p, and A_c, B_c, and C_c:

For the two-dimensional problem

$$\text{ult } p_p = A_p \gamma b + B_p q + C_p c_{\text{thd}} \tag{IV.15'}$$

For the three-dimensional problem

$$\text{ult } p_c = A_c \gamma b_1 + B_c q + C_c c_{\text{thd}} \tag{IV.16'}$$

where γ is the bulk density of the ground above the control level of the foundation, b is the footing width of a continuous foundation, b_1 is the half-width of a square footing area or the radius of a circular footing area, q is the lateral load, and A, B, and C are the bearing capacity coefficients for thawing soils as determined from Table 28 as functions of ϕ_{thd}.

In Table 28, the values of the bearing-capacity coefficients are given only for internal-friction angles $\phi_{\text{thd}} \geqslant 16°$, which correspond to soils with rigid skeletons (loams, sandy loams, sands, etc.); however, for soils with $\phi_{\text{thd}} < 16°$, determination of the critical soil loads can be based only on the equivalent cohesion c^e_{thd}, which can be determined by the Tsytovich ball test method for a given soil density-moisture content, disregarding stress relaxation of thawing soils.

Then, according to relations (IV.11) and (IV.10'), we have for shallow foundations

$$\text{ult } p_{cr} = (\pi + 2)c^e_{\text{thd}} + q \tag{IV.11'}$$

and for deep ones

$$\text{ult } p_{cr} = 8.3 c^e_{\text{thd}} + q \tag{IV.11''}$$

and for the initial critical load

$$\text{init } p_{cr} = \pi c^e_{\text{thd}} + q \tag{IV.10''}$$

If the critical load on the thawing ground was determined from Eqs. (IV.15'), (IV.16') or from (IV.11'), (IV.11''), the standardized (permissible) pressure on the thawing ground will be determined from

$$R^s = km \, (\text{ult } p_{cr}) \tag{X.2}$$

where km is usually put equal to 0.5-0.7.

We note that the initial critical pressure init p_{cr} on the ground can be regarded as a standardized pressure on the thawing ground without introduction of any reducing coefficients and, in addition, that when the pressure imposed by the foundations on the ground is smaller than init p_{cr}, the ground under the footings will be in the compaction phase and the relationships derived earlier, based on the theory of linearly deformable bodies, will be valid for calculation of foundation settling on thawing ground.

Thus, a condition for validity of the equations derived earlier for calculation of foundation settling on thawing ground will be: the pressure p_{eff} imposed on the soil by the foundations must not exceed the initial critical load init p_{cr} on the thawing or thawed soil:

$$p_{ef} \leqslant \text{init } p_{cr} \approx R^s \qquad (X.3)$$

If no experimental data are available on the internal-friction coefficient ϕ_{thd} or on the cohesion c_{thd} of the thawing or thawed soils for determination of the design loads on the ground (standard pressure R^s), use may be made (but only for structures of classes III and IV) of Table 14 in SNiP II-B.1-62, where the standard pressures are given for unfrozen sandy soils as functions of their density, composition, and moisture content and for clayey soils (loams, sandy loams, and clays) as functions of their natural porosities and consistency ranges (fluidity indices).

Concerning thawed bases, their bearing capacity, which is at a minimum immediately after thawing, will increase gradually as pore water is squeezed out and the soil is compacted under its own weight.

In the case of thixotropic clayey soils, new bonds form after thawing, and the ground comes to have a certain structure that increases its bearing capacity somewhat above the value at the time of thawing.

For sandy soils, on the other hand, and especially for coarse clastic ground (rubble, pebbly, gravelly, and others) with rigid skeletons and large pores, water will be pressed out during thawing, and the bearing capacities of these soils will vary little with the passage of time, although, as we stated earlier, the looseness acquired during freezing will cause them to settle much more than the same soils not subjected to freezing and thawing.

The bearing capacities of thawed ground is estimated by the same general methods that are used for unfrozen ground (from the values of its shear resistance in the thawed state); for preliminary calculations, use may be made of the standard resistances according to Table 14 of SNiP II-B.1-62 as functions of the density of sandy soils and the consistency of clayey soils.

3. Foundation Design For Ultimate Deformation (Settling) of Thawing Bases

The ultimate deformation (settling) calculations for foundations on thawed bases include the following determinations:

1) preliminary determination of foundation dimensions and construction;

2) determination of the depth of thaw in the permafrost under the structure;

3) calculation of the settling and design of the foundations for the ultimate settling of the thawing bases.

The preliminary determination of foundation dimensions is based on a known footing depth (which depends on the thickness of the active layer, i.e., the layer of the annual freeze and thaw) and the known standard compressive strength R_{thd}^s of the thawed soil. Determination of the latter characteristic was discussed in the preceding section.

The footing depths of foundations designed according to Principle II with consideration of base thaw may be not less than the thickness $h_{a.1}$ of the active layer, i.e., the layer of the annual freeze and thaw:

$$H \geqslant h_{a.1} \tag{X.4}$$

Here the depth $h_{a.1}$ must be at least as great as the thawing depth of the particular ground, as determined from the results of multiyear observations of soil thawing depths on open areas with snow and vegetation removed, or as determined tentatively from seasonal thaw charts for the ground of the permafrost region (see Figs. 17 and 18). In addition, the freezing depth must be checked by Stefan's equation:

$$h_{a.1} \geqslant H_f^s = \sqrt{\frac{2\lambda_f \theta_2 t_2}{q_2}} \tag{X.5}$$

where H_f^s is the standard (according to the SNiP) freezing depth of the soil, γ_f is the thermal conductivity of the frozen ground taken from Table 10 of SNiP II-B.6-66, in kcal(m)(hr)(deg), θ_2 is the average air temperature during the period of subfreezing temperatures (its absolute value) in °C, t_2 is the duration of the period with negative air temperatures in hr, $q_2 = \zeta(W_d - W_u)\gamma_{sk(f)} - 0.5\, C_f'\theta_2$ is the amount of heat necessary to freeze a unit volume of soil, ζ is the latent heat of freezing of water [$\zeta = 80,000$ kcal/ton, $\gamma_{sk\,(f)}$ is the bulk density of the frozen ground skeleton, which equals $\gamma_{sk\,(f)} = \gamma_f/(1 + W_d)$];

$$C_f' = \frac{1}{W_d}\,[C_f(W_d - W_u) + C_t W_u]$$

is the heat capacity of the soil [Eq. (IX.8)] with consideration of the state change of the water (C_f is the heat capacity of the frozen ground at temperatures of $-10°C$ and lower and C_t is the heat capacity of the thawed soil, Table 10 of the SNiP).

The footing area of the foundation F in m^2, which is needed to calculate the settling of foundations on thawing soils, is determined from the equation known from the "Bases and Foundations" course.*

*See Ref. 32, Chap. IV.

Under a centered load

$$F \geqslant \frac{N}{R_{\text{thd}}^s - \gamma_f h_f \kappa_{\text{fill}}} \tag{X.6}$$

where γ_f is the bulk weight of the foundation masonry, h_f is the foundation depth of placement measured from the basement ceiling to the footing, and k_{fill} is the excavation backfill factor, which is usually assumed equal to 0.8-0.9.

Under off-center loads, i.e., when the force N is applied with an eccentricity e, we have for a foundation of length l and width b (based on the equation for nonuniform compression)

$$l \geqslant \frac{N}{bR_{\text{thd}}^s} \left(1 \pm \frac{6e}{b} \right) \tag{X.7}$$

Specifying the foundation width b, we use relation (X.7) to determine the foundation length l required for strength and then the foundation area $F = bl$.

Having data on the foundation placement depth h_f and its footing area $F = bl$, we design the foundation, i.e., we take transitional dimensions from the foundation area F to the area of the pillar or wall resting on the foundation, i.e., to F_w; here, of course, $F_w \ll F$.

After design of the foundation, it is necessary to recompute all loads for a foundation with the particular dimensions and to check the pressure imposed on the ground; here, so that it will be possible later to use the equations based on the theory of linearly deformable bodies for calculation of the foundation settling on the thawing ground, it is necessary to observe the inequality

$$p_{\text{eff}} \leqslant R^s = \text{init } p_{\text{cr}} \tag{X.8}$$

where init p_{cr} is the initial critical pressure on the ground, at which nonlinear zones of the ultimate state can have only very small values, which are determined, for example, by expression (X.1) or be totally absent [expression (IV.13)], and p_{eff} is the pressure from the superstructure imposed on the ground at the footing level.

Determination of the permafrost thaw depth under structures (when they are built with consideration of base thawing) is absolutely necessary for calculation of foundation settling on thawing bases, which was considered in detail in its theoretical aspect in Chap. VIII, based chiefly on G. V. Porkhayev's studies; here, however, we shall discuss practical methods to determine the ground thaw depths under structures built on permafrost in accordance with Principle II and will present various auxiliary working tables and graphs.

To predict the settling of frozen soils on thawing, it is necessary to know (in addition to the deformation characteristics of the thawing ground) the thaw depths in the permafrost under the structure at various time intervals from the beginning use of the heated structure and the ultimate depth of the thaw bowl, since we must calculate both the settling of the ground in the thawing process and the settling due to secondary compaction of the ground (for clayey and other disperse soils) after maximum thaw depth.

For individual (column and continuous) foundations, it is often sufficient to determine the maximum thaw depth $h_{t.c}$ under the center of the building or structure and the thaw depth $h_{t.m}$ at its margin and then to use them to compute both the total settling of the foundations and the differential settling between the individual foundations at various time intervals from the beginning of thawing.

If it is desired to refine the calculations and determine the ground thawing depth at other points on the profile of the building or structure, for example at distances $0.4B$ and $0.25B$ (where B is the width of the building or structure), as well as the tilt of the foundation on the thawing soils, the reader is referred directly to the original sources,* the more so since depths of thaw at other points on the profile of the structure (and not only at the center and margin) are calculated from analogous formulas, but from slightly different graphs.

In order to make use of prepared tables and graphs of the rather complex functions in calculating the thaw depth of permafrost under buildings and structures, it is first of all necessary, as we noted in Chap. VIII, to determine the parameters [Eqs. (VIII.19), (VIII.20), and (VIII.21)]:

$$I = \frac{\lambda_t \theta_r t}{qB^2}$$

$$\alpha = \frac{\lambda_t R_0}{B}$$

$$\beta = -\frac{\lambda_f \theta_0}{\lambda_t \theta_r}$$

Then the ground thaw depths (in meters), measured from the surface under the floor of the building's first story, after a time t under the center $h_{t.c}$ and margin $h_{t.m}$ (notations ours—N.T.) of the building are determined from Porkhayev's equations:

$$h_{t.c} = k_I(\xi_c - k_c)B \qquad (X.9)$$

where k_I is a correction factor that is numerically equal to the ratio of the thaw depth in the three-dimensional solution of the problem to the thawing depth in the two-dimensional solution (Table 45), and ξ_c and k_c are coefficients that can be computed from Tables 46 and 47 as functions of the parameters I, α, and β or can be determined from the nomograms of Fig. 160.

*See Refs. 6 and 9, Chap. I.

Table 45. Values of the coefficient k_I

I	Coefficient β									
	0	0.4	0.8	1.2	2.0	0	0.4	0.8	1.2	2.0
	$L/B = 1$					$L/B = 2$				
0.10	1.00	0.93	0.87	0.83	0.80	1.00	1.00	0.99	0.97	0.96
0.25	0.95	0.85	0.78	0.74	0.70	1.00	0.97	0.92	0.89	0.88
0.50	0.94	0.78	0.68	0.66	0.70	0.99	0.95	0.88	0.86	0.88
1.00	0.92	0.70	0.63	0.66	0.70	0.97	0.90	0.84	0.86	0.88
1.50	0.90	0.64	0.63	0.66	0.70	0.96	0.87	0.84	0.86	0.88

If there is no heat insulation on the surface of the soil in the basement, the parameter $\alpha = 0$. Then the thawing depth under the margin of the structure will be

$$h_{\text{t.m}} = k_I \xi_k B \qquad (X.10)$$

where ξ_k is the coefficient of the relative thawing depth under the margin of

Table 46. Relative ground thaw depths ξ_c under the center of building as functions of I, β, and α

I	Coefficient β								
	0	0.4	0.8	1.2	1.6	2.0	0	0.4	0.8
	$\alpha = 0$						$\alpha = 0.4$		
0.00	0.00	0.00	0.00	0.00	0.00	0.00	0.00	0.00	0.00
0.05	0.30	0.27	0.26	0.24	0.22	0.21	0.12	0.08	0.05
0.10	0.40	0.37	0.33	0.30	0.27	0.24	0.17	0.14	0.09
0.20	0.55	0.48	0.42	0.37	0.32	0.27	0.26	0.20	0.13
0.40	0.72	0.60	0.51	0.40	0.34	0.27	0.40	0.28	0.17
0.80	0.92	0.73	0.56	0.42	0.34	0.27	0.56	0.37	0.18
1.60	1.12	0.87	0.59	0.42	0.34	0.27	0.74	0.46	0.18
3.20	1.25	0.97	0.60	0.42	0.34	0.27	0.86	0.46	0.18
	$\alpha = 0.2$						$\alpha = 0.6$		
0.00	0.00	0.00	0.00	0.00	0.00	0.00	0.00	0.00	0.00
0.05	0.17	0.14	0.12	0.10	0.07	0.06	0.06	0.04	0.00
0.10	0.25	0.21	0.17	0.15	0.11	0.07	0.09	0.06	0.00
0.20	0.37	0.30	0.24	0.19	0.12	0.07	0.17	0.10	0.00
0.40	0.53	0.41	0.31	0.21	0.13	0.07	0.26	0.14	0.00
0.80	0.73	0.52	0.35	0.22	0.13	0.07	0.30	0.20	0.00
1.60	0.91	0.67	0.39	0.22	0.13	0.07	0.42	0.20	0.00
3.20	1.04	0.73	0.36	0.22	0.13	0.07	0.56	0.20	0.00

Table 47. Values of the coefficient k_c

	Values of I											
α	0.02	0.05	0.1	0.2	0.3	0.4	0.6	0.8	1.0	2.5	2.0	2.5
0.01	0.016	0.026	0.030	0.030	0.028	0.028	0.028	0.028	0.028	0.026	0.026	0.026
0.03	0.028	0.046	0.056	0.060	0.060	0.060	0.060	0.056	0.056	0.056	0.056	0.056
0.05	0.040	0.060	0.072	0.080	0.082	0.084	0.082	0.080	0.080	0.080	0.080	0.080
0.10	0.060	0.088	0.106	0.120	0.124	0.126	0.126	0.126	0.128	0.128	0.126	0.126
0.15	0.074	0.108	0.136	0.156	0.164	0.166	0.168	0.168	0.168	0.168	0.166	0.166
0.20	0.092	0.130	0.160	0.182	0.190	0.194	0.198	0.200	0.204	0.208	0.208	0.208
0.25	0.100	0.144	0.180	0.208	0.222	0.230	0.234	0.244	0.248	0.258	0.260	0.262
0.30	0.104	0.152	0.196	0.232	0.256	0.264	0.276	0.296	0.292	0.300	0.306	0.308
0.35	0.120	0.168	0.212	0.264	0.288	0.302	0.318	0.328	0.336	0.344	0.346	0.348
0.40	0.124	0.180	0.232	0.286	0.316	0.336	0.356	0.368	0.376	0.382	0.386	0.368
0.45	0.138	0.190	0.248	0.314	0.348	0.372	0.394	0.406	0.414	0.424	0.428	0.430
0.50	0.142	0.204	0.268	0.338	0.378	0.402	0.426	0.440	0.448	0.460	0.468	0.470
0.55	0.154	0.220	0.286	0.360	0.404	0.430	0.460	0.480	0.488	—	—	—
0.60	0.168	0.240	0.310	0.386	0.432	0.458	0.492	—	—	—	—	—

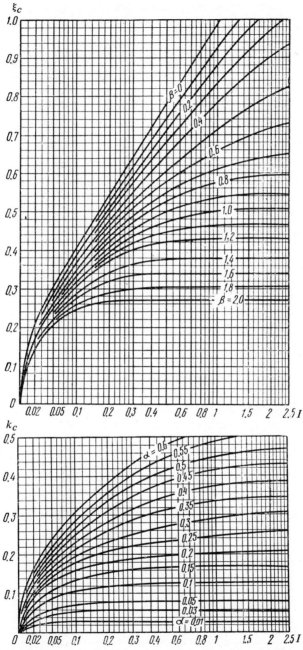

Fig. 160. G. V. Porkhayev nomograms for determination of
the coefficients ξ_c and k_c in calculation of ground thaw depth
under the center of a heated building or structure at any given
time t.

the building or structure, determined as a function of the parameters I, α and β of Table 48 or from the plot of Fig. 161.

But if $\alpha \neq 0$, the soil thaw depth under the margin (at 0.95 width) of a heated building or structure is calculated from the equation

$$h'_{t.m} = k_I (\xi_c - k_c - 0.1\beta \sqrt{I}) B \qquad (X.11)$$

where the coefficient k_c is determined from Table 49 or found from a graph on Fig. 161 as a function of I and α.

The maximum stabilized thaw depths of permafrost under heated buildings and structures (under the center max $h_{t.c}$ and under the margin max $h_{t.m}$) are determined from the relations

$$\max h_{t.c} = k_{II} \xi_{c.u} B \qquad (X.12)$$

$$\max h_{t.m} = k_{II} \xi_{m.u} B \qquad (X.13)$$

where k_{II} is the conversion factor from the plane problem to the three-dimensional problem (determined from Table 50); $\xi_{c.u}$ and $\xi_{m.u}$ are coefficients of influence determined from the nomogram of Fig. 162.

Table 48. Values of the coefficient ξ_m

I	Coefficient β								
	0.0	0.4	0.8	1.2	1.6	0.0	0.05	0.1	0.2
	$\alpha = 0$					$\alpha = 0.4$			
0.00	0.00	0.00	0.00	0.00	0.00	0.00	0.00	0.00	0.00
0.05	0.16	0.12	0.09	0.07	0.06	0.04	0.02	0.01	0.00
0.10	0.23	0.17	0.12	0.08	0.07	0.09	0.06	0.04	0.03
0.20	0.31	0.21	0.16	0.09	0.07	0.11	0.09	0.07	0.05
0.40	0.43	0.31	0.20	0.10	0.07	0.16	0.14	0.12	0.10
0.80	0.61	0.40	0.23	0.10	0.07	0.28	0.25	0.21	0.16
1.60	0.82	0.52	0.25	0.10	0.07	0.44	0.40	0.35	0.23
3.20	1.10	0.64	0.27	0.10	0.07	0.65	0.56	0.47	0.28

I	Coefficient β								
	0.0	0.1	0.2	0.3	0.4	0.0	0.05	0.1	0.2
	$\alpha = 0.2$					$\alpha = 0.6$			
0.00	0.00	0.00	0.00	0.00	0.00	0.00	0.00	0.00	0.00
0.05	0.06	0.03	0.02	0.01	0.00	0.02	0.01	0.01	0.00
0.10	0.10	0.06	0.05	0.04	0.00	0.05	0.03	0.02	0.00
0.20	0.15	0.11	0.08	0.06	0.00	0.07	0.06	0.04	0.00
0.40	0.24	0.19	0.15	0.12	0.00	0.13	0.11	0.10	0.00
0.80	0.41	0.33	0.28	0.20	0.00	0.24	0.21	0.17	0.00
1.60	0.61	0.52	0.45	0.37	0.00	0.35	0.31	0.27	0.00
3.20	0.89	0.73	0.55	0.44	0.00	0.45	0.38	0.30	0.00

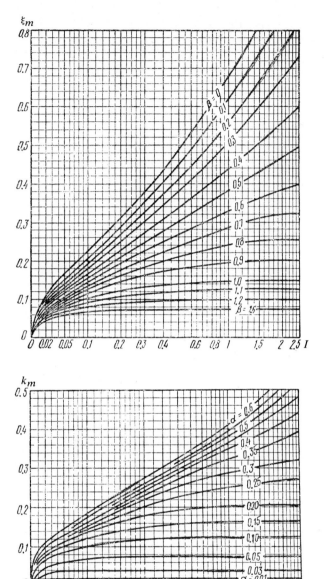

Fig. 161. G. V. Porkhayev's nomograms for determination of the coefficients ξ_m and k_m in calculation of ground thaw depth under the margin of a heated building or structure.

Usually only the E. V. Porkhayev nomograms (Fig. 162) are used in calculating the coefficients $\xi_{c.u}$ and $\xi_{m.u}$, which determine the maximum stabilized ground thaw depths under heated structures. However, to avoid accidental errors, this author recommends that the influence coefficients

Table 49. Values of the coefficient k_m

α	Values of I											
	0.02	0.05	0.1	0.2	0.3	0.4	0.6	0.8	1.0	1.5	2.0	2.5
0.01	0.014	0.020	0.020	0.020	0.020	0.020	0.020	0.020	0.020	0.020	0.020	0.020
0.03	0.030	0.036	0.040	0.040	0.040	0.040	0.040	0.040	0.040	0.040	0.040	0.040
0.05	0.048	0.060	0.070	0.072	0.074	0.076	0.076	0.076	0.074	0.074	0.072	0.072
0.10	0.068	0.086	0.102	0.112	0.118	0.120	0.122	0.124	0.124	0.124	0.126	0.126
0.15	0.080	0.102	0.122	0.140	0.150	0.156	0.158	0.160	0.162	0.164	0.166	0.168
0.20	0.088	0.114	0.138	0.166	0.180	0.188	0.196	0.200	0.202	0.204	0.204	0.206
0.25	0.094	0.118	0.146	0.180	0.202	0.218	0.236	0.244	0.252	0.262	0.268	0.272
0.30	0.096	0.122	0.152	0.192	0.220	0.240	0.264	0.278	0.288	0.304	0.314	0.320
0.35	0.098	0.126	0.156	0.200	0.232	0.256	0.286	0.308	0.324	0.348	0.374	0.394
0.40	0.100	0.130	0.162	0.210	0.244	0.268	0.304	0.328	0.348	0.386	0.412	0.440
0.50	0.104	0.136	0.174	0.226	0.262	0.288	0.318	0.356	0.380	0.434	0.480	—
0.60	0.110	0.146	0.186	0.240	0.276	0.302	0.346	0.408	0.472	—	—	—

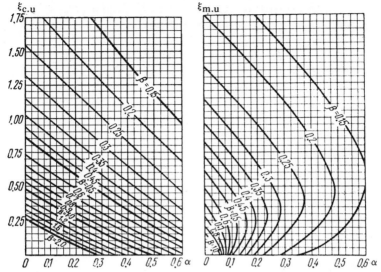

Fig. 162. Porkhayev nomograms for determination of the coefficients $\xi_{c.u}$ and $\xi_{m.u}$ in calculation of the maximum depth of thawing of frozen ground under structures.

always be determined by two methods: analytically (or from tabulated solutions as in Tables 47, 49, and others) and by the use of the nomograms.

S. V. Tomirdiaro's equations can be used in analytic determination of the working coefficients for the maximum (ultimate) thaw depths of permafrost:

For the maximum thaw depth under the center of the building or structure

$$\xi_{c.u} = \frac{1}{2}\left[\cotan\frac{\pi\beta}{2(1+\beta)} - \alpha\right] \qquad (X.14)$$

For the maximum thaw depth under the margin of a building or structure (more precisely, at the distance $0.95B$)

$$\xi_{m.u} = \frac{1}{2}\left[\cotan\frac{\pi\beta}{1+\beta} + \sqrt{\cotan^2\frac{\pi\beta}{1+\beta} + 0.1} - \alpha\right] \qquad (X.15)$$

Table 50. Values of the coefficient k_{II}

L/B	Value of β				
	0.2	0.4	0.8	1.2	2.0
1	0.45	0.56	0.63	0.66	0.70
2	0.62	0.74	0.84	0.86	0.88
3	0.72	0.84	0.91	0.93	0.96
4	1.00	1.00	1.00	1.00	1.00

Together with expressions (X.12) and (X.13), relations (X.14) and (X.15) offer a very simple way of calculating the depth and configuration of the maximum (ultimate) ground thaw bowl under a structure.

M. D. Golovko[3] has prepared extremely convenient diagrams for the determination of the thaw bowl configuration; they were constructed on the basis of numerous determinations that he made on the EHDA electro-integrator (Fig. 163).

The left-side graphs in Fig. 163 indicate the outlines of the thaw bowls in the bases of buildings and structures erected on permafrost for the half-width of the building B depending from the reduced ground layer thickness S, equivalent to the floor of the building in regard to its thermal resistance R_0, and the temperature parameter β'; here

$$S = \lambda_t R_0 \qquad\qquad (X.16)$$

$$\beta' = \frac{\theta'_A - \theta_{\text{thd}}}{\theta'_A - \theta_0} \qquad\qquad (X.17)$$

where θ'_A is the equivalent indoor air temperature

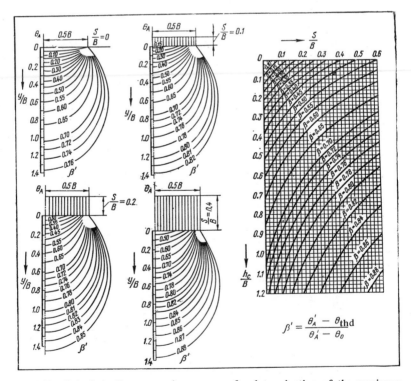

Fig. 163. Golovko's diagrams and nomogram for determination of the maximum thaw bowl under buildings erected on permafrost.

$$\theta'_A = \frac{\lambda_t}{\lambda_f} \theta_A$$

θ_A is the actual average indoor air temperature, θ_0 is the temperature of the permafrost at the depth of zero temperature-variation amplitude, and θ_{thd} is the thawing temperature of the frozen soils, which can be assumed equal to $0°$ for nearly all soils (the only exceptions are highly dispersed clays).

Knowing S/B and β', the ratio max h_t/B is determined from the nomogram of Fig. 163 and used to find the maximum depth h_t of the thaw bowl.

These analytic relationships for the thaw depth in soil bases (tabulated and nomographed) from Porkhayev's solutions and Golovko's integrator experiments are, as was shown in Porkhayev's report, in close conformity to the values observed in nature for buildings and structures on permafrost.

Calculation of settling and design of foundations considering ultimate settling of bases. If the thaw depth h_t under the center of the building and under its margin has been determined from the thermal-engineering calculation set forth above for a given time t, and if we know (from a preliminary calculation) the effective pressure p_{eff} imposed by the structure on the soil and the footing-area dimensions $F = bl$ and the deformability characteristics of the thawing base soils (thawing coefficient $\bar{A}$ and thaw-compacting coefficient $\bar{a}$), we can determine the settling s_{thd} corresponding to any time interval t and the stabilized state of the thaw bowl from the expressions derived in Chap. VI.

Thus, in the absence of distinct ice interlayers, we determine the settling from Eq. (VI.9)

$$s_{thd} = \sum_1^n \bar{A}_i h_i + \sum_1^n \bar{a}_i h_i p_{eff\ i}$$

where

$$p_{eff\ i} = \sigma_z + \gamma(H + z)$$

But when easily measurable ice interlayers of thickness Δ_i are present in the thawing ground, the settling of the foundation will generally be determined by Eq. (VI.13) for the case of the three-dimensional problem (transmission of the external pressure to the ground through a footing area of definite dimensions $F = bl$):

$$s_{thd} = \sum_1^n \bar{A}_i(h_i - \Delta_{eff\ i}) + \sum_1^n \bar{a}_i F_{\gamma i} + \frac{p}{2h_e} \sum_1^n \bar{a}_i h_i z_i + \sum_1^n \Delta_{eff\ i}$$

It will be recalled that the thickness h_e of the equivalent ground layer whose settling is equal in magnitude to that of a foundation of the given dimensions is determined from Eq. (V.23):

$$h_e = A\omega_m b$$

In the case of more uniform thawing frozen soils without pronounced ice interlayers, the expression for the ultimate settling of the foundation on the thawing soils assumes the simpler form [Eq. (VI.14)]:

$$s_{thd} = \bar{A}h_t + \frac{\bar{a}\gamma}{2}(h_t^2 + 2Hh_t) + \bar{a}h_t p_{eff}\left(1 - \frac{h_t}{4h_e}\right)$$

This relation determines the total stabilized settling of the foundations in the case of water-permeable thawing soils (gravelly, sandy, and similar types), for which it can be assumed that the settling time is practically equal to the thaw time of the soil.

The time variation of the settling of clayey soils, which are incompletely compacted in the thawing and load application process (as a result of the low water permeability, which slows down the extrusion of melt water out of the pores in the soil) will, as was shown in Chap. VI, consist of three terms: the thaw settling s_{1t}, the compaction settling during thaw s_{2t}, and the secondary compaction settling after thawing s_{3t}, i.e., according to Eq. (VI.26):

$$s_t = s_{1t} + s_{2t} + s_{3t}$$

where the thaw settling $s_{1t} = \bar{A}h_t$, and the compaction settling [Eq. (VI.28)] is

$$s_{2t} = \bar{a}\left(\chi_1 h_t p_{eff} + \chi_2 \frac{\gamma' h_t^2}{2}\right)$$

The coefficients χ_1 and χ_2 are determined* from Table 39 as functions of r_{thd}.

In the case of action of a localized load from the foundations of a structure, whose influence on settling is taken into account by the equivalent-layer method, we shall have the following expression [Eq. (VI.37)] for the total settling of the foundations as the frozen soils in the base thaw:

$$s_{2t} = \bar{a}\left(\chi_1 h_t p_{eff} + \chi_2 \frac{\gamma' h_t^2}{2} - \frac{1}{2}\chi_1 \frac{p_{eff} h_t^2}{2h_e}\right)$$

Relations (VI.14), (VI.28), and (VI.37) correspond to the total stabilized thaw settling of the soils and, in the case of soils with considerable water permeability (gravelly, pebbly, sandy, and similar types), for which it can be assumed that the settling time is practically equal to the thawing time, they will also determine the settling time variation of the thawing soils.

But for clayey soils, since they are not fully compacted during thawing (due to very low permeability to water), the settling will, as we pointed out in Chap. VI, also depend on the degree of their consolidation; it will be determined by a more complex relation.

*See Chap. VI, Sec. 6.

Combining Eqs. (VI.34), (VI.35), and (VI.36) into a single general relation for determination of the secondary compaction settling s_{3t} of the soils after the thawing depth has reached the constant value for the particular conditions (the maximum value h_t) and recognizing that consolidation of clayey soils will proceed in time in proportion to the degree of consolidation under its own weight $U_3^{(\gamma h)}$ and the action of the external load $U_3^{(p)}$, we have

$$s_{3t} = [\bar{a}(1 - \chi_1)h_{t\infty}p]\ U_3^{(p)} + \left[\frac{1}{2}\ \bar{a}(1 - \chi_2)\gamma'h_{t\infty}^2\right]\ U_3^{(\gamma h)} \quad \text{(X.18)}$$

where $U_3^{(p)}$ and $U_3^{(\gamma h)}$ are the degrees of consolidation of the clayey soils after completion of thawing under the action of the external load p and their own weight γh, determined (based on the solution of the corresponding differential equation of consolidation) from Table 40 as functions of the parameters

$$r_{\text{thd}} = \frac{\beta t}{2\sqrt{c_m}} \qquad \text{and} \qquad N = \frac{\pi^2 c_m}{4h_{t\infty}^2}\ (t - t_{\text{thd}})$$

(where t_{thd} is the time to complete thawing of the soil and βt is the thermal coefficient).

We note that the first term in Eq. (X.18) corresponds to the case of a continuous uniformly distributed load.

In the case of a localized load (from structural foundations with a definite footing area), the first term of Eq. (X.18) assumes a different form, since the magnitude and decay of the settling s_{3t} in time must be derived from the local load by the equivalent soil layer method or by elementary summation with calculating compressive stresses at various points in the compressed zone under the foundation.

Using the Tsytovich equivalent-layer method and assuming that the equivalent plot of the compacting pressure against depth is triangular with a base equal to the external load p and a height equal to twice the equivalent soil layer $h_a = 2h_e$, we have

$$s_{3t} = \frac{h_a\bar{a}_m p}{2}\ (1 - \chi_1)\left\{1 - \frac{16}{\pi^2}\left[\left(1 - \frac{2}{\pi}\right)e^{-N} + \frac{1}{9}\left(1 + \frac{2}{3\pi}\right)e^{-2N} - \cdots\right]\right\}$$

$$\text{(X.19)}$$

where

$$N = \frac{\pi^2 c_m}{4h_a^2}\ t'$$

with $t' = t - t_{\text{thd}}$ and

$$c_m = \frac{k}{\bar{a}_m\gamma_B}$$

These formulas should be used to predict both the total stabilized settling of foundations on thawing soils and the time variation of the settling.

As a result, we obtain the design settling des s of the foundations, which must be smaller than the SNiP limit, i.e.,

$$\text{des } s \leqslant \text{ult } s \qquad (X.20)$$

Example calculation of ground-thaw depth under a structure (to be built on permafrost with consideration of thawing) and determination of foundation settling on thawing soils. We should note first of all that this problem is complicated, and that we shall solve it for the most part by the use of the tabulated influence-function values and nomograms given above; this, of course, gives approximate rather than exact theoretical values, but it does not introduce substantial errors into the design procedure; nevertheless, the situation must be remembered when the calculated results are put to use.

Given: the floor-area dimensions of the heated area with width $B = 10$ m; length $L = 20$ m; indoor temperature $\theta_r = +15°C$; thermal resistance of floor (insulation) $R_0 = 1.0$ (m^2)(hr)(deg)/kcal; soil below foundation footings a sandy loam with bulk density $\gamma = 1.8$ ton/m^3, moisture content $W_d = 20\% = 0.20$, and an unfrozen-water content (at the temperature of the frozen soil) $W_u = 5\% = 0.05$; the temperature of the permafrost at a depth of 10 m is $\theta_0 = -4°C$; the depth of the seasonal thaw $H = 1.5$ m.

It is necessary to determine the thaw depth of the ground under the structure after one year, five years, and 50 years of use, and the settling of the foundations for footing areas of 1.2×20 meters, an effective pressure $p_{\text{eff}} = 0.1$ kg/cm^2, and the thaw deformability characteristics of the ground $\bar{A} = 0.2$ and $\bar{a} = 0.01$ cm^2/kg.

We determine the thaw depth of the soils under the foundation footing at the specified time intervals.

First of all, we find the heat of thawing of frozen sandy loam from the equation

$$q = \zeta(W_d - W_u)\gamma_{\text{sk}} \text{ (f)} = \zeta(W_d - W_u)\frac{\gamma}{1 + W_d}$$

$$= 80,000 \, (0.20 - 0.05)\frac{1.8}{1 + 0.20} = 20,000 \text{ kcal/m}^3$$

From Table 10 of SNiP II-B.6-66, we find for sandy loam with $\gamma = 1.8$ ton/m^3 and $W_d = 0.20$ the thermal conductivities $\lambda_t = 1.34$ and $\lambda_f = 1.52$ kcal/(m)(hr)(deg).

We determine the influence functions (parameters) needed for calculation of the thaw depth of the ground under the structure:

$$\alpha = \frac{\lambda_t R_0}{B} = \frac{1.34 \times 1.0}{10.0} \approx 0.13$$

$$\beta = -\frac{\theta_f\theta_0}{\theta_t\theta_r} = -\frac{1.52\,(-4.0)}{1.34 \times 15} \approx 0.30$$

$$I = \frac{\lambda_t\theta_r}{qB^2}\,t = \frac{1.34 \times 15}{20{,}000 \times 10^2}\,t \approx 0.00001t$$

at $t = 1$ yr $=$ 8,760 $I_1 = 0.00001 \times$ 8,760 $= 0.09$

at $t = 5$ yr $=$ 43,800 $I_5 = 0.00001 \times$ 43,800 ≈ 0.44

and at $t = 50$ yr $= 438{,}000$ $I_{50} = 0.00001 \times 438{,}000 \approx 4.4$

Having the parameters α, β, and I, we determine the thaw depth of the ground after one year of building use: for $I_1 = 0.09$ and $\beta = 0.30$, we find in Table 45 that $k_I = 0.98$ and determine the thawing depth [Eq. (X.9)]:

$$h_{t1} = k_I(\xi_c - k_c)B$$

From the nomogram of Fig. 160, we find for ξ_c and k_c for $I_1 = 0.09$, $\beta = 0.30$, and $\alpha = 0.13$, and check them against Tables 46 and 47: $\xi_c = 0.38$ and $k_c = 0.19$. Then

$$h_{t1} = 0.98\,(0.38 - 0.19) \times 10 = 1.86 \text{ m}$$

In exactly the same way, we shall have:

At $t = 5$ years, $I_5 = 0.44$, $\beta = 0.30$, $\alpha = 0.13$, and $L/B = 2$; from the tables and nomograms we find $k_1 = 0.96$, $\xi_c = 0.66$, and $k_c = 0.27$. Then

$$h_{t5} = 0.96\,(0.66 - 0.27) \times 10 = 3.74 \text{ m}$$

For $t = 50$ years, $I_{50} = 4.4$, $\beta = 0.30$, and $L/B = 2$; the tables and nomograms give $k_I = 0.94$, $\xi_c = 1$, and $k_c = 0.31$. Then

$$h_{t50} = 0.94\,(1.0 - 0.31) \times 10 \approx 6.49 \text{ m}$$

We determine the maximum stabilized depth of the thaw bowl [from formula (X.12)]:

$$\max h_{\text{t.c}} = k_{II}\xi_{\text{c.u}}\,B$$

For $\alpha = 0.13$ and $\beta = 0.30$, the nomogram of Fig. 162 gives $\xi_{\text{c.u}} = 1.10$; for $L/B = 2$ and $\beta = 0.3$, we find (from Table 50):

$$k_{II} = \frac{0.62 + 0.74}{2} = 0.68$$

Then

$$\max h_{\text{t.c}} = h_{t\infty} = 0.68 \times 1.1 \times 10 = 7.48 \text{ m}$$

Now we determine the settling for a foundation of the given dimensions — 1.2×20 m — under an effective pressure on the ground $p_0 = 1$ kg/cm^2.

In calculating the settling, it will be more convenient to state all dimensions in cm and the pressure in kg/cm^2.

First of all, we find the thickness of the equivalent soil layer for the foundation: at $l/b = 20/1.2 \geqslant 10$ and $\mu_r = 0.3$ (for sandy loam), Table 37 gives $A_{\omega m} = 2.77$. Then the thickness h_e of the equivalent soil layer, whose settling is exactly equal to that of a foundation of the given dimensions according to relation (V.23)

$$h_e = A_{\omega m} b = 2.77 \times 1.2 = 3.32 \text{ m} = 332 \text{ cm}$$

Having the deformability characteristics of the thawing sandy loam $(\bar{A}, \bar{a})$ and the thawing depth of the ground h_t, we determine the settling of the foundation from Eq. (VI.14):

$$s_{thd} = \bar{A}h_t + \frac{\bar{a}\gamma}{2}(h_t^2 + 2Hh_t) + \bar{a}h_t p_{eff}\left(1 - \frac{h_t}{4h_e}\right)$$

Considering that according to the specifications $\bar{A} = 0.02$, $\bar{a} = 0.01$ cm^2/kg, $\gamma = 1.8$ g/cm$^3 = 0.0018$ kg/cm^3 at $h_{t=1} = 1.86$ m $= 186$ cm

$$s_1 = 0.02 \times 186 + \frac{0.01 \times 0.0018}{2}(186^2 + 2 \times 150 \times 186)$$

$$+ 0.01 \times 186 \times 1\left(1 - \frac{186}{4 \times 332}\right) = 3.72 + 0.41 + 1.60 = 5.73 \text{ cm}$$

The settling after five years with $h_t = 3.74$ m $= 374$ cm is

$$s_5 = 0.02 \times 374 + \frac{0.01 \times 0.0018}{2}(374^2 + 2 \times 150 \times 374)$$

$$+ 0.01 \times 374 \times 1\left(1 - \frac{374}{4 \times 332}\right) = 7.48 + 2.27 + 3.74 \times 0.72 = 12.44 \text{ cm}$$

The settling corresponding to a time $t = 50$ years is

$$s_{50} = 0.02 \times 649 + \frac{0.01 \times 0.0018}{2}(649^2 + 2 \times 150 \times 649)$$

$$+ 0.01 \times 649 \times 1\left(1 - \frac{649}{4 \times 332}\right) = 12.98 + 5.54 + 3.30 \approx 21.8 \text{ cm}$$

We note that if the thawing depth $h_t \geqslant 2h_e$, the last term in the settling equation assumes a constant value $\bar{a}h_e p_{eff}$, to which a foundation settling of $0.01 \times 332 \times 1 - 3.32$ cm will correspond in this case.

We determine the settling at maximum development of the thawing dish, i.e., at $h_{t\infty} = 7.48$ m $= 748$ cm:

$$s = \bar{A}h_{t\infty} + \frac{\bar{a}\gamma}{2}(h_{t\infty}^2 + 2Hh_{t\infty}) + \bar{a}h_e p_{eff} = 0.02 \times 748$$

$$+ 0.000009 (748^2 + 2 \times 150 \times 748) + 3.32$$

$$= 16.0 + 7.1 + 3.3 = 26.4 \text{ cm}$$

The performed calculation shows that under the given conditions, the settling of the foundations as the permafrost soils thaw under them will continue for a long time, reaching a value larger than 26 cm (more than the limit permitted by the norms); this calls for redesign of the foundations or special design measures to absorb this settling of the foundations and superstructures (partitioning of the building into separate compartments, use of rigid reinforced belts, etc.) or else it will be necessary to specify preconstruction thawing and strengthening of the base soils.

It is also important to note that most of the settling is due to the thaw settling $\bar{A}h_t$ and a relatively small fraction to the load on the foundation.

If the design settling is larger than the ultimate settling (i.e., des $s \geqslant$ ult s), it would appear advisable to thaw the permafrost base soils before construction; this will eliminate the thaw settling $\bar{A}h_t$ and provide compaction of the thawed soil under its own weight.

If the permanently frozen ground has a low water permeability (is clayey), prediction of the thaw settling under the foundations of structures built on it in accordance with Principle II should not be limited to calculation of the total stabilized settling of the foundations by Eq. (VI.14); in this case, it is also necessary to compute the compaction settling and the secondary compaction of the soils in time, i.e., to use relation (X.18) to predict the settling.

The example calculation given above, as well as other similar specific predictions of foundation settling on thawing soils indicate that foundations can be placed with allowance for the thawing of frozen soil bases only if the soils are with low ice contents (preferably with $\bar{A} \leqslant 0.02$) and low compressibility (as discussed above), such as coarse sandy, gravelly, pebbly, etc. However, it is necessary to predict the settling and to design the foundations with allowance for less than the maximum settling even for these types of soils when they are in the frozen state prior to construction.

It must be mentioned once again that before any structure is built on permafrost and the construction principle is chosen (with preservation of the frozen state of the base soils or with consideration of thawing during use of the structures), it is necessary to make an experimental determination of the thaw coefficient $\bar{A}$, which is responsible for most of the thaw settling of the frozen soils; as we pointed out above, this can be done in an extremely simple field experiment in which the soils are thawed at the footing level of the future foundation columns in a wide pit (with shallow thawing, $h_t \leqslant b_0/2$, where b_0 is the width of the pit) and measurement of the thaw depth and thermal settling (without external loads) of the thawed soil layer. Knowledge of the amount of thermal settling $\bar{A}h_t$ of the soil at the maximum thaw depth under the structure will give a correct reference point for design of the foundations.

4. Consideration of Combined Work of Foundations
and Thawing Bases

When structures are built on permafrost with consideration of thawing of their bases, the question arises as to the combined work of the deformable foundations and the base, which settles nonuniformly over the length of the foundations.

The use of the known method for calculation of flexible foundations as beams and plates resting on an elastic base is complicated by a number of circumstances in this case.

First, the thawing base cannot be regarded as ideally elastic, since the settling of the thawing ground will be largely permanent and inelastic with components that do not depend on the external pressure, and the ground will exhibit a certain elasticity only under repeated force application (e.g., the vibrations of machinery).

When the ground thaws under rigid (compared to the thawing base) foundations, which transmit a constant load to the ground, elastic deformations will be manifested only to a very small degree, and the permanent ones (for the most part thawing and thaw-compaction strains) will be of dominant importance.

Second, thawing of soils under foundations occurs nonuniformly: the greatest thaw depth will be observed under the center of the building and the smallest thaw depth under its margins; further, settling of the foundations will depend on the deformation properties of the thawing soils, which differ for various areas of the soil base.

For this reason, the data presently available for calculation of flexible foundations on thawing bases are not yet adequate to permit confident application of the known calculation methods for beams and plates on an elastic base.

At the same time, field experiments carried out specifically for this purpose by the scientific division of "Fundamentproekt"[4] on model foundations on thawing bases to investigate the work of flexible foundations indicated that the reaction (resistance) of the thawing base is proportional to a certain degree to the compaction settling of the soil.

The layer of thawing frozen sand used in the experiments was of variable thickness, which was achieved by artificial regulation of thaw depth in the middle of the excavation. The experiments were performed with model continuous foundations in the form of steel channels 2.1 meters in length, resting directly on seven square 30 × 30-cm plates of the same width as a continuous-foundation model, allowing stepwise reaction pressure distribution of the thawing base.

Daily measurements of the contact pressures of the thawing soil were made throughout the experiment (using dynamometers); also measured were the settling of the plates under the model foundation (using dial gauges) and the thawing depth, by pressing a steel rod (feeler) into the soil.

Some of the results of the "Fundamentproekt" experiments with two model foundations on thawing sand are shown in Fig. 164, where the upper curves indicate the thawing depth h_t of the soil under the model foundation, the middle curves the settling of the thawing soil base $s_{thd} = y$, and the lower plots the reactions p_x of the thawing base.

These experiments are of definite interest as a justification for the premises upon which flexible foundations are designed for thawing bases, but they do, of course, require repetition with thawing soils with a variety of properties (composition, ice content, inhomogeneity, etc.) so that the results can be used as a basis for development of a general design method for elastic foundations on nonuniformly thawing inhomogeneous bases.

It was found on reduction of Kolesov's measurements that the reaction of the thawing base does not depend on the total settling, but only on the difference between the total settling of the frozen soil on thawing and the thaw settling, i.e., the settling due to ice inclusions, which is, generally speaking, independent of the external pressure.

Here the variation of the reaction pressure p_x can be described by the equation

$$p_x = \frac{k_{thd}y}{h(x)} - \frac{k_{thd}y_0}{h(x)} \qquad (X.21)$$

where p_x is the reaction pressure of the thawing soil in kg/cm^2, which varies over the length of the foundation beam x, k_{thd} is the compliance of the thawing base in kg/cm^2, $y = s_{thd}$ is the settling of the soil during thawing under load in cm, $h(x)$ (h_t in our notation — N.T.) is the variable thawing depth of the soil under the foundation in cm, and y_0 is the settling in cm due to melting of ice inclusions.

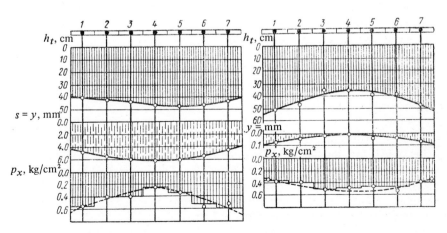

Fig. 164. Settling s, reaction p_x, and thawing depth h_t of soil along two continuous foundations (according to A. A. Kolesov's experiments).

We note that examination of the variables in relation (X.21) indicates that the compliance $k_{thd} = 1/\bar{a}$, where $\bar{a}$ is the reduced (relative) coefficient of thaw compaction; $y_0/h(x)$ is the relative thaw settling (e_{thd} in our notation), which is equal to the "thawing coefficient" $\bar{A}$.

Equation (X.21) can then be rewritten

$$p_x = \frac{s_{thd}}{\bar{a}h_t} - \frac{\bar{A}}{\bar{a}} \tag{X.21'}$$

which also follows from the well-known earlier basic equation for the settling of thawing soils [Eq. (VI.3)], namely:

$$s_{thd} = \bar{A}h_t + \bar{a}h_t p_x$$

solution of which for p_x gives expression (X.21').

However, neither Eq. (VI.III) nor Eq. (X.21') takes account of the compaction of the thawing soil under its own weight, a quantity that can be neglected only for small thaw depths, as mentioned in Chap. VI, which gives equations that take account of the compaction of the thawing soil under its own weight [Eqs. (VI.9) to (VI.14) and others].

The proposal to use simplified equation (X.21) or (X.21') as a starting point for calculation of elastic foundations on thawing bases, which are also assumed to be ideally elastic (e.g., the use of the familiar engineering method of B. N. Zhemochkin for calculation of elastic beams on elastic bases)[5] must presently be regarded as inadequately justified, since a number of assumptions are made in application of this method to the present case (e.g., vertical constancy of compaction settlement, etc.) and certain points remain unclear.*

Evidently the development of a sufficiently accurate method for calculation of flexible foundations on thawing bases will be possible only if more general calculation methods are used (e.g., V. Z. Vlasov's variational method,[6] a system of linear equations solved by the initial-parameter method, or others). Here, the basic task in calculation of flexible foundations on thawing bases remains the determination of the thawing base's reactions with consideration of the inhomogeneity of the soils along the length of the foundation and the compressibility variation of the thawing soil bed with increasing thawing depth.

If the reaction pressures of the thawing ground over the length of the foundation beam are known, they are applied to the foundation from below (Fig. 165) and the applied pressures are regarded as forces external with respect to the foundation beam. The rest of the calculation proceeds in accordance with the rules of structural mechanics: the maximum values of the bending moments max M_x and the shear forces max Q_x are determined and used to calculate the dimensions of the foundation beam.

---

*See Ref. 6, Chap. I.

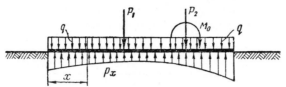

Fig. 165. Design diagram of rigid foundation based on thawing ground.

For short (sufficiently rigid) foundations (with length-to-height ratios smaller than 7–10), it can be assumed that bending of the foundation beam will have practically no effect on the settling of thawing soils. The following approximate device can then be employed for preliminary calculations. The ground thawing depth h_t is calculated by Eqs. (X.9) to (X.17) for various points on the base of the foundation beam, and Eq. (VI.14) is used with the known values of the design characteristics (thaw coefficient $\bar{A}$, relative thaw compaction coefficient $\bar{a}$, base depth of foundation H, average pressure on the soil p, and bulk density of fill γ) to determine the settling s_i at various points along the length of the foundation, as well as the maximum s_{max} and average s_{av} settling of the base. Then, considering that the reactions of the thawing base are to a certain degree proportional to the sum of the average settling and the settling difference between specific points of the base, we can note the approximation for short foundation beams

$$p_x \approx [s_{av} + (s_{max} - s_i)]c_{el} \qquad (X.22)$$

where c_{el} is the compliance of the compressible base, determined from the condition that the sum of the base reactions be equal to the sum of the external forces in kg/cm³.

The compliance of the thawing base can be calculated from the expression

$$c_{el} \approx \frac{\sum P_i + pL}{s_{av}bL} \qquad (X.23)$$

where $\sum P_i$ is the sum of the vertical concentrated forces applied to the foundation beam, p is the uniformly distributed load on the beam, and L and b are the length and width of the foundation beam.

Knowing the average compliance c_{el} of the thawing base in kg/cm³, we use relation (X.22) to calculate the reaction pressures p_x, apply them to the foundation beam from below, and use the rules of structural mechanics to find the design values of the maximum bending moment max M_x and the maximum shear force max Q_x due to the action of all forces on the foundation beam (including the base reactions), which form the basis for selection of a foundation cross section that meets strength conditions.

We note that the character of pressure distribution obtained by the above approximate procedure is consistent with the results of direct observations and measurements.

However, it is evident that the entire calculation procedure for flexible foundations on thawing bases requires further improvements and development.

5. Design of Foundations to Be Built on Permafrost with Consideration of Base Thawing

As described above, the thawing of permanently frozen ground under structures always results in settling of their foundations in a nonuniform fashion. For this reason, the erection of structures on disperse high ice content soils using them in the thawing state (even with allowance for thaw settling) is not recommended because of the inevitable extensive settling and local slump of the stratum. Adequately strong and stable structures can be built under thawing-permafrost conditions only on relatively incompressible competent ground (pebbly, gravelly, coarse sandy, and similar types), whose settling upon thaw under the foundations of the structures may be less than the ultimate values, and, of course, on strong bedrock.

The following evolves from the analysis of combined work of the structures and the thawing ground:

1) Differential settling of structure foundations always occurs on frozen stratum during thaw because of the varying depths of thaw under different parts of the structure and ground inhomogeneity.

2) The most rapid settling of foundations on thawing ground is observed early in the life of the structure, when the thaw rate of the soils under the structure is highest; this makes it advantageous to thaw the ground to a certain depth prior to placing the foundations as a measure that reduces the subsequent settling and its irregularity.

3) The most expedient foundation system for structures to be built on thawing ground with consideration of thaw settling will consist of rigid, massive foundations that are capable of absorbing extensive irregular settling and redistributing the reaction pressures in the base.

In designing and building structures on thawing soils, it is necessary to use settling joints to partition the entire structure into individual rigid compartments that can settle independently of one another. Devices used to ensure rigidity of these compartments include continuous foundation slabs under the entire compartment, or a system of rigid reinforced-concrete strip foundations, or, finally, box-section foundations, which are capable of withstanding substantial nonuniform settling of the thawing base.

In the present case, that of structures built on thawing bases, the foundation bottom level is placed slightly deeper than the seasonal ground thaw depth; measures must then be taken to deal with frost heaving of soils.

Figure 166 shows diagrams of a rigid box-section foundation and a foundation consisting of rigid intersecting reinforced-concrete strips on which the foundation columns supporting the wall beams rest.

For continuous foundation walls (concrete or rubble-concrete) it is advisable to use steel reinforcement in the foundation slab and pour a reinforced-concrete binder over the top.

Reinforced-concrete stiffening flanges should also be designed into brick-walled buildings; they are placed at the level of the top face of the foundation (one version is the reinforced-concrete binder) and at window level of the next-to-top story. In the latter case, the stiffening flanges must be burdened from above to ensure that they will work together with the wall masonry.

Figure 167 illustrates strengthening of reinforced-concrete stiffening flanges at corners and crossings of walls.

Thus, when structures are built on thawing soils with extensive differential settling (but less than the maximum permissible amounts), it is necessary to use a system of rigid, massive (reinforced-concrete) foundations designed to take substantial nonuniform thaw settling.

If bedrock can be reached at an acceptable depth (on the order of 4–8 meters or less), it should always be given preference over other bases for structures to be built under permafrost conditions. Deep column foundations are used in such cases, and if strong incompressible bedrock is encountered just below the surface (in practice, no deeper than 4 meters), all of the disperse soil above the bedrock is replaced by sand-and-gravel fill.

Thus, Fig. 168 shows two schemes for deep foundations under structures built permitting thawing of the

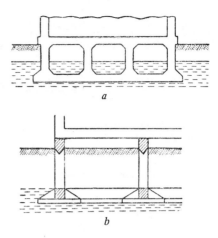

Fig. 166. Diagrams of rigid reinforced-concrete box-section foundation (a) and a foundation in the form of crossing reinforced concrete strips (b) for structures to be built on permafrost considering thaw.

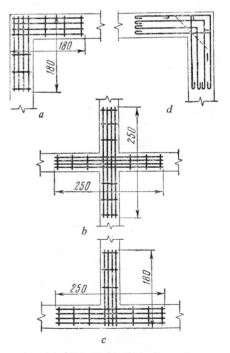

Fig. 167. Examples showing strengthening of corners and intersections of walls with welded steel mesh (a, b, c) and with single rods (d), after V. V. Dokuchayev.

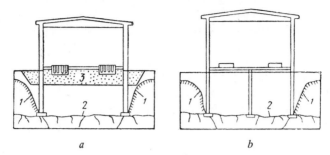

Fig. 168. Examples of design solutions for foundations of structures to be built on permafrost with consideration of their thawing: (a) deep foundations under walls, with partial replacement of soil under floors and under light-machinery foundations; (b) deep foundations under walls with machinery placed on false floors; (1) boundary of the thaw bowl; (2) bedrock; (3) compacted sand-and-gravel fill.

frozen base soils. The foundations of the structures shown schematically in the figure rest on competent bedrock, and the floors and machinery are placed either on a compacted sand-and-gravel pad with concrete admixture or on shallow foundations (Fig. 168a), depending on the expected settling and on operational requirements, or else the equipment is mounted on an appropriately designed false floor (Fig. 168b). Sometimes deep foundations based on bedrock are also used under machinery.

When bedrock is encountered near the surface (at depths of less than 4-5 meters), it is economical to replace the soil and build small structures on shallow foundations based on a well-drained gravel-and-peastone pad. An

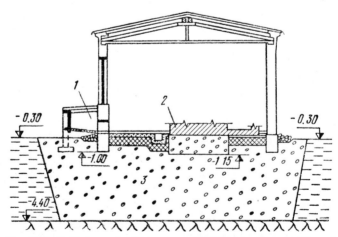

Fig. 169. Boilerhouse building on gravel-and-peastone pad: (1) coal bin; (2) foundation under boiler; (3) gravel-and-peastone fill; (4) bedrock.

example of this solution, used for a small boilerhouse built in the Arctic permafrost zone, appears in Fig. 169.[7]

We should note that the properties of the bedrock must be investigated thoroughly by excavating and using hot-plunger load tests in order to avoid the mistakes that occur very frequently under permafrost conditions, since bedrock is strongly weathered in the overwhelming majority of cases under permafrost conditions and settles perceptibly on thawing, a fact that must be taken into account.

References

1. N. I. Saltykov, Foundation Design by the Structural Method, *Merzlotovedeniye*, no. 1, 1946.
2. N. N. Maslov, Osnovy mekhaniki gruntov i inzhenernoy geologii (Fundamentals of Soil Mechanics and Engineering Geology), *Izd-vo Vysshaya Shkola*, 1968.
3. M. D. Golovko, Metod rascheta chashi protaivaniya v osnovaniyakh zdaniy, vozvodi-mykh na mnogoletnemerzlykh gruntakh (A Method for Calculation of the Thaw Bowl for Buildings to be Erected on Permafrost Soils), A Publication of the V. S. Luk'-yanov Laboratory of Hydraulic and Electrical Analogies, TsNIIS Minstroya, 1958.
4. A. A. Kolesov, Flexible Foundations on Thawing Ground. Advanced Experience Abstract Bulletin, *Stroitel'stvo* (Construction) Series V, no. 4(50), IBTI, 1970.
5. B. N. Zhemochkin and A. P. Sinitsyn, Prakticheskiye metody rascheta fundamentnykh balok na uprugom osnovanii (Practical Methods for the Design of Foundation Beams and Plates on an Elastic Base), Gosstroyizdat, 1962.
6. N. A. Tsytovich, ed., *Trudy MISI*, no. 14, Gosstroyizdat, 1956.
7. Dokuchayev, V. V.: "Osnovaniya i fundamenty na vechnomerzlykh gruntakh" (Bases and Foundations on Permanently Frozen Ground), Gosstroyizdat, 1963.

Chapter XI

PRECONSTRUCTION THAWING AND EARTHWORK SPECIFICS UNDER PERMAFROST CONDITIONS

1. Applicability Conditions of Preconstruction Thawing in Construction on Permafrost

The technique of building on a soil base that has been prethawed to an appropriate depth has recently been coming into more extensive practical use since it has a number of unquestionable advantages under certain conditions over other methods of construction on permanently frozen ground.

This method was first proposed by Prof. A. V. Liverovskiy,[1] and certain economic validations of preconstruction thawing of permafrost base soils for the erection of structures and the physical aspects of the structural processes that take place in the thawing ground were set forth by V. F. Zhukov.[2]

Practical construction experience on permafrost gained in the USSR indicates that two methods are currently most promising: preservation of a frozen state of the base soils with winter basement ventilation (see Chap. IX) and the preconstruction-thawing method. Thus, for example, according to T. A. Aydla, preconstruction thawing of heaved coarse-skeleton permafrost in the Far East (in mountainous terrain) makes construction easier and ensures stability of the structures regardless of unforeseen changes in the soil-temperature regime during operation, without requiring special preventive maintenance.[3]

Let us consider under what conditions it would be most advantageous to use preconstruction thawing of permafrost.

If a calculation indicates that the settling of the frozen soils under the structure will exceed the permissible limit upon thawing, it is difficult to design the foundations and the super-structures in such a way that they will meet strength and stability conditions, and it is necessary to take expensive measures such as partitioning the structure into small rigid compartments resting on solid foundation slabs or measures that increase the rigidity of individual parts of the structures, which is often difficult and uneconomical. In the case under consideration, i.e., when the settling of the thawing base soils after formation of the thaw bowl will exceed the maxima, it is necessary to prethaw the base soils in order to reduce the base settling by the necessary amount before the structure is put up.

Depending on the properties of the permafrost (its composition, ice content, texture, etc.), prethawing of the base soils may produce different effects, and this must be taken into account in the design work.

It is most advantageous to prethaw permafrost in the southern regions of the permafrost zone where thin (up to 7-10 meters in thickness) highly icy soils overlie low-compressible monolithic bedrock.

The experience of Dal'stroy [Main Construction Administration of the Far East] has shown that it is economical to build by the prethawing method on heaving coarse-skeleton ground, which compacts very quickly on prethawing and has adequate bearing capacity in the thawed state.[3]

Prethawing is also used successfully on sandy soils, whose compaction rate is practically the same as their thaw rate; these soils may have extensive nonuniform thaw settling if not prethawed.

Use of the preconstruction thaw method in the case of disperse, e.g., clayey permafrost gives good results only when the stratum is interbedded by water-permeable (sandy or gravelly) interlayers, which promote rapid drainage of melt water and compaction of the clayey soil, or when simultaneous thawing measures of icy disperse soils are taken to accelerate their consolidation or to reinforce and strengthen them artificially.

Construction on permafrost using the prethawing method has the following advantages:

1) Thaw settling of the permafrost ($\Sigma \bar{A}_i h_i$) is totally eliminated over the entire depth of prethawing, which is usually done to a depth less than that of the stabilized thaw bowl; the settling is often 60-80 percent of the total settling of the frozen ground at the depth of the stabilized thaw bowl.

2) Thawing of permafrost is accompanied by compaction under its own weight and an increase in its bearing capacity as a result of displacement of water from the thawed ground mass;

3) When high ice content ground is thawed (especially soil with laminar texture), its filtering capacity is greatly increased, and this accelerates consolidation (compaction) under its own weight;

4) Hydraulic thawing of coarse-skeleton soils causes intensive self-compaction;

5) For coarse-skeleton soils, preconstruction thawing substantially reduces foundation-construction costs (according to T. A. Aydla, foundation-trench excavation work accounts for about 9 percent of the cost of the foundations when prethawing is used, whereas when it is not the earthworks account for about 59 percent of the cost of column foundations);

6) Prethawing makes it possible to increase the amount of useful space inside the buildings due to the use of heated basements;

7) For high-temperature permafrost, prethawing is the preferred method and (if properly done) guarantees that the structures will give no trouble.

2. Selection of Preconstruction-Thawing Depth

Prior to selecting the prethawing depth h_{pt}, it is necessary to calculate the total stabilized settling of the soils at the maximum (stabilized) depth h_∞ of the thaw bowl or for the depth h_{sl} corresponding to the service life of the building (which is usually put equal to $t_{sl} = 50$-60 years).

If the design settling [e.g., according to (VI.9) or (VI.14)] exceeds the limit stated in SNiP II-B.1-62 i.e., if $s_{des} \geqslant av \ s_{ult}$, it is necessary to prethaw the ground in the base.

The values obtained for the design settling and its components (thermal and compaction) give a point of reference for determination of the necessary prethawing depth in the permafrost, since comparison of the total settling and the compaction settling of the soil under its own weight forms a basis for statement of a first value of the prethawing depth h_{pt1}, which must then be verified by a calculation.

Since the thermal settling usually accounts for a major part of the total settling of the thawing soils, we first specify a tentative value h_{pt1} for the preconstruction-thawing depth below foundation footing level such that the difference between the previously calculated total and thermal settlings will be smaller than the maximum permitted (by the SNiP) for the particular type of structure.

We then calculate the settling from the more exact relationships, which take account of:

1) The compaction settling of the soil layer of thickness h_{pt1} that has been thawed in preconstruction preparation under its own weight and the weight of the overlying soil layer from the surface to footing level H;

2) Thaw settling (which does not depend on pressure) from the prethawing-layer depth h_{pt1} down to the maximum thaw depth h_∞ to be taken into account, i.e., in a layer $h_{m1} = h_\infty - h_{pt1}$;

3) The compaction settling of the thawing (during use of the structure) soil layer h_{m1} under the weight of all overlying layers: H, h_{pt1}, and h_{m1}.

To determine the indicated settling components the following quantity must be known: the zero-load compression curve of the prethawed soil down to depth h_{pt1}, which is used to determine the relative compressibility a_r of the thawed soil or the compression-nonlinearity parameter (at substantial variations in the porosity of the soil) $M' = a'_c \gamma h_{m1}$, where a'_c is the compression coefficient determined from experimental data (see Fig. 121, Chap. VI).

Then the total stabilized settling under the weight of all soil layers above thaw depth will be determined by the expression

$$s^{(\gamma h)} = \frac{\Delta \epsilon'_{max}}{1 + \epsilon'_m} h_{pt1} D + \bar{A}(h_\infty - h_{pt1}) + \frac{\Delta \epsilon_{max}}{1 + \epsilon_m}(h_\infty - h_{pt1})D' \qquad (XI.1)$$

where ϵ'_m and ϵ_m are the average initial void ratios of the thawed and thawing soils, D and D' are the corresponding relative settling values for the thawed and thawing soils under soil weight as determined from expression (VI.45) or from Table 41 (last line) as functions of $M = a'_c \gamma h_{pt1}$ or $M' = a_c \gamma (h_\infty - h_{pt1})$.

When the pressure variations are small (approximately 0.5 to 2.5 kg/cm^2), the porosity can be assumed to depend linearly on the external pressure, and then the compaction settling under the load imposed by the weight of all ground layers above the thaw depth will, in the general case (at average values of the deformability indicator), be given by

$$s'(\gamma h) = \frac{a_0 \gamma'}{2} (h_{pt\,1}^2 + 2Hh_{pt\,1}) + \bar{A}(h_\infty - h_{pt\,1}) + \frac{\bar{a}\gamma' h_{m\,1}}{2} (h_{m\,1} + 2H + 2h_{pt\,1})$$

(XI.2)

where γ' is the bulk weight of the soil less the weight of the water displaced by it.

The settling obtained from formula (XI.1) or (XI.2) is then compared with the SNiP limit and if it exceeds the limit a new prethawing depth $h_{pt\,2}$ or $h_{pt\,3}$ greater than the preceding one is specified and the settling is recalculated, and so forth until the inequality $s'(\gamma h) < s_{ult}$ is satisfied. However, for a given preconstruction thaw depth $h_{pt\,i}$ is finally settled upon, it is necessary to check the settling prediction with consideration of the load from the foundation of the structure, i.e., to determine the design settling

$$s_{des} = s^{(\gamma h)} + s^{(p)}$$

(XI.3)

where $s^{(\gamma h)}$ is the settling due to compaction of the ground under its own weight [determined from Eq. (XI.1) or (XI.2)] and $s^{(p)}$ is the settling of the ground under the load imposed by the foundations of the structure.

According to Chap. VI, the latter is determined by the following expressions:

If compression is nonlinear,

$$s'^{(p)} = \frac{\Delta\epsilon_{max}}{1 + \epsilon_m} h_e D''$$

(XI.4)

where $D'' = f(M = a_c p)$;

In calculations according to the theory of linearly deformable bodies and by the equivalent-layer method [Eqs. (VI.14) and (V.24)] when $h_{pt\,1} < 2h_e$

$$s''^{(p)} = h_{pt\,1} a_r p \left(1 - \frac{h_{pt\,1}}{4h_e}\right)$$

and if $h_{pt\,1} \geqslant 2h_e$

(XI.5)

$$s''^{(p)} = h_e a_r p$$

For preliminary calculations of $s^{(p)}$, it is also possible to use the simplified elementary summation Eq. (V.22).

If the total design settling is now smaller than the limit, i.e., if

$$s_{des} = s^{(\gamma h)} + s^{(p)} < s_{ult}$$

the calculation is terminated and the chosen prethawing depth $h_{pt\,1}$ is accepted.

Thus, the preconstruction thaw depth is determined by trial calculations (by specifying a thaw depth and determining the corresponding settling), and it is also acceptable to use the elementary-summation equations [e.g., Eq. (VI.9) for the thermal and compaction settling, etc.].

Electronic computers facilitate selection of the optimum preconstruction thawing depth.

A number of computer calculations performed by L. N. Khrustalev, et al.,[4] for the Vorkuta region (a large-panel building 12 meters wide with a 40-ton load on the walls, with continuous-foundation varying from 1.6 to 4 meters in width and ground pressures from 2.5 to 1 kg/cm²) indicate that the preconstruction thaw depth in permafrost depends on the deformability indices of the thawing soil: the relative compaction coefficient $\bar{a}$ (cm²/kg) and the thaw coefficient $\bar{A}$ (Fig. 170).

If prethawing is done in advance in such a way that the compaction of the thawed soil under its own weight has been completed, the total settling of the foundations on the partially thawed base (thawed to the depth h_{pt}) will be determined by a simpler expression:

$$s_{des} = \bar{A}(h_\infty - h_{pt}) + \frac{\gamma'(h_\infty - h_{pt})^2}{2}\,\bar{a} + h_{pt}a_r p\left(1 - \frac{h_{pt}}{4h_e}\right) \quad \text{(XI.7)}$$

We note that when $h_{pt} \geqslant 2h_e$, the last term in Eq. (XI.7) assumes a constant value $h_e a_r p$, and that when the thaw depth is equal to the depth of the stabilized (ultimate) thaw bowl, i.e., when $h_{pt} = h_\infty$, the settling of the foundations will be determined by the known expression for the settling of

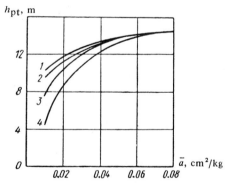

Fig. 170. Preconstruction-thawing depth h_{pt} vs. thaw compaction coefficient $\bar{a}$ at various values of the thawing coefficient $\bar{A}$: (1) at $\bar{A} = 0.050$; (2) at $\bar{A} = 0.040$; (3) $\bar{A} = 0.025$; (4) $\bar{A} = 0.01$ (from results of computer calculations).

bases on unfrozen ground. For laminar stratified soils it is necessary to calculate the average compressibility coefficient from Eq. (V.25).

If the settling calculation for prethawing of permafrost indicates that it will be difficult to satisfy inequality (XI.6), several methods can be used if construction is necessary: (1) the structure can be designed for very high rigidity with partitioning into independently settling compartments, reinforcement with stiffening flanges, etc., although this will be possible only if the expected differential settling is not particularly large; (2) the properties of the thawed ground can be improved by strengthening and reinforcing it (see below in Sec. 5 of this chapter) and, finally, (3) preconstruction thawing can be carried out in advance (before building the structures) and the time necessary for compaction of the soils to less than ultimate settling can then be determined, as discussed in the following section.

3. Determination of Permissible Time for Beginning of Construction After Prethawing of Soils

If the total stabilized settling, determined as the average value from expressions (XI.1) or (XI.2) will exceed the limit at the specified prethawing depth h_{pt}, i.e., if

$$\text{av } s_{\infty}^{(\gamma h)} \geqslant s_{ult}$$

it is not necessary to wait for complete compaction of the thawed soils under their own weight; instead, a construction time that is permissible from the standpoint of thawed-soil compaction can be determined on the premise that the additional settling that can occur after thawing of the soils to the depth h_{pt}, i.e., $s_{add} = s_{\infty}^{(\gamma h)} - s_t$ will be less than or equal to the limit:

$$s_{\infty}^{(\gamma h)} - s_t \leqslant s_{ult}$$

Since the degree of compaction (consolidation) of the soils is, as we know [Eq. (VI.46)], determined by the relation

$$U_{\Delta}^{(\gamma h)} = \frac{s_t}{s_{\infty}^{(\gamma h)}}$$

we substitute $s_t = s_{\infty}^{(\gamma h)} - s_{ult}$ to obtain the following expression for the degree of consolidation of the ground:

$$U_{\Delta}^{(\gamma h)} = \frac{s_{\infty}^{(\gamma h)} - s_{ult}}{s_{\infty}^{(\gamma h)}} \tag{XI.8}$$

where $s_{\infty}^{(\gamma h)}$ is the final compaction settling of the thawed ground under its own weight, calculated from Eq. (XI.1) or (XI.2).

Having determined $U_{\Delta}^{(\gamma h)}$ from Eq. (XI.8) and having calculated the parameter $M = a_c \gamma' h$ from Table 41 (Chap. VI), we find the corresponding value of N.

According to Eq. (VI.47),

$$N = \frac{\pi^2 c_c}{4h_0^2} t_0$$

where

$$c_c = \frac{k_0(1 + \epsilon_m)}{\gamma_w \Delta\epsilon_{max} a_c}$$

is the ground-consolidation coefficient determined from the initial ground filtration coefficient k_0, its average initial void ratio is ϵ_m, the maximum void ratio change is $\Delta\epsilon_{max}$, and the compression-nonlinearity parameter is a_c where γ_w is the specific gravity of the water.

Solving the expression for N to obtain the time t_0, we obtain

$$t_0 = \frac{4h_{pt}^2}{\pi^2 c_c} N \tag{XI.9}$$

Substituting numerical values for the variables in (XI.9), we obtain the permissible time to construction beginning after prethawing of the ground.

4. Practical Methods for Prethawing of Permanently Frozen Ground

A wide variety of factors will influence selection of the technique to be used to prethaw permafrost soils: (1) the preconstruction thaw depth and the volume of permafrost to be thawed; (2) the properties of the permafrost to be thawed (its composition, clay content, ice content, temperature, texture, etc.); (3) the technical feasibility of prethawing operations (availability of the appropriate equipment, heat and power sources, etc.). Also of great importance is the time allowed for the prethawing operations (a few days, several months, or even 2-3 years), and also the necessary degree of compaction and strengthening of the thawed soils to take the load from the new structure.

These circumstances make it necessary to draw up a special plan that takes account of all of the mentioned circumstances before starting the prethawing operations.

If we follow the suggestion by V. F. Zhukov,[5] methods of thawing permafrost can be classified into two types: (1) methods using natural heat and (2) methods based on the use of artificial heat sources.

Methods using natural solar heat are used when there is sufficient time for thawing the soils to a certain depth under exposure only to warm air (during the summer). To speed up thawing, the moss and grass cover is cleared away, the surface is *blackened*, and the future pit is covered with transparent vinyl chloride plastic sheets. It is usually possible to thaw the ground to no more than 3-4 meters in a single season. The pit is filled with water for the winter (if the properties of the thawing ground do not suffer too badly, for example, in cases of sandy and coarse-skeleton soil compositions), to prevent the ground from freezing and to help enlarge the thaw bowl. Thawing of the ground in the excavation by natural air heat is continued during the following summer. According to V. P. Bakakin,[6] permafrost can be thawed to depths up to 7 meters by this method.

Artificial heat sources are more extensively used in practice for preconstruction thawing of permafrost, because they permit a degree of industrialization of the thaw process.

The following are among the methods of thawing permafrost that make use of artificial heat sources:

1) Hydraulic thawing of permafrost with cold and hot water;

2) Steam thawing (open and closed method);

3) Electric heating of the permafrost with alternating current to the necessary depth;

4) Thermochemical methods of thawing permafrost.

According to Dal'stroy data,[3] the most effective hydraulic method for preconstruction thawing of coarse-skeleton ground is the filtration-needle method—thawing by the use of needle filters into which water is forced under pressure. Finely dispersed material is flushed out rapidly by the water, causing self-compaction of the thawed coarse-skeleton ground which is practically complete within two to three weeks. The hydraulic needles are usually spaced 3.5–4 meters apart and sunk to depths of up to 7 meters into the coarse-skeleton ground with the aid of drilling rigs. If drainage or the local terrain relief allows free runoff of the melt water (i.e., if conditions preclude freezing in a closed volume), the coarse-skeleton ground is easily dried after hydraulic thawing and for all practical purposes does not heave. Cost studies have shown that construction on coarse-skeleton ground with preconstruction thawing in hilly terrain and especially on high-temperature permafrost is considerably more advantageous than other methods used under these conditions.

There are two methods of preconstruction steam-thawing of permafrost to the design depth: an open method, in which the steam is released directly into the ground from the end of a pipe (steam needle), and a closed method in which the steam is circulated through a closed system of steam pipes. The open method of steam-thawing is used in most practical cases, although the moisture content of the thawed ground will increase as a result of water vapor condensation.

The steam needles are usually sunk under their own weight or under light percussion into disperse frozen soils. Permafrost is seldom thawed deeper than 10 meters with steam needles, while preconstruction pressure-filtration (needle) hydraulic thawing of permafrost has been used down to a depth of 25 meters (preconstruction thawing of permafrost for expansion of the main building of the Arkagalinsk state regional electric power plant).[7]

Alternating-current electric heating (from 120 to 400 volts and more) can be used to thaw permafrost massifs of various configurations, without increasing the moisture content of the thawed soils and even facilitating drying to a certain degree. The thawing soils compact during electric heating. Thus, according to the Vorkuta Scientific Research Station, preconstruction thawing of the soils under the first test building at Vorkuta produced compaction settling from 25 to 45 cm.[8]

The technology of permafrost thawing by electric heating (Fig. 171)[9] is based on utilization, in accordance with the Joule-Lenz law, of the heat that appears in the massif of frozen ground itself when alternating current is passed through it. During electric heating, compaction of the thawing soils

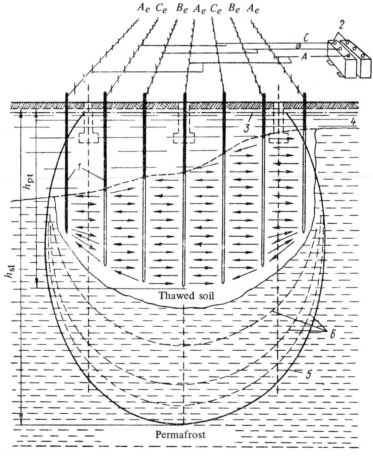

Fig. 171. Process scheme for thawing of permafrost by electric heating: A_e, B_e, C_e are rows of phased electrodes; (1) tubular iron electrodes; (2) step-down transformers; (3) water table; (4) permafrost boundary; (5) contour of ultimate thaw bowl; (6) development of thaw bowl in time.

occurs as a result of the coalescence of mineral interlayers as ice lenses and layers are melted; the water formed in this process runs into drain electrodes.

Alternating current is supplied from the source to perforated electrodes, each of which acts simultaneously as a drain. The electrodes are arranged in parallel rows and the entire space in which electric thawing of the soils occurs is fenced off for safety reasons and posted with warning signs. A voltage of 380 v (that used most frequently at construction sites) is usually used for heating purposes.

Thermochemical methods for preconstruction thawing of permafrost soils include the use of thermite (after V. S. Vylomov), the use of moistened quicklime, which causes not only thawing of the permafrost due to hundreds of degrees of temperature developed in the lime drillpiles, but also substantial compacting of the surrounding soil masses as a result of the significant expansion of the quicklime during slakening, and, finally, a proposal that flameless combustion (oxidation) be used to thaw permafrost, although practical experience with this method is still inadequate.

Special steel needles (Fig. 172) are

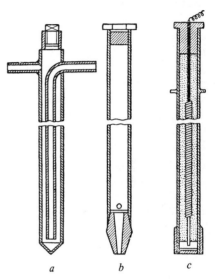

a   *b*   *c*

Fig. 172. Designs of the metallic needles most commonly used to thaw permafrost: (*a*) for hydraulic thawing under pressure; (*b*) for steam heating; (*c*) for electric heating.

used in hydraulic, steam, and electric thawing of permafrost soils.

Aspects of the preconstruction thawing of permafrost that depend on the properties of the soils to be thawed include the following.

When high-ice content ground (in small thicknesses, say 5-7 meters) overlies bedrock, at least the entire bed of high-ice content ground must be thawed. Under permafrost conditions, however, bedrock is exfoliated in the overwhelming majority of cases and interbedded by wedges and lenses of ice, so that there will be substantial thaw settling. Thus it will often be unsafe to order preconstruction thawing only to the small depth at which the upper rock layers are encountered, since it is also necessary to allow for the possible settling of the permanently frozen rocks when they are subsequently thawed. Even in this case, it is mandatory to calculate the total settling of the foundations on the prethawed soil bed with consideration of the thaw settling of the bedrock to the depth of the established (stabilized) thaw bowl under the structure, and to compare the calculated settling with the admissible maximum for the particular type of structure.

In the case of frost-expanded coarse-skeleton ground in mountainous or hilly terrain (assuming good drainage), the expedient prethawing method, as explained above, is hydraulic thawing.

If there are thick layers of sandy and generally well-drained soils, preconstruction thawing will be advantageous in many cases; any of the methods described above may be used to thaw the permafrost to the required depth and, if necessary, the soils may be compacted further after thawing, including by mechanical methods, until they have acquired the necessary bearing capacity and acceptable deformability.

In the presence of disperse soils (silty, clayey, and fine sandy types), it is preferable to prethaw by the electric-heating method with partial strengthening and drying of the soils. In most cases (and especially for icy and very icy soils), it is also necessary to provide additional artificial compacting and strengthening of the soils, since consolidation under their own weight is a very slow process and artificial compacting improves the bearing capacity and deformability of soils of this type. As a rule, it is necessary in the prethawing of finely disperse ground to provide for subsequent compacting and strengthening in order to meet strength requirements (to increase the bearing capacity of the soils) and reduce settling to amounts that do not exceed the permissible limits.

5. Methods for Compaction and Strengthening of Thawed Ground

A distinction should be drawn between compacting and strengthening methods used during preconstruction ground thawing and methods used afterward for further compaction and strengthening of thawed soils.

We shall consider the compaction of thawing ground under its own weight and compaction under its own weight combined with a so-called overburden, i.e., an additional load on the surface of the thawing stratum. We can compute the compaction under gravity and the added external load by the methods of frozen-ground mechanics if we know the compression curves of the thawing ground only at pressures equal to the weight of the soils at soil-sampling depths and the compression curves for the thawed soil after compaction under the weight of the overlying bed and the additional external uniformly distributed pressure.

Methods of this type for ground compaction and strengthening in the preconstruction thawing process also include electroosmotic drying and strengthening of thawing soils, a method that is already in rather extensive use and has been developed in adequate detail.[10] In electroosmosis, the pore water in water-saturated colloidal clayey and silty soils moves under pressure toward the cathode and the effective pressure on the soil skeleton is increased at the anode, first drying the soil, and compacting its skeleton. This method is useful in the case of disperse clayey soils, for which other methods are either totally useless or require a long time to obtain the desired compaction results.

Direct current is used in electroosmotic compaction of thawing clayey soils, in most cases with tubular electrodes of the same type as are used in electric heating. Figure 173 shows a diagram of an electrode arrangement for electroosmotic compaction of thawing ground[9] in which the electrodes of the outer circuit are the cathodes (negatively charged), to which the osmotically attracted pore water moves, while the inner electrodes are anodes (with positive charge); the ground massif to be compacted has several electrode circuits. In the compaction of high ice content thawing soils, all electrodes function

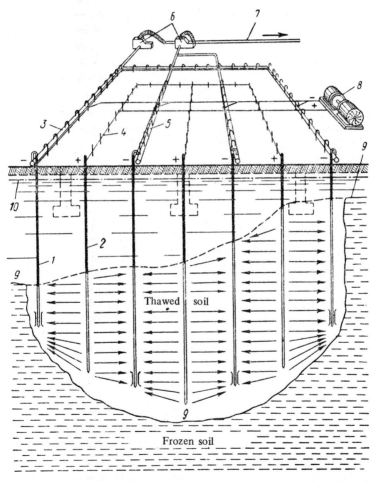

Fig. 173. Diagram of electroosmotic soil compacting in prethawing: (1) cathode needle filter; (2) anode needle filter; (3) connecting hose; (4) electrical conduit; (5) collecting manifold of dewatering system; (6) pumps; (7) water discharge; (8) direct-current source; (9) upper surface of the permafrost; (10) water table.

simultaneously as vertical drains, while the cathodes are also designed in the form of dewatering needle filters.

It has also been found by experiments at Vorkuta (G. D. Potrashkov) that a chemical additive is necessary for uniform reinforcement of soils by electroosmosis (a saturated solution of lime containing 7 percent calcium chloride).

Electroosmotic reinforcement is conducted at maximum vacuum, and measurements are made during compaction to monitor the settling of the ground, water-table level, and the temperature of the thawing soil being compacted.

According to B. A. Rzhanitsyn *et al.*[9] the electroosmotic method is characterized by the following typical indices: time of operations 1–2 months, electric power per square meter of base 0.1–0.3 kw, power consumption 60–80 kw-hr per cubic meter of electroosmotically treated thawing ground.

Electroosmotic compacting of thawing ground was used successfully by the NIIosnovaniy for test sections of residential buildings in Vorkuta, where the permafrost was prethawed only under sections A of the buildings (Fig. 174),[11] which were separated from the rest of the building by a settling joint and permafrost was originally encountered at highly irregular depths. After thawing of the massif of permafrost to the depth required according to calculation and after electroosmotic compaction, the settling of the buildings was equalized; electroosmosis and water level lowering produced an additional compaction settling of about 10–15% from the total. The apartment buildings erected on the improved prethawed base settled a total of not more than 30 mm and are in good condition.

After the preconstruction ground thawing and compaction during the thaw process under their own weight or by electroosmosis, it is necessary to check by calculation the strength and deformability of the thawed soils.

These calculations are carried out in accordance with the familiar rules and equations of general soil mechanics, based on the design characteristics: strength (internal-friction angle ϕ and cohesion c in kg/cm^2) and deformation properties (relative compressibility a_r in cm^2/kg or the total strain modulus E_t in kg/cm^2) established as a result of tests on specimens of thawed soils with undisturbed natural structure.

If the calculations should indicate that the bearing capacity of the ground is not sufficient to take the load from the projected structure, and the settling

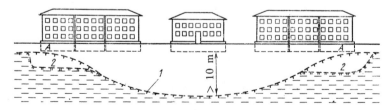

Fig. 174. Surface of permafrost under experimental building at Vorkuta: (1) natural state; (2) after prethawing under building sections A.

prediction indicates that settling will exceed the permissible maximum, it will be necessary to design and implement measures for further compaction and strengthening of the ground.

The measures for postcompaction and strengthening of already thawed ground are the same as those used for unfrozen ground and described in various specialized sources and the pertinent sections of the general textbook on bases and foundations.*

Without dwelling in detail on a description of the various methods used to compact and strengthen thawed ground, we shall list only the most important ones, with an indication of their ranges of usefulness for the compaction and strengthening of weak thawed soils.

Mechanical methods: Tamping and vibrocompaction, usually deep, sometimes with the addition of surfactants. Tamping should be used only on low-moisture ground (with less than 70 percent water saturation), and clayey soils compact well under tamping to a depth of about 1.5-2 meters at moisture contents that do not exceed the plastic limit by more than 3 percent; while, deep compaction with vibrators compacts fine sandy soils to depths of about 10 meters its use requires spot testing.

Physical methods: Compaction by overburdening, by the use of vertical sand drainage, compacting by lowering the water table, etc. The most effective way of accelerating the consolidation of weak water-saturated clayey soils is, in the author's view, vertical sand drainage, by which the consolidation of such soils can be accelerated by several tens of times. Vertical drainage can also be recommended as a means of accelerating compaction and necessary strength, development in weak thawed silty and clayey soils, either by placing special sand drains 40-50 cm in diameter at distances of 2-3 meters from one another or by the use of casing pipe of the type used for electric thawing of permafrost if simultaneously with pipe extraction the wells are filled with coarse sand. The latter method, of course, still requires improvement in practice. In any event, the author regards vertical drainage (with sand drains, sand wells, cardboard drains, etc.) as the most effective physical method for speeding up the compaction of weak water-saturated ground.

Another possible method of compacting weak soils might be to press them out by increasing the weight of the soil together with a temporary lowering of the water table (by pumping); it would be based on the established fact that the swelling of a given soil (on removal of the load) is much smaller than its compressibility (under load).

Chemical methods. Two-solution and single-solution methods are now used only for ground with sufficient water permeability: the two-solution method at filtration coefficients from 2 to 80 m/day, and the single-solution method from 1.2 to 2 m/day. Less water-permeable soils resist injection with chemical strengthening reagents.

*See Ref. 32, Chap. IV.

Electrochemical method: This consists of electroosmosis simultaneously with injection of a strengthening reagent such as water glass. Electroosmosis is used for clayey soils with low water permeability; the action of the direct current on the envelope of water loosely bound by the surfaces of the mineral particles renders it mobile (in the direction toward the cathode) and increases the permeability of the soil which can be exploited for injection of various chemical strengthening reagents into the ground during electroosmosis.

We should note once again that alternating-current electric heating of permafrost, discussed above, is naturally accompanied by a whole series of physicochemical effects on the thawing ground that promote its compaction and strengthening.

Thus, a given method of postcompaction and strengthening of the thawed ground will be chosen depending on its properties as well as the technical facilities of the building organization.

6. Special Aspects of Earthworks Under Permafrost Conditions

Trenching and excavating operations in frozen ground, the placement of fill, and other earthmoving operations and the laying of foundations for structures have features that distinguish them from construction under ordinary conditions in unfrozen ground. The chief distinction of these operations consists in the fact that the builders must reckon with the thermal state of the soils and their properties in the frozen and thawed states.

Depending on the method by which it is planned to construct the building—retention of the frozen state of the ground in the base, considering gradual thaw, or preliminary thawing—different requirements will be imposed on the earthwork and foundation-laying operations.

Construction with preservation of the frozen state in the base soils requires measures to preserve the natural state. If local disturbances occur during the foundation work it is necessary to ensure restoration of the ground's frozen state.

In this case, the engineering preparation of the construction site and building of the approach roads will be subject to special requirements, consisting chiefly in doing everything possible to preserve the natural plant and moss cover and the temperature regime of the soils; all work done to prepare the area, draining, staking it out, etc. must be carried out strictly according to the plan.

The best time to do this work and perform the excavations, especially in swampy terrain, will be the winter, when it is easiest to preserve the frozen state of the ground in the bases of the structures.

However, it is more economical to plan the earthwork in such a way that it will be possible to do much of it in the layer of the summer thaw, i.e., to time the earthwork for the fall, when the ground will be thawed to the greatest depth. This is all the more advantageous in that the subsequent winter frost makes it possible to keep the frozen base soils undisturbed.

We should also note that in construction in swampy terrain or when placing the supports of bridges over small streams, it is advisable to do the excavation work by "promorozka" (after A. V. Liberovskiy), which consists in gradual natural freezing of the water-saturated soils and freeze up of taliks with subsequent water-free work in the frozen soils.

Now that major construction has been developed substantially under permafrost conditions, industrial methods of foundation building and earthwork are coming into increasing use in practice in the sinking of shafts and trenches for column and strip foundations of industrial and residential buildings and structures. Pile foundations have a wide use as a result of the increased capacity of the piledrivers, use of the vibration method and powerful drilling equipment, and the development of special techniques.

The first plans and calculations for pile foundations to take vertical and horizontal forces under permafrost conditions were apparently made by Gipromez (State All-Union Institute for the Planning of Metallurgical Plants) for the shops of the Petrovsk-Zabaykal'skiy metallurgical plant, presently under construction.

In 1928 the author published[12] closed solutions of the bending problem of a pile embedded partly or completely in permafrost under the action of horizontal forces, treating the piles as rods in an elastic medium subject to the theory of local elastic deformations. V. V. Dokuchayev[13] also presents a similar calculation, but in a somewhat different interpretation using Zhemochkin's diagrams for the bending of a rod in an elastic medium. Without dwelling on these calculations, since they pertain to the structural mechanics of elastic systems, we shall consider the performance of operations involved in placing piles and pile foundations under permafrost conditions.

Back in 1936–1937, in designs for surface structures for the Vorkuta coal basin in the region of high-temperature permafrost, Lengiproshakht called for foundation erection on wooden piles.

Later, reinforced-concrete piles came in use; in Yakutsk they were for the most part lowered with steam needles;[14] at Noril'sk (in hard-frozen soils) they were placed in drilled wells (M. V. Kim, G. N. Maksimov, et al.,[15] and at Vorkuta (under conditions of high-temperature permafrost) they were driven with diesel-powered pile drivers and powerful vibrators.[16]

The steam-needle method of pile installing (Fig. 175) into thawed wells steamed in permafrost was one of the first methods used for pile operations back in the early 1920's; under certain conditions, this method is still used successfully today (in Yakutiya and other areas). It is economical and efficient (two men can run one steam boiler, servicing 7–8 needles), but it has a disadvantage in that it introduces a great amount of heat into the permafrost, so that a long time (sometimes several months) is required for back freeze of the steamed volume of soil, and the use of the steam-needle method has come

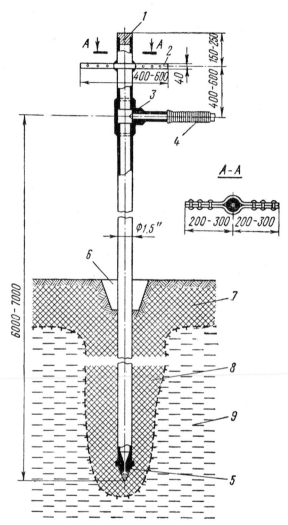

Fig. 175. Design of steam needle for thawing permafrost:
(1) plug; (2) handle; (3) tee; (4) flexible hose; (5) tip; (6)
guide pit in ground for positioning of steam needle; (7)
active-layer zone; (8) thawed zone in permafrost; (9) permafrost stratum.

to be limited for the most part to areas with low-temperature permafrost.
However, during freezing of the steamed soil around the driven pile, water
migrates toward the freezing front, i.e., away from the piles and toward the
permafrost, and a frozen soil of more massive texture forms around the pile
(Fig. 176),[17] with the moisture content of the frozen ground no higher, as
indicated by direct experiments, than that of the surrounding permafrost, but

interlayers (films) of ice are often formed on the surface of the pile, lowering the strength of its adfreezing to the soil.

S. G. Tsvetkova's experimental and analytical studies enabled her to prepare convenient graphs of the time to recovery of the frozen state of a talik around a pile steamed into permafrost as a function of a composite indicator θ.

According to these plots (Fig. 177), the time to recover the frozen states of the steamed ground around a pile is determined as a function of the indicator

$$\theta = \frac{i_r W_v (r_0^2 - r_1^2)}{\theta_f \lambda_f} \qquad (XI.10)$$

where i_r is the relative ice content of the frozen ground, W_v is the total volumetric moisture content of the ground (kg/cm^3), r_0 is the radius of the pile in meters, r_1 is the radius of the (steamed) cylindrical talik in meters, θ_f is the temperature of the permafrost, and λ_f is the thermal conductivity of the frozen ground in kcal/(m^2)(hr)(deg).

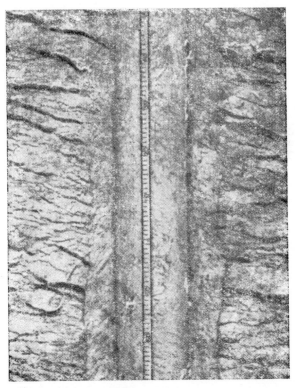

Fig. 176. Structure of frozen ground around a pile driven by steaming.

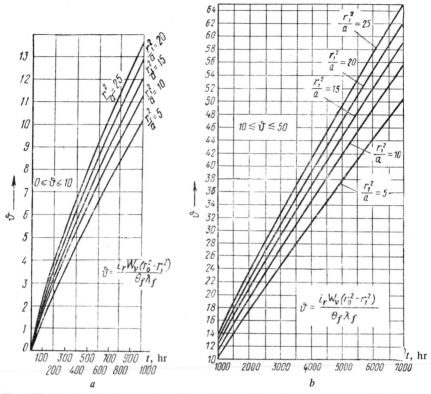

Fig. 177. Tsvetkova's diagrams for determination of the time to recovery of the frozen state in the ground around a pile: (*a*) for values of θ from 0 to 13; (*b*) θ from 10 to 64 (*a* is the thermal diffusivity of the soil).

We note that piles can be placed in permafrost with the aid of steam needles only if the soil contains no gravelly or, in general, coarse-skeleton mineral particles, and no rocks and boulders, since coarse-skeleton particles settle to the bottom of the steamed well and prevent further penetration.

In coarse fragment and detrital clayey soils and in all low-temperature permanently frozen soils (hard-frozen), it is better to place the piles in predrilled wells; if the diameter of the well exceeds that of the piles, the wells are filled with drilling mud (after Kim, Maksimov, *et al.*) before the piles are lowered into them, but if the piles are larger in diameter than the wells, they are driven into the wells with a diesel pile driver or vibration sinker (according to Vyalov, Targulyan, and others).

In the former case, i.e., when mud is used, it is necessary to add more than one-half part of sand (by volume) to the drilling mud; then the bearing capacity of the piles increases to at least 20 percent above that calculated by the SNiP figure after the soil has frozen around them, whereas if the well is filled with mud alone (fine drilling cuttings), the bearing capacity of the piles is slightly below the desired value for the particular soil conditions.

In the case of high-temperature (plastically frozen) soils, reinforced-concrete piles are driven directly into the permafrost with diesel pile drivers, vibration sinkers, and vibrohammers. Field tests made by V. N. Yeroshenko on piles driven directly into plastically frozen ground of the Vorkuta region indicated that the bearing capacity of driven floating piles is approximately twice the value calculated from the standard SNiP data for the same ground type. According to the same experiments, the bearing capacity reaches 100 tons per pile with only 2.8 cm of settling for 35 × 35-cm pile pillars resting on bedrock talus of the Vorkuta region.

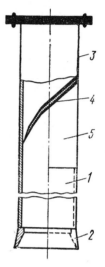

An improvement on the direct driving of piles into permafrost must be credited to L. P. Markizov[18] who used a "tubular leader" (Fig. 178) in the form of a hollow tube with a special cutting tip,[19] with the aid of which a layer of frozen active-layer up to 2 meters in thickness or more can be penetrated in a few minutes, after which the pile is driven into the resulting well by a conventional method, using a diesel driver or vibration sinker. This improvement has greatly reduced the cost of pile-driving operations and insured a year-round operation.

Fig. 178. Tubular leader for advanced sinking of wells in seasonally frozen soils before piledriving. (1) hollow pipe; (2) special tip; (3) cap; (4) deflector to empty soil; (5) hole for escape of soil.

Let us cite certain examples of the use of pile foundations in the construction of buildings under permafrost conditions. Figure 179 shows a sectional drawing of a substation building put up on reinforced-concrete piles frozen into pre-drilled wells on a site composed of broken bedrock with ice interlayers; Fig. 180 is a section through the lower part of a large-panel residential building with a winter-ventilated basement of the standard series, with load bearing outer and transverse walls on reinforced-concrete piles placed in rows about 2.3 to 3.4 meters apart, with a design bearing capacity of 40 to 50 tons per pile.

Figure 181 is a sectional drawing of a storehouse building at Dudinka, in which the reinforced-concrete piles function simultaneously as the frame of the building. The interior of the storehouse is unheated, and the permafrost remains frozen with no special heat-removing provisions under the harsh climatic conditions.[15]

As Markizov[18] showed, the tubular leader can be used successfully in excavating frozen ground not only in the seasonal frozen layer, but also in permafrost; it speeds up loosening of the frozen ground and lowers the cost of

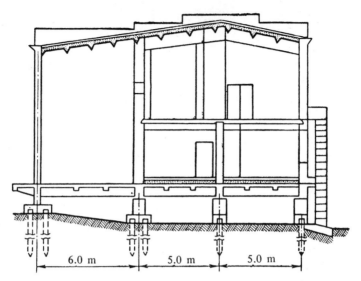

Fig. 179. Sectional drawing of substation built on permafrost on pile foundations.

this operation, and the ground can then be worked easily with a power shovel (Fig. 182). This method is used successfully under permafrost conditions not only for digging strip-foundation trenches, but also in the excavation of cellars, laying pipelines, and other earthwork operations.

It must also be noted here that mechanical breaking of frozen ground cuts costs by 50 percent as compared to drilling and blasting operations, but that research on the cutting of frozen ground must be referred to in selecting the cutting elements of power shovels and other earth-moving machines (see Sec. 5 of Chap. VI).[20]

In the construction of dams from local materials and in building other hydraulic-engineering structures under permafrost conditions, earth-moving and

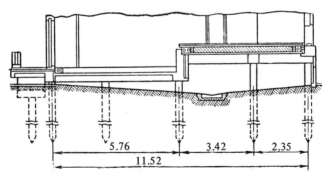

Fig. 180. Section through pile foundations of large-panel residential building erected on permafrost.

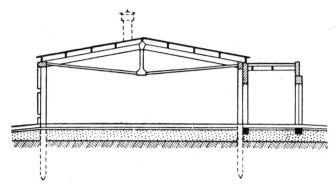

Fig. 181. Unheated frame storehouse on pile foundations.

excavating operations become particularly important; most important among these are the operations involved in building the antifiltration cores of rock-filled dams, which are the types preferred at medium and high water heads (more than 20 meters).

During construction of the Vilyuy rock-fill dam, for example (see Fig. 131), the placement of a water-impermeable core consisting of 580,000 cubic meters of local gritty clayey loams in the winter with temperatures as low as $-50°C$ was a technically very complex problem. For the first time in Soviet hydraulic-engineering practice, a progressive technology for the earthworks involved in placing antifiltration cores on rock-filled dams was elaborated under the conditions of an exceptionally harsh climate and permafrost by the Vilyuy State Electric Power Plant Construction Trust with the participation of the VNIIG (All-Union Scientific Research Institute of Hydraulic Engineering), the Leningrad Polytechnic Institute, and the MISI (Moscow Construction Engineering Institute).[21]

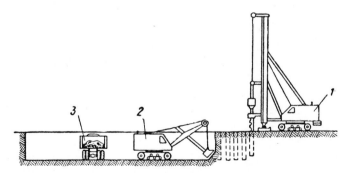

Fig. 182. Trench excavation in frozen ground with a tubular leader:
(1) piledriver working in the seasonally frozen layer with a tubular leader; (2) power shovel; (3) dump truck for loading soil.

In the construction of the Vilyuy dam, the naturally encountered loams had a filtration coefficient (according to LPI tests) $k = i \times 10^{-4}$ cm/sec, which, according to the calculations, would be adequate with appropriate treatment to ensure water impermeability of the core, the more so since these loams had excellent strength-property indicators (after consolidation and draining, an internal-friction angle $\phi \approx 26°$ and a cohesion $c \approx 0.1$ kg/cm²).

In consideration of the harsh climate (short summer period) and the availability of only a thin layer of topsoil loams that would be suitable for the antifiltration core only after appropriate treatment, the Vilyuy Trust was forced to develop a method for pretreatment of the loams and stockpiling them in very large winter-storage heaps (16-18 meters high and containing up to 200,000 cubic meters each).

These operations were carried out in accordance with the diagram shown in Fig. 183; stripping of the loams was started on a large area at the end of April, and by mid-June they had been worked to depths of up to 1 meter. The soil collected by the bulldozers was left in windrows for up to three weeks, during which it dried to some extent in the open air.

However, drying of clayey soils is not alone sufficient to permit their use as an antifiltration material; it is necessary to work them into an optimum soil mix, which can be done by two methods: by mixing layers of topsoil loam and layers of bedrock eluvium in the pits (as was done by the Vilyuy Trust), or by addition of large-fragment mixes (gravel, crushed stone) from other deposits when the soils are dumped in the winter-storage heaps (by the technology developed by the Khantaysk State Electric Power Plant Construction Trust jointly with MISI).[22]

Heaping of large volumes of loam for winter storage required electric heating of the peripheral zones of the piles and salting the piles (to increase heating efficiency).

Figure 184 shows a diagram of the process of salting the loam in the windrows and in the peripheral zones of piles on flat areas of the pit reserved for salinization by uniform spreading of calcium chloride over the area to be salted in amounts of 20-30 kg/m². After 15-20 days, working of the flat was begun by bulldozing the soil into windrows, from which it was delivered to the piles for dumping in a layer 2-2.5 meters deep.

The amounts of salt needed for both the preparation of the loams and their compaction were determined for the most part on the basis of MISI studies (by S. B. Ukhov and Ya. A. Kronik),[23] and the properties of the clayey soils after artificial salinization by the procedure described by B. I. Dalmatov and V. S. Lastochkin.[24]

To control frost heaving of the clayey soils, to improve their antifiltration capacity, and ensure long-term antiheave stabilization in hydraulic-engineering structures for operation under harsh climatic conditions, the MISI (Ya. A. Kronik, *et al.*) proposed, and the Vilyuy Trust implemented, a complex

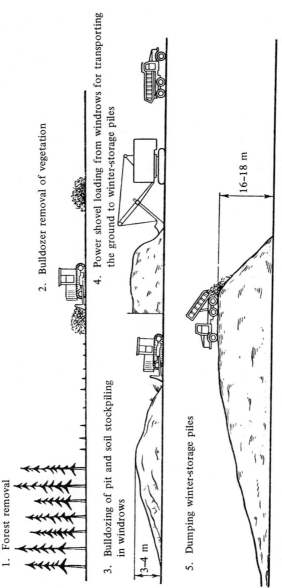

1. Forest removal

2. Bulldozer removal of vegetation

3. Bulldozing of pit and soil stockpiling in windrows

3-4 m

4. Power shovel loading from windrows for transporting the ground to winter-storage piles

5. Dumping winter-storage piles

16–18 m

Fig. 183. Diagram of loam preparation in pit according to Vilyuy State Electric Power Plant Trust technology.

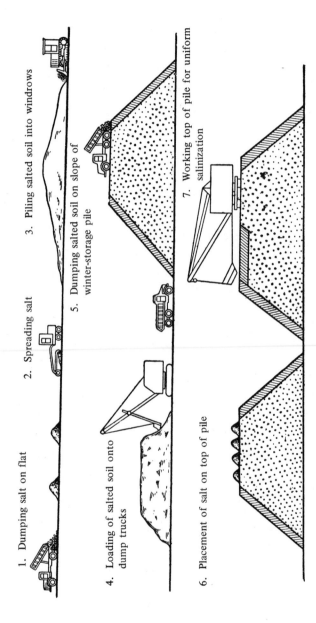

1. Dumping salt on flat

2. Spreading salt

3. Piling salted soil into windrows

4. Loading of salted soil onto dump trucks

5. Dumping salted soil on slope of winter-storage pile

6. Placement of salt on top of pile

7. Working top of pile for uniform salinization

Fig. 184. Diagram illustrating salinization of the peripheral zone of loam heap according to Vilyuy Trust data.

method for antiheave soil melioration* that includes: salinization of the soils, overcompaction of the soils (to a density higher than that required by the standards for unfrozen soils) while they are placed into the body of the core, and capping the parts of the earth structure to be frozen with a nonfrost susceptible material.

In addition, the loam piles were also covered for winter storage with a layer of foamed ice, the recipe for which was developed by the Leningrad Polytechnic Institute around the PO-6 foam generator. Figure 185 shows a diagram of a mobile unit mounted on a tank trailer after a proposal of the LPI and VNIIG. However, it should be noted that extensive use of this method in practice has been prevented by the difficulty of removing the foamed ice from the piles when the stockpiled loams were excavated in the winter.

The loam was laid into the dam cove after special preparation of the dumping area (it was heated to + 5 and even to + 10°C by a special turbojet unit), with immediate laying of the next layer of loam in a thickness of about 0.5 m and strong compacting by driving heavily loaded MAZ-525 dump trucks over it repeatedly (10–15 times). It was also mandatory to provide salt treatment of the contact zones onto which the loam was spread.

For the most part on the basis of MISI studies (S. B. Ukhov, Ya. A. Kronik), the Vilyuy Trust Laboratory (Yu. N. Myuznikov, *et al.*) prepared a nomogram for determination of the required amount of salt (sodium chloride or calcium chloride has usually been used) as a function of the depth of salinization, soil temperature, and the required thickness of the loam layer for treatment with the salt solution (Fig. 186).

*See Sec. 7 of Chap. II.

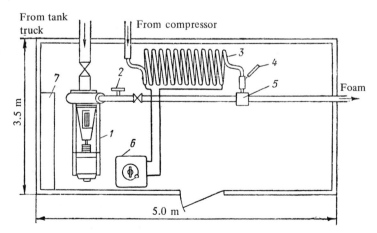

Fig. 185. Diagram of LPI-VNIIG ice foam machine: (1) pump unit; (2) pressure gauge; (3) air preheater; (4) thermometer; (5) air mixer; (6) transformer; (7) control panel.

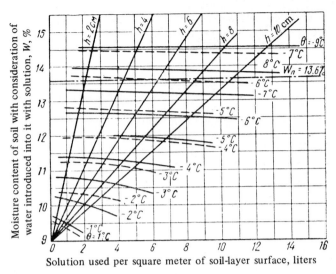

Solution used per square meter of soil-layer surface, liters

Fig. 186. Freezing temperature and gravimetric moisture content of soils as functions of amounts of NaCl and $CaCl_2$ solutions and depth of salinization: (——) freezing line of $CaCl_2$ soil solutions; (– – –) same, NaCl ($W_{t.m}$ is the total moisture capacity of the soil at γ_{sk} = 2.1 g/cm³).

Field observations made at the site of loam placement into the dam core showed that the actual amount of salt solution used (according to Ye. N. Batenchuk, G. F. Biyanov, *et al.*) is close to the calculated amount, averaging 2 to 4 liters per square meter of treated loam area; the depth of salt penetration averages about 5 cm without special raking.

We note in conclusion that the combined heave prevention measures described above, eliminate frost heave in winter fill placement operations, and ensure long-term heave stability of the earthwork parts of hydraulic-engineering structures during operation under the harsh conditions of the Far North and produce adequate antifiltration properties in these soils. This has permitted the use of these soils in the construction of the Vilyuy and Khantaysk dams, where they were used after preparation on the site and proved quite effective, both technically and economically.

Thus, there is sufficient basis for recommending combined anti-heave melioration of cohesive soils for more extensive use in construction practice, provided, of course, that experimental test work is done in other permafrost regions.

References

1. A. V. Liverovskiy and K. D. Morozov, Stroitel'stvo v usloviyakh vechnoy merzloty (Construction Under Permafrost Conditions), Lenstroyizdat, 1941.

2. V. F. Zhukov, Predpostroyechnoye protaivaniye mnogoletnemerzlykh gornykh porod pri vozvedenii na nikh sooruzheniy (Preconstruction Thawing of Permafrost Rock for Erection of Structures), USSR Academy of Sciences Press, 1958.
3. T. A. Aydla, Introduction of a Preconstruction Thawing Technique, *Trudy VNII-1*, vol. XXII, Magadan, 1963.
4. V. F. Zhukov, L. N. Khrustalev, and V. M. Vodolazkin, Selection of Optimum Preconstruction Thawing Depth in the Ground Under a Building (Using a Computer), "Trudy VI soveshchaniya-seminara po stroitel'stvu v surovykh klimaticheskikh usloviyakh" (Proceedings of Sixth Conference-Seminar on Construction Under Harsh Climatic Conditions), Krasnoyarsk, 1970.
5. V. F. Zhukov, Predpostroyechnoye protaivaniye mnogoletnemerzlykh gornykh porod pri vozvedenii na nikh sooruzheniy (Preconstruction Thawing of Permafrost Rock for Erection of Structures), chap. IV, sec. 2, USSR Academy of Sciences Press, 1958.
6. V. P. Bakakin, Opyt upravleniya teploobmenom deyatel'nogo sloya merzlykh gornykh porod v tselyakh povysheniya effektivnosti ikh razrabotki (Experience in the Control of Heat Exchange in the Active Layer of Frozen Soils with the Object of Increasing Soil-Working Efficiency), USSR Academy of Sciences Press, 1954.
7. T. A. Aydla and Yu. K. Trincher, Design and Implementation of Hydraulic Preconstruction Thawing of Base Soils under the Main Building of the Arkagalinsk State Regional Electric Power Plant, Trudy VI Vsesoyuznogo soveshchaniya po stroitel'stvu v surovykh klimatecheskikh usloviyakh (Proceedings of Sixth All-Union Conference on Construction under Harsh Climatic Conditions), vol. V, no. 2, Krasnoyarsk, 1970.
8. Collection "Teoriya i praktika merzlotovedeniya v stroitel'stve" (Permafrostological Theory and Practice in Construction), Komi Branch, USSR Academy of Sciences, and NIIosnovaniy, Nauka Press, 1965.
9. B. A. Rzhanitsyn, B. P. Gorbunov, *et al.* Preconstruction Thawing and Compaction of Permafrost, in collection "Doklady na Mezhdunarodnoy konferentsii po merzlotovedeniyu" (Papers at International Conference on Permafrostology), USSR Academy of Sciences Press, 1963.
10. (1) B. A. Rzhanitsyn, Electrochemical Reinforcement of Clayey Soils, in collection "Soveshchaniye po zakrepleniyu gruntov i gornykh porod" (Conference on Reinforcement of Soils and Rocks), USSR Academy of Sciences Press, 1941; (2) G. M. Lomidze and A. V. Netushil, Elektroosmoticheskoye vodoponizheniye (Electroosmotic Dewatering), Gosenergoizdat, 1958; (3) Ye. P. Kudryavtsev, Calculations for Electrical Compaction of Clayey Soils, Nauchnyye doklady vysshey shkoly, *Energetika*, 1959, no. 2; (4) G. N. Zhinkin, Elektrokhimicheskoye zakrepleniye gruntov v stroitel'stve (Electrochemical Ground Reinforcement in Construction), Stroyizdat, 1966.
11. Preconstruction Thawing of Foundation Bases, in collection "Teoriya i praktika merzlotovedeniya v stroitel'stve" (Permafrostological Theory and Practice in Construction), Nauka Press, 1955.
12. Tsytovich, N. A.: Design of Foundations for Structures to be Built on Permafrost, *Trudy Gipromeza*, no. 2, 1928.
13. V. V. Dokuchayev, Raschet fundamentov na vechnomerzlykh gruntakh po predel'nym sostoyaniyam (Design of Foundations on Permafrost for Extreme States), Stroyizdat, 1968.
14. P. I. Mel'nikov, S. S. Vyalov, *et al.*, Pile Foundations on Permafrost, in collection "Doklady na mezhdunarodnoy konferentsii po merzlotovedeniyu" (Papers at International Conference on Geocryology), N. A. Tsytovich (ed.), USSR Academy of Sciences Press, 1963.
15. (1) M. V. Kim, G. N. Maksimov, *et al.*, Vozvedeniye fundamentov v usloviyakh vechnomerzlykh gruntov (Erection of Foundations Under Permafrost Conditions), Gosstroyizdat, 1962; (2) G. N. Maksimov, Cooling of High-Temperature Permafrost

in Pile Foundation Construction, "Osnovaniya, fundamenty i mekhanika gruntov," no. 1, 1968.

16. S. S. Vyalov, Yu. O. Targulyan, and D. P. Vysotskiy, Interaction of Frozen Soil with Piles and Pipes in Vibration Sinking, "Materialy VIII Vsessoyuznogo soveshchaniya po geokriologii" (Materials of Eighth All-Union Conference on Geocryology), no. 5, Yukutsk, 1966.

17. According to experiments of S. G. Tsvetkova. Author's abstract of candidate's dissertation "Skorost' smerzaniya svay, zabitykh v vechnomerzlyye grunty" (Adfreezing Rate of Piles Driven into Permafrost), MISI, 1954.

18. (1) L. P. Markizov, Organizatsiya truda po razrabotke merzlykh gruntov (Labor Organization for Working of Frozen Soils), *Izd-vo Komi*, Syktyvkar, 1970; (2) L. P. Markizov, Working of Frozen Soils at Vorkuta, "Osnovaniya, fundamenty i mekhanika gruntov", no. 5, 1970.

19. D. P. Vysotskiy and V. P. Vlokh, Soviet Patent No. 209,312, *Byull. Komiteta po delam izobreteniy*, no. 4, 1968.

20. See also A. N. Zelenin, Osnovy razrusheniya gruntov mekhanicheskimi sposobami (Fundamentals of Mechanical Disintegration Methods of Soils), *Izd-vo Mashinostroyeniye*, 1968.

21. Ye. N. Batenchuk, G. F. Biyanov, *et al.*, Placement of Cohesive Soils in the Far North in Winter, in series: "Biblioteka gidrotekhnika i gidroenergetika" (Library of Hydraulic Engineering and Hydroelectric Power), no. 2, *Izd-vo Energiya*, 1968.

22. (1) N. A. Tsytovich and Ya. A. Kronik, Control of Moisture-Content Regime in Cohesive Soils in Dam Building in the North, Collection: "Temperaturno-vlazhnostnyy rezhim plotin iz mestnykh materialov v surovykh klimaticheskikh uslovyakh Vostochnoy Sibiri i Kraynego severa" (Temperature-Moisture Regime in Dams Built from Local Materials under the Harsh Climatic Conditions of Eastern Siberia and the Far North), VNIIG, 1969; (2) Nersesova, Z. A., Heave of Silty Loams and Physicochemical Countermeasures Against It, *Trudy NIITransstroy*, no. 62, 1967.

23. (1) S. B. Ukhov, Artificial Salinization of Loamy Ground for Wintertime Construction, "Izvestiya vuzov, Stroitel'stvo i arkhitektura," no. 1, 1959; (2) Ukhov, S. B., Physicochemical Prevention of the Frost Heaving of Soils, Materials of Eighth All-Union Conference on Geocryology, no. 8, Yakutsk, 1966; (3) Ya. Kronik, Antiheave Melioration of Clayey Soils of the Far North in Dam Building, dissertation, MISI, 1970.

24. B. I. Dalmatov and V. S. Lastochkin, Iskusstvennoye zasoleniye gruntov v stroitel'stve (Artificial Ground Salinization in Construction), Stroyizdat, 1966.

BIBLIOGRAPHY

Aydla, T. A.: Introduction of a Preconstruction Thawing Technique, Trudy VNII-1, vol. XXII, Magadan, 1963.

Andrianov, P. I.: (1) Koeffitsienty rasshireniya gruntov pri zamerzanii (Expansion Coefficients of Soils on Freezing); (2) Temperatury zamerzaniya gruntov (Ground Freezing Temperatures), SOPS AN SSSR, 1936.

Bakakin, V. P.: Opyt upravleniya teploobmenom deyatel'nogo sloya merzlykh gornykh porod v tselyakh povysheniya effektivnosti ikh razrabotki (Experience in the Control of Heat Exchange in the Active Layer of Frozen Soils with the Object of Increasing Excavation Efficiency), USSR Academy of Sciences Press, 1954.

Batenchuk, Ye. N., G. Ya. Biyanov, et al.: Placement of Cohesive Soils in the Far North in Winter, in series "Biblioteka gidrotekhnika i gidroenergetika," (Library of Hydraulic and Hydroelectric Engineer), no. 2, Izd-vo Energiya, 1968.

Berezantsev, V. G.: Resistance of Soils to Local Loads at Constant Negative Temperature, Collection 1, "Materialy po laboratornym issledovaniyam merzlykh gruntov pod rukovodstvom N. A. Tsytovicha" (Materials from Laboratory Studies on Frozen Soils Under the Supervision of N. A. Tsytovich), USSR Academy of Sciences Press, 1953.

Berezantsev, V. G.: Raschet osnovaniy sooruzheniy (Calculation of Bases for Structures), Stroyizdat, 1970.

Bliznyak, Ye. V.: Design and Construction of Dams Under Permafrost Conditions, Gidrotekhnicheskoye Stroitel'stvo, no. 9, 1937.

Bogoslovskiy, P. A.: Calculation of Multiyear Temperature Variations in Earth Dams Based on Frozen Ground, Trudy Gor'kovskogo ISI, no. 28, 1957, and no. 29, 1958; also Nauchnyye Doklady Vysshey shkoly, Stroitel'stvo, no. 1, 1958; Izvestiya vysshikh uchebnykh zavedeniy, no. 5, 1958; nos. 11-12, 1963.

Bogoslovskiy, P. A., A. V. Stotsenko, et al.: Dams in the Permafrost Region, "Doklady na Mezhdunarodnoy konferentsii po Merzlotovedeniyu" (Papers at International Conference on Permafrostology), N. A. Tsytovich (ed.), USSR Academy of Sciences Press, 1963.

Bokiy, G. B.: Crystal-Chemical Aspects of the Behavior of Water in Frozen Clayey Soils, Vestnik MGU, Geologiya, no. 1, 1961.

Bondarev, P. D.: Deformatsii zdaniy v rayone Vorkuty, ikh prichiny i metody predotvrashcheniya (Deformation of Buildings in the Vorkuta Region; Causes and Methods of Prevention), USSR Academy of Sciences Press, 1957.

Bozhenova, A. P.: Significance of Osmotic Forces in the Moisture-Migration Process, Collection 3, "Materialy po laboratornym issledovaniyam merzlykh gruntov pod rukovodstvom N. A. Tsytovicha," USSR Academy of Sciences Press, 1957.

Bozhenova, A. P. and F. G. Bakulin: Experimental Studies of Moisture-Migration Mechanisms in Freezing Soils, Collection 3, "Materialy po laboratornym issledovaniyam merzlykh gruntov," USSR Academy of Sciences Press, 1957.

Brodskaya, A. G.: Szhimayemost' merzlykh gruntov (Compressibility of Frozen Ground), USSR Academy of Sciences Press, 1962.

Bykov, N. I. and P. N. Kapterev: Vechnaya merzlota i stroitel'stvo na ney (Permafrost and Construction on it), Transzheldorizdat, 1940.

Chernyshev, N. Ya.: Deformation of Wooden Bridges due to Heaving of Frozen Soil, "Zheleznodorozhnoye delo," nos. 1-2, 1928.

References to foreign literature and to separate articles by Soviet specialists that are for the most part of specialized interest are in the text references.

Collection "Prochnost' i polzuchest' merzlykh gruntov" (Strength and Creep of Frozen Soils), Siberian Division, Institute of Geocryology, USSR Academy of Sciences, USSR Academy of Sciences Press, 1963.

Collection: "Teoriya i praktika merzlotovedeniya v stroitel'stve" (Permafrostological Theory and Practice in Construction), Komi Branch of the USSR Academy of Sciences and NIIosnovaniy, Nauka Press, 1965.

Dalmatov, B. I.: Vozdeystviye moroznogo pucheniya gruntov na fundamenty sooruzheniy (Effects of Frost Heave of Soils on Foundations of Structures), Gosstroyizdat, 1957.

Dalmatov, B. I. and V. S. Lastochkin: Iskusstvennoye zasoleniye gruntov v stroitel'stve (Artificial Ground Salinization in Construction), Stroyizdat, 1966.

Dodin, V. Z.: Sooruzheniye kanalov podzemnykh kommunikatsiy (Construction of Channels for Underground Pipelines), Stroyizdat, 1965.

Dokuchayev, V. V.: Osnovaniya i fundamenty na vechnomerzlykh gruntakh (Bases and Foundations on Permanently Frozen Ground), Gosstroyizdat, 1963.

Dokuchayev, V. V.: Raschet fundamentov na vechnomerzlykh gruntakh po predel'nym sostoyaniyam (Design of Foundations on Permafrost Soils for Extreme States), Stroyizdat, 1968.

Dostavalov, B. N. and V. A. Kudryavtsev: Obshcheye merzlotovedeniye (General Permafrostology), Izd-vo MGU, 1967.

Dorman, Ya. A.: Iskusstvennoye zamorazhivaniye gruntov pri stroitel'stve metropolitenov (Artificial Freezing of Soils in Subway Construction), Izd-vo Transport, 1971.

Dubnov, Yu. D.: Experience in the Use of Potassium Chloride as a Countermeasure Against Structure Foundations heave, Sb. trudov VNII Transstroya, no. 62, 1967.

Fedorov, N. F.: Sanitary-Engineering Pipelines and Structures, in book by Yu. Ya. Velli, V. V. Dokuchayev, and N. F. Fedorov, "Zdaniya i sooruzheniya na Kraynem Severe" (Buildings and Structures in the Far North), Gosstroyizdat, 1963.

Fedosov, A. Ye.: Mechanical Processes in Soils During Freezing of Their Liquid Phases, Trudy IGI AN SSSR, no. 4, 1940.

Fel'dman, G. M.: Migratsiya vlagi v gruntakh pri promerzanii, Sb. "Teplofizika promerzayushchikh i protaivayushchikh gruntov" (Migration of Moisture in Soils During Freezing, in collection: "Thermophysics of Freezing and Thawing Soils"), Nauka Press, 1964.

Gol'dshteyn, M. N.: Deformations of an Earth Dam and Bases of Structures During Freezing and Thawing, Trudy NIIZhTa, no. 16, Transzheldorizdat, 1948.

Golovko, M. D.: Metod rascheta chashi protaivaniya v usloviyakh zdaniy, vozvodimykh na mnogoletnemerzlykh gruntakh. Laboratoriya gidravliki i elektrogidrodinamicheskikh analogiy im. V. S. Luk'yanova (A Method for Calculation of Thaw Bowls for Buildings to be Erected on Permafrost. V. S. Luk'yanov Laboratory of Hydraulics and Electrohydraulic Analogies), Izd-vo TsNIIM Minstroya, 1958.

Grandilevskiy, V. N.: Use of a Finite-Difference Method for Solution of Three-Dimensional Problems in Nonstationary Heat Conduction, Trudy GISI, no. 37, 1961.

Grechishchev, S. Ye.: Creep of Frozen Soils in the Complex Stressed State, in collection "Prochnost' i polzuchest' merzlykh gruntov" (Strength and Creep of Frozen Ground), Izd-vo Sib. Otd. AN SSSR, 1963.

Kachurin, S. P.: Termokarst na territorii SSSR (Thermokarst on the USSR Territory), USSR Academy of Sciences Press, 1963.

Khakimov, Kh. R.: Voprosy teorii i praktiki iskusstvennogo zamorazhivaniya gruntov (Problems in the Theory and Practice of Artificial Soil Freezing), USSR Academy of Sciences Press, 1957.

Khakimov, Kh. R.: Zamorazhivaniye gruntov v stroitel'nykh tselyakh (Freezing of Soils for Construction Purposes), Gosstroyizdat, 1962.

Khrustalev, L. N.: Influence of Land Development on Temperature Regime of Permafrost, in collection: "Teoriya i praktika merzlotovedeniya v stroitel'stve" (Permafrostological Theory and Practice in Construction), Nauka Press, 1965.

Kim, M. V., *et al.*: Vozvedeniye fundamentov v usloviyakh vechnomerzlykh gruntov (iz opyta Noril'skogo rayona) (Erection of Foundations Under Permafrost Conditions), Gosstroyizdat. 1962.

Kiselev, M. F.: K raschetu osadok fundamentov na ottaivayushchikh gruntakh (Calculation of Foundation Settling on Thawing Soils), Gosstroyizdat, 1957.

Konovalov, A. A.: Solution of Certain Thermal Problems in the Control of Ground Temperature Regimes, Trudy VI Soveshchaniya-seminara po obmenu opytom stroitel'-stva v surovykh klimaticheskikh usloviyakh (Transactions of the Sixth Conference— Seminar for Exchange of Experience in Construction under Harsh Climatic Conditions), Krasnoyarsk, 1970.

Kovner, S. S.: A Problem of Heat Conduction, *Geofizika*, vol. III, no. 1, 1933.

Kronik, Ya. A.: Heave Prevention by Salinization of Loams During Winter Placement of the Core of the Vilyuy State Electric Power Station Dam, "Ekspress-informatsiya OES," series "Stroitel'stvo gidroelektrostantsiy," no. 8(236), 1968.

Kronik, Ya. A., S. B. Ukhov, and N. A. Tsytovich: Artificial Ground Salinization to prevent Frost Heaving, *Osnovaniya, fundamenty i mekhanika gruntov*, no. 1, 1969.

Kudryavtsev, V. A.: The Frost Survey as the Basic Form of Frost Research, in collection "Merzlotnyye issledovaniya" (Frost Research), no. 1, Izd-vo MGU, 1961.

Kulikov, Yu. G. and N. A. Peretrukhin: Determination of Normal Heaving Forces, "Trudy VNII transporta," *Izd-vo Transport*, 1967.

Lapkin, G. I.: Raschet osadok sooruzheniy na ottaivayushchikh gruntakh po metodu kontaktnykh davleniy (Calculation of Structure Settling on Thawing Soils by the Contact-Pressure Method), Stroyizdat, 1947.

Lebedev, A. F.: Pochvennyye i gruntovyye vody (Soil and Ground Water), USSR Academy of Sciences Press, 1937.

Liverovskiy, A. V. and K. D. Morozov: Stroitel'stvo v usloviyakh vechnoy merzloty (Construction under Permafrost Conditions), Lenstroyizdat, 1941.

Lomize, G. M. and A. V. Netushil: Elektroosmoticheskoye vodoponizheniye (Electro-osmotic Dewatering), Gosenergoizdat, 1958.

Lukin, G. O.: Observations of the Temperature Regime in the Bases of Buildings in Yakutsk, in collection: "Fundamenty sooruzheniy na merzlykh gruntakh v Yakutii" (Foundations of Structures on Frozen Soils in Yakutiya), Nauka Press, 1968.

Luk'yanov, V. S.: Tekhnicheskiye raschety na gidravlicheskikh priborakh (Engineering Calculations on Hydraulic Analog Devices), Transzheldorizdat, 1937.

Lykov, A. V.: Teoriya teploprovodnosti (The Theory of Heat Conduction), Gosstroyizdat, 1954.

Lykov, A. V. and Yu. A. Mikhaylov: Teoriya perenosa energii i veshchestva (The Theory of Energy and Mass Transport), Minsk, *Izd-vo AN BSSR*, 1959.

Maksimov, G. N.: Artificial Air Cooling in the Placement of Pile Foundations in Permafrost, NIIOSP Collection No. 55, Stroyizdat, 1964.

Maksimov, G. N.: Cooling of High-Temperature Permafrost in Pile Foundation Construction, *Osnovaniya, fundamenty i mekhanika gruntov*, no. 1, 1968.

Malyshev, M. V.: Calculation of the Settling of Foundations on Thawing Soil, *Osnovaniya, fundamenty i mekhanika gruntov*, no. 4, 1966.

Markizov, L. P.: Labor Organization for Frozen Ground Excavation, Syktyvkar, *Izd-vo KOMI*, 1970.

Mel'nikov, P. I., S. S. Vyalov, *et al.*: Pile Foundations on Permafrost, "Doklady na Mezhdunarodnoy konferentsii po merzlotovedeniyu," pod red. N. A. Tsytovicha, USSR Academy of Sciences Press, 1963.

Mel'nikov, P. I., S. Ye. Grechishchev, *et al.*: Fundamenty sooruzheniy na merzlykh gruntakh v Yakutii (Foundations of Structures on Frozen Ground in Yakutiya), Siberian Division, USSR Academy of Sciences, Nauka Press, 1968.

Meyster, L. A.: Methods of Engineering-Geocryological Research, In book: "Osnovy geokriologii" (Fundamentals of Geocryology), part II, chap. XII, USSR Academy of Sciences Press, 1969.

Moiseyev, I. S.: Calculation of the Temperature Regime in Earth Dams in the Permafrost Region, *Trudy NISI*, no. 29, 1959.

Nersesova, Z. A.: Melting of Ice in Soils at Negative Temperatures, *Doklady AN SSSR*, vol. 4, no. 3, 1951.

Nersesova, Z. A.: Influence of Exchange Cations on Water Migration and Soil Heaving during Freezing, Collection 4, "Issledovaniya po fizike i mekhanike merzlykh gruntov" (Studies in the Physics and Mechanics of Frozen Ground), USSR Academy of Sciences Press, 1961.

Nersesova, Z. A.: Heaving of Silty Loams on Freezing, and Physico-chemical Counter-measures Against It, "Trudy VNII transportnogo stroitel'stva," no. 62, *Izd-vo Transport*, 1967.

Ornatskiy, N. V.: Planning of Antiheaving Measures, in DORNII collection: "Reguliravaniye vodyanogo rezhima," [Control of Ground Water Regime], Dorizdat, 1946.

Orlov, V. O.: Kriogennoye pucheniye tonkodispersnykh gruntov (Cryogenic Heaving of Fine-Particle Soils), USSR Academy of Sciences Press, 1962.

Osnovy geokriologii (merzlotovedeniya) (Fundamentals of Geocryology (Permafrostology)), Collected Publication by Staff Members of the USSR Academy of Sciences Institute of Permafrostology, Parts I and II, USSR Academy of Sciences Press, 1959.

Pchelintsev, A. M.: Stroyeniye i fiziko-mekhanicheskiye svoystva merzlykh gruntov (Structure and Physicomechanical Properties of Frozen Soils), Nauka Press, 1964.

Pekarskaya, N. K.: Shear Resistance of Permanently Frozen Ground with Various Textures and Ice Contents, in collection "Issledovaniya po fizike i mekhanike merzlykh gruntov," USSR Academy of Sciences Press, 1961.

Pekarskaya, N. K.: Prochnost' merzlykh gruntov pri sdvige i yeye zavisimost' ot tekstury (Shear Strength of Frozen Soils and its Dependence on Texture), USSR Academy of Sciences Press, 1963.

Pekarskaya, N. K.: Short-Term Resistance of Frozen Soils to Uniaxial Compression and Tension, "Materialy VIII Vsesoyuznogo soveshchaniya po geokriologii," no. 5, Yakutsk, 1966.

Peretrukhin, N. A.: Frost-Heaving Forces of Foundations, "Trudy VNII transportnogo stroitel'stva," no. 62, *Izd-vo Transport*, 1967.

Poltev, I. F.: Osnovy merzlotnoy s'emki (Fundamentals of Frost Survey), *Izd-vo MGU*, 1963.

Popov, A. I.: Merzlotnyye yavleniya v zemnoy kore (kriolitologiya) (Frost Phenomena in the Earth's Crust (Cryolithology)), *Izd-vo MGU*, 1967.

Porkhayev, G. V.: Temperature Fields in the Bases of Structures, "Doklady na Mezhdunarodnoy konferentsii po merzlotovedeniyu" N. A. Tsytovich, (ed.), USSR Academy of Sciences Press, 1963.

Porkhayev, G. V. and V. K. Shchelokov: Variations of Ground-Temperature Regime During Land Development, in book: "Teplofizika promerzayushchikh i protaivayushchikh gruntov" (Thermophysics of Freezing and Thawing Soils), G. V. Porkhayev, (ed.), Nauka Press, 1964.

Porkhayev, G. V.: Calculation of Temperature Regime in Bases of Buildings and Structures with Year-Round Ventilated Basements, in collection "Fundamenty sooruzheniy na merzlykh gruntakh v Yakutii," Nauka Press, 1968.

Porkhayev, G. V.: Teplovoye vzaimodeystviye zdaniy i sooruzheniy s vechnomerzlymi gruntami (Thermal Interaction of Buildings and Structures with Permanently Frozen Ground), Nauka Press, 1970.

Porkhayev, G. V. and V. K. Shchelokov: Prediction of Temperature-Regime Variations in Permafrost Rocks after Land Development, NIIOSP, 1971.

Posobiye k SNiP II-B.6-66. Posobiye po proektirovaniyu osnovaniy i fundamentov zdaniy i sooruzheniy na vechnomerzlykh gruntakh (Attachment to SNiP II-B.6-66. Handbook of the Design of Bases and Foundations for Buildings and Structures on Permafrost), compiled under the supervision of S. S. Vyalov and G. F. Porkhayev, Stroyizdat, 1969.

Puzakov. N. A.: Vodno-teplovoy rezhim zemlyanogo polotna avtomobil'nykh dorog (Moisture and Temperature Regime in the Earth Fill of Highways), Avtotransizdat, 1960.

Rzhanitsyn, B. A., B. P. Gorbunov, et al.: Preconstruction Thawing and Compaction of Permafrost, "Doklady na Mezhdunarodnoy konferentsii po merzlotovedeniyu," USSR Academy of Sciences Press, 1963.

Saltykov, N. I.: Fundamenty zdaniy v rayone Bol'she-zemel'noy tundry, "Trudy Instituta merzlotovedeniya" (Foundations of Buildings in the Region of the Bol'shezemel'skaya Tundra, Transactions of the Institute of Geocryology), USSR Academy of Sciences Press, 1947.

Saltykov, N. I.: Teoreticheskiye osnovy proektirovaniya fundamentov na ottaivayushchem osnovanii (Theoretical Fundamentals of the Design of Foundations on Thawing Bases), USSR Academy of Sciences Press, 1952.

Saltykov, N. I.: Thermal-Engineering Calculations for Above-Ground Refrigeration Devices, in book: "Osnovy geokriologii," chap. II, Inzhenernaya geokriologiya (Fundamentals of Geocryology, Part II, Engineering Geocryology), USSR Academy of Sciences Press, 1959.

Saltykov, N. I. and N. N. Saltykova: Thermal-Engineering Designs of Ventilated Basements, in collection: "Issledovaniye vechnoy merzloty v Yakutskoy respublike" (An Investigation of Permafrost in the Republic of Yakutiya), no. 2, USSR Academy of Sciences Press, 1950.

Savel'yev, B. A.: Stroyeniye, sostav i svoystva ledyanogo pokrova morskikh i presnykh vodoyemov (Structure, Composition, and Properties of the Ice Cover of Maritime and Fresh Water Bodies), Izd-vo MGU, 1963.

Sheykov, M. L.: Shear Resistance of Frozen Soils, "Laboratornyye issledovaniya mekhanicheskikh svoystv merzlykh gruntov pod rukovodstvom N. A. Tsytovicha," Collections 1 and 2, USSR Academy of Sciences Press, 1936.

Shumskiy, P. A.: Osnovy strukturnogo l'dovedeniya (Fundamentals of Structural Glaciology), USSR Academy of Sciences Press, 1955.

Shusherina, Ye. P.: A Method of Determining Thawing and Compaction Coefficients of Frozen Soils during Thawing, Collections 1 and 2, "Materialy po laboratornym issledovaniyam merzlykh gruntov pod rukovodstvom N. A. Tsytovicha," USSR Academy of Sciences Press, 1953, 1954.

Shusherina, Ye. P.: Coefficient of Transverse Deformation and Volume Deformations of Frozen Soils in the Process of Creep, Sb. MGU, no. V, 1969.

Shusherina, Ye. P. and Yu. P. Bobkov: Influence of the Moisture Content of Frozen Soils on their Strength, Collection: "Merzlotnyÿe issledovaniya" (Frost Studies), no. IX, Izd-vo MGU, 1969.

Sokolovskiy, V. V.: Statika sypuchey sredy (Statics of Disperse Media), Gostekhizdat, 1954.

Stotsenko, A. V.: Special Problems in Major Hydraulic-Engineering Construction in the Permafrost Zone, Materialy VII Mezhvedomstvennogo soveshchaniya po merzlotovedeniyu, USSR Academy of Sciences Press, 1959.

Sumgin, M. I.: Vechnaya merzlota pochvy v predelakh USSR (Permafrost in the USSR), 2nd edition, USSR Academy of Sciences, 1937.

Ter-Martirosyan, Z. G. and N. A. Tsytovich: Secondary Consolidation of Clays, Osnovaniya, fundamenty i mekhanika gruntov, no. 5, 1965.

Terzaghi, K.: Structural Soil Mechanics, Translation edited by N. M. Gersevanov, Gosstroyizdat, 1933.

Tomirdiaro, S. V.: Teplovyye raschety osnovaniy v rayonakh vechnoy merzloty (Thermal Calculations for Bases in Permafrost Regions), *Izd-vo SVKNII*, Magadan, 1963.

Trupak, N. G.: Spetsial'nyye sposoby provedeniya gornykh vyrabotok (Special Mine Excavation Methods), Ugletekhizdat, 1951.

Tsvetkova, S. G., L. A. Brattsev, *et al.*: Sovremennoye sostoyaniye geokriologicheskikh issledovaniy za rubezhom (The Present Status of Geocryological Research Abroad), Izd. VSEGINGEO, 1966.

Tsytovich, N. A.: Design of Foundations for Structures to be Built on Permafrost, Nauchno-issledovatel'skiye raboty Gipromeza, no. 2, 1928.

Tsytovich, N. A.: Permafrost as a Base for Structures, "Materialy KYePS," No. 80, cb. "Vechnaya merzlota" (Materials of the Permanent Commission for the Study of the Natural Productive Forces of the USSR, No. 80, collection: "Permafrost"), USSR Academy of Sciences Press, 1930.

Tsytovich, N. A.: Lectures on the Design of Foundations Built Under Permafrost Conditions, Izd. Leningradskogo instituta sooruzheniy, 1933.

Tsytovich, N. A., I. S. Vologdina, *et al.*: Laboratornyye issledovaniya mekhanicheskikh svoystv merzlykh gruntov (Laboratory Studies of the Mechanical Properties of Frozen Soils), Collections I and II, SOPS AN SSSR, 1936.

Tsytovich, N. A. and M. I. Sumgin: Osnovaniya mekhaniki merzlykh gruntov (Fundamentals of the Mechanics of Frozen Soils), USSR Academy of Sciences Press, 1937.

Tsytovich, N. A.: Issledovaniye deformatsiy merzlykh gruntov (A Study of the Deformations of Frozen Soils), (part II of doctoral thesis), Leningrad, 1940.

Tsytovich, N. A.: (1) A Study of the Elastic and Plastic Deformations of Frozen Soils, Laboratornyye issledovaniya mekhanicheskikh svoystv merzlykh gruntov (Laboratory Studies of the Mechanical Properties of Frozen Soils), Collection I, USSR Academy of Sciences Press, 1936; (2) Certain Mechanical Properties of Permafrost in Yakutiya, "Trudy KOVM AN SSSR," vol. X, 1940.

Tsytovich, N. A.: Raschet osadok fundamentov (Calculation of Foundation Settling), Stroyizdat, 1941.

Tsytovich, N. A.: Toward a Theory of the Equilibrium State of Water in Frozen Soils, *Izvestiya AN SSSR*, Seriya "Geograficheskaya i geofizicheskaya," nos. 5-6, vol. IX, 1945.

Tsytovich, N. A.: Unfrozen Water in Unconsolidated Rocks, Izvestiya AN SSSR, Seriya "Geologicheskaya," no. 3, 1947.

Tsytovich, N. A., N. I. Saltykov, *et al.*: Fundamenty elektrostantsii na vechnoy merzlote (Foundations of Electric Power Stations on Permafrost), USSR Academy of Sciences Press, 1947.

Tsytovich, N. A.: Printsipy mekhaniki merzlykh gruntov (Principles of the Mechanics of Frozen Soils), USSR Academy of Sciences Press, 1952.

Tsytovich, N. A.: Influence of Freezing Conditions on the Porosity of Water-Saturated Sands, in collection: "Voprosy geologii Azii" (Problems of the Geology of Asia), vol. 2, USSR Academy of Sciences Press, 1955.

Tsytovich, N. A.: Problems in the Theory of Soil Mechanics for Construction (Paper at the 12 October 1955 session of the Hungarian Academy of Sciences), Proceedings of the Hungarian Academy of Sciences, vol. XIX, nos. 1-3, 1956.

Tsytovich, N. A.: A Study of the Adfreezing Forces of Cohesive Clayey Soils by the Ball-Test Method, Proceedings of the Czechoslovak Academy of Sciences, vol. V, No. 3, Prague, 1956.

Tsytovich, N. A. and Z. A. Nersesova: Physical Phenomena and Processes in Freezing, Frozen, and Thawing Soils, Collection 3, "Materialy po laboratornym issledovaniyam merzlykh gruntov pod rukovodstvom N. A. Tsytovicha" (Report on Laboratory Studies of Frozen Soils under the Supervision of N. A. Tsytovich), USSR Academy of Sciences Press, 1957.

Tsytovich, N. A.: Osnovaniya i fundamenty na merzlykh gruntakh (Bases and Foundations on Frozen Soils), USSR Academy of Sciences Press, 1958.

Tsytovich, N. A., I. N. Votyakov, and V. D. Ponomarev: Metodicheskiye rekomendatsii po issledovaniyu osadok ottaivayushchikh gruntov (Methodological Recommendations for Investigation of the Settling of Thawing Soils), USSR Academy of Sciences Press, 1961.

Tsytovich, N. A. and Kh. R. Khakimov: Use of Artificial Soil Freezing in Construction and Mining, in collection: "Doklady k V Mezhdunarodnomu kongressu po mekhanike gruntov," N. A. Tsytovich, (ed.), Stroyizdat, 1961.

Tsytovich, N. A.: Instability of the Mechanical Properties of Frozen and Thawing Soils, Proceedings of the First International Conference on Geocryology, USA, 1963.

Tsytovich, N. A., Yu. K. Zaretskiy, et al.: (1) Consolidation of Thawing Soils, "Doklady k VI Mezhdunarodnomu kongressu po mekhanike gruntov" (Papers at Sixth International Congress on Soil Mechanics), Stroyizdat, 1965; (2) Sb. NIIosnovaniy, no. 56, Gosstroyizdat, 1966.

Tsytovich, N. A., Yu. K. Zaretskiy, et al.: Prediction of the Settling of Thawing Soils as a Function of Time, "Materialy V Vsesoyuznogo soveshchaniya po obmenu opytom stroitel'stva v surovykh klimaticheskikh usloviyakh" (Materials of Fifth All-Union Conference for Exchange of Construction Experience under Harsh Climatic Conditions), no. 5, Krasnoyarsk, 1968.

Tsytovich, N. A., Ukhov, S. B., and Ya. A. Kronik: Combined Physicochemical Frost Heaving Countermeasures in Fill Soils, "Doklady k Dunaysko-Evropeyskoy konferentsii po mekhanike gruntov v dorozhnom stroitel'stve" (Papers at Danube-European Conference on Soil Mechanics and Highway Construction), (May, 1968, Vienna, Austria), Izd. NIIosnovaniy, 1968.

Tsytovich, N. A., V. G. Berezantsev, et al.: "Osnovaniya i fundamenty" (Bases and Foundations), Izd-vo Vysshaya shkola, 1970.

Tsytovich, N. A. and I. I. Cherkasov: Determination of the Coefficients of Compressibility of Soils from Plunger-Indentation Tests, Osnovaniya, fundamenty i mekhanika gruntov, no. 6, 1970.

Tsytovich, N. A., N. V. Ukhova, and S. B. Ukhov: Prognoz temperaturnoy ustoychivosti plotin iz mestnykh materialov na vechnomerzlykh osnovaniyakh (Prediction of the Temperature Stability of Dams Built from Local Materials on Permafrost Bases), Stroyizdat, 1972.

Tsytovich, N. A.: Mekhanika gruntov (Soil Mechanics), fourth edition, Stroyizdat, 1963.

Tsytovich, N. A.: Mekhanika gruntov (Kratkiy kurs) (Soil Mechanics (A Short Course)), Izd-vo Vysshaya shkola, 1968.

Tulayev, A. Ya.: Obzor literaturnykh rabot, posvyashchennykh izucheniyu puchin i meram bor'by s nimi (Survey of Literature on Frost Heaving and Countermeasures Against It), no. II, Dorizdat, 1941.

Turovskaya, Ya. A.: O vliyanii deformatsiy na strukturu glinistykh gruntov (Influence of Deformation on Structure of Clayey Soils), Collection No. 4, LIIZhT, 1957.

Tyutyunov, I. A.: Migratsiya vody v torfogleyevoy pochve i periody zamerzaniya i zamerzshego yeye sostoyaniya v usloviyakh neglubokogo zaleganiya vechnoy merzloty (Migration of Water in Peat-Gley Soil, and the Durations of its Freezing and Frozen States under Conditions of Shallow Depth Permafrost), USSR Academy of Sciences Press, 1951.

Tyutyunov, I. A. and Z. A. Nersesova: Priroda migratsii vody v gruntakh pri promerzanii i osnovy fiziko-khimicheskikh priyemov bor'by s pucheniyem (The Nature of Water Migration in Soils During Freezing, and Fundamentals of Physicochemical Countermeasures Against Heaving), USSR Academy of Sciences Press, 1963.

Ukhov, S. B.: Artificial Salinization of Loamy Soils for Wintertime Construction, Isvestiya vusov, Stroitel'stvo i arkhitekhtura, no. 1, 1959.

Ukhov, S. B.: A Physicochemical Countermeasure against Frost Heaving of Soils, Materialy VIII Vsesoyuznogo soveshchaniya po geokriologii, no. 8, Yakutsk, 1966.

Ukhov, N. V.: Consideration of Convective Heat Transfer during Thawing of Water-Saturated Soil, Materialy VIII Vsesoyuznogo Mezhvedomstvennogo soveshchaniya po geokriologii (merzlotovedenniyu), (Materials of Eighth All-Union Interdepartmental Conference on Geocryology (Permafrostology)), no. 4, Yakutsk, 1966.

Ukhova, N. V.: A Study of the Nonstationary Temperature Regime in Frozen Dams, dissertation, *MISI*, 1967.

Ushkalov, V. P.: Glubina i skorost' ottaivaniya merzlogo osnovaniya (Depth and Rate of Thawing of a Frozen Base), Gosstroyizdat, 1962.

Velli, Yu. Ya., V. V. Dokuchayev, and N. F. Fedorov: Zdaniya i sooruzheniya na Kraynem Severe (spravochnoye posobiye) (Buildings and Structures in the Far North (A Reference Guide)), Gosstroyizdat, 1963.

Veynberg, B. P. Led (Ice), Gostekhteorizdat, 1940.

Vologdina, I. S.: Adfreeze Forces of Frozen Ground to Wood and Concrete, Collection 1, "Laboratornyye issledovaniya mekhanicheskikh svoystv merzlykh gruntov," USSR Academy of Sciences Press, 1936.

Voytkovskiy, K. F.: Raschet sooruzheniy iz l'da i snega (Design of Ice and Snow Structures), USSR Academy of Sciences Press, 1959.

Vyalov, S. S.: Reologicheskiye svoystva i nesushchaya sposobnost' merzlykh gruntov (Rheological Properties and Bearing Capacity of Frozen Soils), USSR Academy of Sciences Press, 1959.

Vyalov, S. S. and N. A. Tsytovich: Estimation of the Bearing Capacity of Cohesive Soils from the Impression Depth of a Spherical Die, *Doklady AN SSSR*, vol. III, no. 6, 1956.

Vyalov, S. S., S. E. Gorodetskiy *et al.*: Prochnost' i polzuchest' merzlykh gruntov i raschety ledogruntovykh ograzhdeniy (Strength and Creep of Frozen Soils, and Calculations for Frozen Ground Barriers), USSR Academy of Sciences Press, 1962.

Vyalov, S. S., Yu. O. Targulyan, *et al.*. Interaction of Frozen Ground with Piles and Pipes in Vibration Sinking, "Materialy VIII Vsesoyuznogo soveshchaniya po geokriologiy" (Materials of the Eighth All-Union Conference on Geocryology), no. 5, Yakutsk, 1966.

Vyalov, S. S., S. E. Gorodetskiy, *et al.*: "Metodika opredeleniya kharakteristiki polzuchesti, dlitel'noy prochnosti i szhimayemosti merzlykh gruntov" (A Method for Determination of the Creep, Long-Term-Strength, and Compressibility Characteristics of Frozen Soils), Nauka Press, 1966.

Vyalov, S. S., P. I. Mel'nikov, *et al.*: Merzlotovedeniye i opyt stroitel'stva na vechnomerzlykh gruntakh v SShA i Kanade (American and Canadian Permafrostology and Construction Experience on Permafrost), Stroyizdat, 1968.

Vyalov, S. S., N. K. Pekarskaya, and R. V. Maksimyak: Physical Nature of Deformation and Breakdown Processes of Clayey Soils, *Osnovaniya, fundamenty i mekhanika gruntov*, no. 1, 1970.

Yegerev, K. Ye.: An Electrical Method of Determining Tangential Reactions Distributed Over the Lateral Surface of a Loaded Pile Frozen into the Soil, *Trudy instituta Merzlotovedeniya AN SSSR*, vol. XIV, 1958.

Zaretskiy, Yu. K.: Strength and Creep Calculations for the Walls of Mine Shafts Sunk with the Aid of Artificial Freezing, in collection "Prochnost' i polzuchest' merzlykh gruntov i raschety ledogruntovykh ograzhdeniy" (Strength and Creep of Frozen Ground and Calculations for Frozen Ground Barriers), USSR Academy of Sciences Press, 1962.

Zaretskiy, Yu. K.: Calculations of Thaw Settling of Soils, *Osnovaniya, fundamenty i mekhanika gruntov*, no. 3, 1968.

Zaretskiy, Yu. K.: Problems of the Theory of Creep and Consolidation of Soils and Their Practical Applications, doctoral thesis, NIIOSP, 1971.

Zaretskiy, Yu. K.: Rheological Properties of Plastically Frozen Ground, *Osnovaniya, fundamenty i mekhanika gruntov*, no. 2, 1972.

Zaretskiy, Yu. K. and S. S. Vyalov: Problems in the Structural Mechanics of Clayey Soils, *Osnovaniya, fundamenty i mekhanika gruntov*, no. 3, 1971.

Zelenin, A. N.: Osnovy razrusheniya gruntov mekhanicheskimi sposobami (Fundamentals of the Mechanical Disintegration of Soils), *Izd-vo Mashinostroyeniye*, 1968.

Zhemochkin, B. N. and A. P. Sinitsyn: Prakticheskiye metody rascheta fundamentnykh balok i plit na uprugom osnovanii (Practical Methods for the Design of Foundation Beams and Plates on an Elastic Base), Gosstroyizdat, 1962.

Zhinkin, G. N.: Elektrokhimicheskoye zakrepleniye gruntov v stroitel'stve (Electro-chemical Reinforcement of Soils in Construction), Stroyizdat, 1966.

Zhukov, V. F.: Predpostroyechnoye protaivaniye mnogoletnemerzlykh gornykh porod pri vozvedenii na nikh sooruzheniy (Preconstruction Thawing of Permafrost Rock for Erection of Structures), USSR Academy of Sciences Press, 1958.

INDEX